Springer Laboratory

Manuals in Polymer Science

For further volumes:
http://www.springer.com/series/3721

Harald Pasch • Bernd Trathnigg

Multidimensional HPLC of Polymers

Springer

Harald Pasch
University of Stellenbosch
Department of Chemistry
and Polymer Science
Stellenbosch
South Africa

Bernd Trathnigg
Karl-Franzens-University Graz
Institute of Chemistry
Graz
Austria

ISSN 0945-6074
ISSN 2196-1174 (electronic)
ISBN 978-3-642-36079-4
ISBN 978-3-642-36080-0 (eBook)
DOI 10.1007/978-3-642-36080-0
Springer Heidelberg New York Dordrecht London

Library of Congress Control Number: 2013935702

Printed on acid-free paper

Springer is a part of Springer Science+Business Media (www.springer.com)

Preface

Liquid chromatography is one of the workhorses in the analysis of polymers. When it comes to the determination of the molar mass distribution, there is no other technique that can compare with size exclusion chromatography in terms of accuracy and reliability. However, size exclusion chromatography separates according to the size of the macromolecules and not molar mass, and has its limits when complex polymer systems must be analyzed.

Complex polymer systems such as random, block and graft copolymers, polymer blends, telechelics and macromonomers necessitated liquid chromatography to be used not only for molar mass determinations but also for evaluation of the chemical heterogeneity and the functionality type distribution. Powerful methods for this task include liquid adsorption chromatography and liquid chromatography at the critical point of adsorption. To address multiple distributions of molecular parameters in complex polymers, multidimensional fractionation and analysis techniques have been developed, most prominently comprehensive two-dimensional liquid chromatography.

A number of text books on liquid chromatography of polymers have been published over the years covering the fundamentals of the different techniques. In our textbook 'HPLC of Polymers' we addressed the experimental aspects of the different techniques, in particular in interaction chromatography, and gave detailed instructions for conducting experiments using the diverse techniques. Since experiment is always the proof of the theory, we intended to give an introduction into liquid chromatography by proposing a number of more or less simple experiments. 'HPLC of Polymers' was published in 1998 and became quite popular as a basis for teaching polymer chromatography courses.

Over the last 15 years polymer chromatography advanced significantly with comprehensive 2D-LC becoming more and more a 'routine' method. New coupling techniques have been developed including LC-NMR and LC-MALDI-TOF. It became even possible to analyse complex polyolefins by high-temperature 2D-LC. Considering these new developments we thought that the time had come to update and extend the previous 'HPLC of Polymers' book by focusing on multidimensional separation methods. The result of our efforts is the present textbook

'Multidimensional HPLC of Polymers' that intends to review the state of the art in polymer chromatography and to summarize the developments in the field during the last 15 years.

Similar to the previous textbook, this laboratory manual is written for beginners as well as for experienced chromatographers. The subject of the book is the description of the experimental approach to the analysis of complex polymers. It summarizes important applications in liquid chromatography of polymers with emphasis on multidimensional experiments. The theoretical background, equipment, experimental procedures and applications are discussed for each separation technique. It will enable polymer chemists, physicists and material scientists, as well as students of polymer and analytical sciences to optimize experimental conditions for a specific separation problem. The main benefit for the reader is that a great variety in instrumentation, separation procedures and applications is given, making it possible to solve simple as well as sophisticated separation tasks.

This book is dedicated to friends and colleagues that contributed (directly or indirectly) to this book by pioneering HPLC, cross-fractionation and multidimensional chromatography of polymers, among others Steve Balke (Canada), Gottfried Glöckner (Germany) and Sadao Mori (Japan), Boris Belenkii, Victor Evreinov and Alexander Gorshkov (Russia), Yefim Brun (USA), Peter Kilz, Helmut Much and Günter Schulz (Germany), Taihyun Chang (Korea), Peter Schoenmakers (The Netherlands), Tibor Macko and Wolf Hiller (Germany).

Stellenbosch, South Africa — Harald Pasch
Graz, Austria — Bernd Trathnigg
March 2013

Contents

Symbols and Abbreviations

ε	Potential of interaction
η	Viscosity
λ	Wavelength
δ	Solubility parameter
ΔG	Gibbs free energy
ΔG_m	Gibbs free energy of mixing
ΔH	Interaction enthalpy
ΔH_m	Mixing enthalpy
η_o	Viscosity of a solvent
η_{rel}	Relative viscosity
ΔS	Conformational entropy
ΔS_m	Mixing entropy
η_{sp}	Specific viscosity
$[\eta]$	Intrinsic viscosity, Staudinger index
2D-LC	Two-dimensional liquid chromatography
a	Mark-Houwink exponent
ACN	Acetonitrile
AH	Adipic acid-hexane diol polyester
APCI	Atmospheric pressure chemical ionization
API	Atmospheric pressure ionization
aPP	Atactic polypropylene
BPA	Bisphenol A
BR	Butyl rubber
c	Concentration
CAP	Critical adsorption point
CCD	Chemical composition distribution
CE	Capillary electrophoresis
CEC	Capillary electrokinetic chromatography
CEF	Crystallization elution fractionation
CH	Cyclohexane
CRYSTAF	Crystallization fractionation
2D	Two-dimensional
D	Pore diameter
DAD	Diode-array detector
DHB	Dihydroxy benzoic acid

DMAC	Dimethyl acetamide
d-PS	Deuterated polystyrene
D-RI	Dual density-refractive index detection
DVB	Divinyl benzene
EGMBE	Ethyleneglycol monobutylether
ELSD	Evaporative light scattering detector
EO	Ethylene oxide
EP	Ethylene-propylene copolymer
EPDM	Ethylene-propylene-diene rubber
ESI	Electrospray ionization
EVA	Ethylene-vinylacetate copolymer
f	Molecular functionality
f_n	Number-average functionality
f_p	Practical functionality
f_w	Weight-average functionality
FAE	Fatty alcohol ethoxylate
FFF	Field flow fractionation
FID	Fast induction decay
FTD	Functionality type distribution
FTIR	Fourier transform infrared spectroscopy
GC-MS	Gas chromatography coupled to mass spectrometry
GPC	Gel permeation chromatography
HDC	Hydrodynamic chromatography
HDPE	High density polyethylene
HEMA	Hydroxyethyl methacrylate
HFiP	Hexafluoro isopropanol
HPLC	High performance liquid chromatography
HPPLC	High performance precipitation liquid chromatography
HT-LC	High temperature liquid chromatography
IC	Interaction chromatography
iPP	Isotactic polypropylene
IR	Infrared
IVD	Intrinsic viscosity distribution
K	Constant factor in the Mark-Houwink equation
K^*	Optical constant in light scattering
K_d	Distribution coefficient
K_{LAC}	Distribution coefficient of adsorption
K_{SEC}	Distribution coefficient of size exclusion
LA	Lactide
LAC	Liquid adsorption chromatography
LALLS	Low angle laser light scattering
LC	Liquid chromatography
LCCC	Liquid chromatography at critical conditions
LDPE	Low density polyethylene
LLDPE	Linear low density polyethylene

LS	Light scattering
M	Molar mass
MAA	Methacrylic acid
MALDI-MS	Matrix-assisted laser desorption/ionization mass spectrometry
MALLS	Multi angle laser light scattering
MEK	Methyl ethyl ketone
MeOH	Methanol
m_i	Mass of species *i*
M_{eq}	Equivalent molar mass
MMA	Methyl methacrylate
MMD	Molar mass distribution
MS	Mass spectrometry
M_n	Number-average molar mass
M_o	Molar mass of repeat unit
M_v	Viscosity-average molar mass
M_w	Weight-average molar mass
n	Degree of polymerization
NELC	Non-exclusion liquid chromatography
n_i	Number of species *i*
NMR	Nuclear magnetic resonance
NP	Normal phase
P	Degree of polymerization
P_n	Number-average degree of polymerization
P_w	Weight-average degree of polymerization
$P(\Theta)$	Scattered light angular dependence
PA	Polyamide
PB	Polybutadiene
PBA	Poly(butylene adipate)
PBSe	Poly(butylene sebacate)
PBSu	Poly(butylene succinate)
PCL	Polycaprolactone
PDMA	Poly(decyl methacrylate)
PDMS	Poly(dimethyl siloxane)
PE	Polyethylene
PEG	Poly(ethylene glycol)
PEMA	Poly(ethyl methacrylate)
PEO	Poly(ethylene oxide)
PI	Polyisoprene
PLA	Poly(L-lactide)
PMA	Poly(methyl acrylate)
PMMA	Poly(methyl methacrylate)
PnBMA	Poly(*n*-butyl methacrylate)
PO	Propylene oxide
PP	Polypropylene
PPG	Poly(propylene glycol)

PPO	Poly(propylene oxide)
PS	Polystyrene
PtBMA	Poly(t-butyl methacrylate)
PVAc	Poly(vinyl acetate)
PVC	Polyvinyl chloride
R (Θ)	Intensity of scattered light
R, R_g	Radius of gyration
RALLS	Right angle laser light scattering
RGD	Radius of gyration distribution
RI	Refractive index
RP	Reversed phase
RT	Retention time
S/N	Signal-to-noise ratio
SAN	Styrene-acrylonitrile copolymer
SBA	Styrene-butyl acrylate copolymer
SBR	Styrene-butadiene rubber
SCB	Short chain branching
SDV	Styrene-divinylbenzene copolymer
SEC	Size exclusion chromatography
SFC	Supercritical fluid chromatography
sPP	Syndiotactic polypropylene
T	Temperature
TCB	1,2,4-Trichlorobenzene
TGIC	Temperature gradient interaction chromatography
THF	Tetrahydrofuran
TLC	Thin layer chromatography
ToF	Time of flight
TREF	Temperature rising elution fractionation
UPLC	Ultrahigh pressure liquid chromatography
UV	Ultraviolet
V_e	Elution volume
V_i	Interparticle volume
V_p	Pore volume
V_R	Retention volume
V_{stat}	Volume of the stationary phase
VISC	Viscometer
w_i	Weight fraction

1 Introduction

Synthetic polymers are highly complex multicomponent materials. They are composed of macromolecules varying in chain length, chemical composition, and architecture. By definition, complex polymers are heterogeneous in more than one distributed property (for example, linear copolymers are distributed in molar mass and chemical composition).

Properties typically considered important to polymer performance in products may be very diverse and can be divided into simple and distributed properties. Simple properties are the total weight of polymer present, the residual monomer or oligomer content, total weight of microgels or aggregates, and properties that depend only on these measures, such as conversion in the polymerization reaction, monomer composition and average copolymer composition. For other properties, different molecules in the same polymer will have different values of the property. These properties are termed distributed properties, the most important of them in polymer chemistry being the molar mass distribution, the distribution of chemical compositions, the distribution of sequence lengths, the distribution of functional groups and the distribution of molecular topologies.

The end-use application of polymers is most frequently determined not only by their chemical identity but more importantly by the distributions of the key physical and physico-chemical parameters. This is similarly true for synthetic and for biopolymers, for technical polymers used as construction materials and for specialty polymers used in drug delivery and tissue engineering. Adequate understanding and monitoring of polymer distributions helps to improve polymer performance and to predict long term behaviour.

Separation science is an important tool for the determination of polymer distributions. A summary of different separation methods, the accessible macromolecular parameters and representative end-use properties are summarized in Table 1.1. The principles and details of the different separation methods will be discussed in the forthcoming chapters and typical applications will be presented.

H. Pasch and B. Trathnigg, *Multidimensional HPLC of Polymers*, Springer Laboratory,
DOI 10.1007/978-3-642-36080-0_1,

Table 1.1 Polymer distributions, end-use properties and separation methods

Polymer distribution	End-use properties	Separation methods[a]
Molar mass	Elongation, tensile strength, adhesion	SEC, FFF, HDC, TGIC, CEC, SFC
Chemical composition	Toughness, biodegradability, morphology, solubility	Gradient HPLC, TGIC, CEC, LCCC
Long-chain branching	Shear strength, tack, peel, crystallinity	SEC-MALLS, SEC-VISC
Short-chain branching	Haze, stress-crack resistance, crystallinity	SEC-FTIR, SEC-NMR, TREF, CRYSTAF, HPLC
Topology	Flow, viscosity, diffusion	SEC-MALLS-VISC
Tacticity	Crystallinity, toughness, solubility	SEC-NMR, TGIC, LCCC
Copolymer sequence	Miscibility, flexibility, haze	SEC-spectroscopy, gradient HPLC, LCCC, 2D-LC
Polyelectrolyte charge	Flocculation, complexation, transport	SEC-conductivity, CEC

Adapted with permission from [1]. Copyright (2004) American Chemical Society
[a]*SEC* size exclusion chromatography, *FFF* field flow fractionation, *HDC* hydrodynamic chromatography, *TGIC* temperature gradient interaction chromatography, *CEC* capillary electrokinetic chromatography, *SFC* supercritical fluid chromatography, *HPLC* high performance liquid chromatography, *LCCC* liquid chromatography at critical conditions, *MALLS* multi-angle laser light scattering, *VISC* viscometry, *TREF* temperature rising elution fractionation, *CRYSTAF* crystallization fractionation, *2D-LC* two-dimensional liquid chromatography

1.1 Molecular Heterogeneity of Complex Polymers

In general, the molecular structure of a macromolecule is described by its size, its chemical structure, and its architecture. The chemical structure characterizes the constitution of the macromolecule, its configuration and its conformation. For a complete description of the constitution, the chemical composition of the polymer chain and the chain ends must be known. In addition to the type and quantity of the repeat units, their sequence of incorporation must be described (alternating, random, or block in the case of copolymers). Macromolecules of the same chemical composition can still have different constitutions due to constitutional isomerism (1,2- vs. 1,4-coupling of butadiene, head-to-tail vs. head-to-head coupling, linear vs. branched molecules). Configurational isomers have the same constitution but different steric patterns (cis- vs. trans-configuration; isotactic, syndiotactic and atactic sequences in a polymer chain). Conformational heterogeneity is the result of the ability of fragments of the polymer chain to rotate around single bonds. Depending on the size of these fragments, interactions between different fragments, and a certain energy barrier, more or less stable conformations may be obtained for the same macromolecule (rod-like vs. coil conformation).

Depending on the composition of the monomer feed and the polymerization procedure, different types of heterogeneities may become important. For example, in the synthesis of tailor-made polymers telechelics or macromonomers are frequently used. These oligomers or polymers usually contain functional groups at the

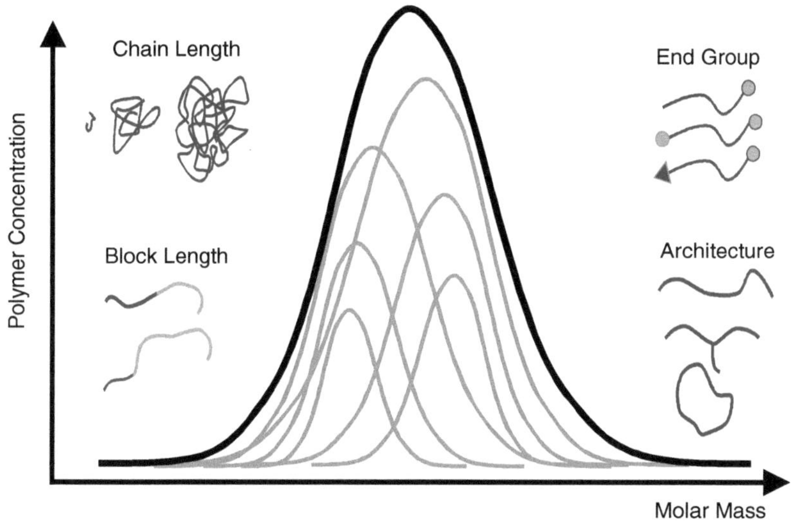

Fig. 1.1 Schematic representation of the molecular heterogeneity of complex polymers (reprinted from [2] with permission of Springer Science + Business Media)

polymer chain end. Depending on the preparation procedure, they can have a different number of functional endgroups, i.e., they can be mono-, bifunctional, etc. In addition, polymers can have different architectures, i.e. they can be branched (star- or comb-like), or cyclic.

The structural complexity of synthetic polymers can be described using the concept of molecular heterogeneity, see Fig. 1.1, meaning the different aspects of molar mass distribution (MMD), distribution in chemical composition (CCD, e.g. block length distribution), functionality type (e.g. endgroup) distribution (FTD) and molecular architecture distribution (MAD). They can be superimposed one on another, i.e. bifunctional molecules can be linear or branched, linear molecules can be mono- or bifunctional, copolymers can be block or graft copolymers etc. In order to characterize complex polymers it is necessary to know the molar mass distribution within each other type of heterogeneity.

All synthetic polymers are disperse or heterogeneous in terms of molar mass. The *molar mass distribution* originates from randomness of the polymerization process. In the daily routine synthetic polymers are often characterized by average molar masses, considering the frequencies (numbers) of macromolecules of a certain molar mass M_i in the sample. Most frequently used are the *number-average* molar mass M_{n}, expressing the amount of species in terms of number of moles n_i, and the *weight-average* molar mass M_{w}, considering the mass m_i of the species. As m_i is related to n_i via $m_i = n_i\, M_i$, and for a single species $M_i = M_{\mathrm{o}}\, P_{\mathrm{i}}$, the molar mass averages may be expressed as average degrees of polymerization, where P_{n} is the number-average and P_{w} is the weight-average degree of polymerization.

$$M_n = \sum n_i M_i / \sum n_i = P_n M_o, \tag{1.1}$$

$$M_w = \sum w_i M_i / \sum w_i = \sum n_i M_i^2 / \sum n_i M_i = P_w M_o. \tag{1.2}$$

Molar masses of polymers may be determined by different methods, SEC being the most important [1–7]. The difference between number and weight average molar masses gives a first estimate of the width of the MMD. The broader the distribution, the larger is the difference between M_n and M_w. The ratio of M_w/M_n is a measure of the breadth of the molar mass and is termed the dispersity.

When two or more monomers of different chemical structures are involved in a polymerization reaction, instead of a chemically homogeneous homopolymer in most cases a chemically heterogeneous copolymer is formed. Depending on the reactivity of the monomers and their sequence of incorporation into the polymer chain, macromolecules can be formed which differ significantly in composition (meaning the amounts of repeat units A, B etc. in the copolymer), and the sequence distribution. With respect to sequence distribution, copolymers can be classified as alternating, random, block and graft copolymers.

Chemical heterogeneity is a consequence of CCD and can be presented as an integral or differential distribution curve of composition vs. molar mass. Consider a random copolymer obtained in a homogeneous reaction from a mixture of A and B monomers. Even under such favourable conditions the resulting macromolecules will differ in chemical structure. There are differences in the sequence of the A and B monomers along the macromolecules, differences in the average chemical composition of the copolymer molecules formed at any instant of the polymerization (instantaneous heterogeneity), and differences due to the depletion of the reaction mixture in one of the monomers.

The sequence distribution of a copolymer chain may be characterized by the number-average lengths of uninterrupted sequences of A and B units in this chain. The average sequence lengths can be measured by physical or chemical methods. The former methods (FTIR or NMR analyses) usually measure the percentage of A and B units inside of triades, pentades etc. whereas the latter methods measure the percentage of A–A, A–B, and B–B linkages. Macromolecules of random copolymers, even if identical in chain length and composition (and thus also in the average sequence length), still offer a great variety with respect to the order of individual sequences in the molecules. Thus, a copolymer sample contains a tremendous number of constituents. In terms of liquid chromatography (LC), a sample of this kind is an extremely complex mixture; it is difficult to separate by size exclusion or interaction chromatography [1].

In addition to the sequence distribution, conversion heterogeneity has to be considered when analyzing copolymers. Only in special cases is the composition of a copolymer identical to the composition of the monomer batch. These cases are azeotropic copolymers or systems whose monomer reactivity ratios equal 1. In general, the instantaneous composition of a copolymer differs from the composition of the monomer mixture, which causes depletion of the batch of the monomer that is preferably incorporated. Thus, subsequent portions of a copolymer sample are polymerized from mixtures of various compositions and this gives rise to additional chemical heterogeneity. Accordingly, when discussing the CCD of copolymers, *sequence distribution*, *instantaneous heterogeneity* and *conversion heterogeneity* must be considered.

Oligomers and polymers with reactive functional groups have been used extensively to prepare a great variety of polymeric materials. In many cases, the behaviour and reactivity of these functional homopolymers is largely dependent on the nature and the number of functional groups. In a number of important applications the functional groups are located at the end of the polymer chain. Macromolecules with terminal functional groups are usually termed "telechelics". Special cases are "macromonomers" which contain one polymerizable endgroup. They can be used a starting materials for the synthesis of graft polymers, combs or brushes.

Molecular functionality, f, of a telechelic polymer is described as the number of functional groups per molecule. Macromolecules with the same structure of the polymer chain may differ in the number and the nature of the functional groups. When functional homopolymers are synthesized, functionally defective molecules are formed in addition to macromolecules of required functionality, For example, if a target functionality of $f = 2$ is required, then generally in the normal case species with $f = 1, f = 0$ or higher functionalities are formed as well [3], which may result in a decreased or increased reactivity, cross-linking density, surface activity etc. Each functionality fraction has its own molar mass distribution. Therefore, for a complete description of the molecular structure of a functional homopolymer, the determination of the molar mass distribution and the functionality type distribution is required.

Typically, functionality is quantitatively described as number-average functionality, f_n, where f_n is the ratio of the total number of functional groups to the total number of molecules in the system, i.e. the average number of functional groups per initial molecule. The f_n value provides information on the average functionality but does not characterize the functional dispersity. An average functionality of 1 may be simulated by equal amounts of non-functional and difunctional species, and is therefore ambiguous. The characterization of the width of the functionality type distribution is more informative. In analogy to the average molar masses, number-average and weight-average functionalities may be introduced,

$$f_{\mathrm{n}} = \sum n_i f_i / \sum n_i, \tag{1.3}$$

$$f_{\mathrm{w}} = \sum w_i f_i / \sum w_i = \sum n_i f_i^2 / \sum n_i f_i, \tag{1.4}$$

where n_i is the number of molecules of functionality f_i, and $w_i = n_i f_i$. For the description of the functional dispersity the term f_w/f_n may be used. For polymers containing only one type of molecule, $f_w/f_n = 1$ is obtained. In the case of a distribution of molecules of different functionality $f_w/f_n > 1$ is obtained. In the characterization of polymers, a separation according to different distributions is required. This can be achieved using different modes of liquid chromatography.

Using the traditional methods of polymer analysis, such as infrared spectroscopy or nuclear magnetic resonance, one can determine the type of monomers or functional groups present in the sample. However, the determination of functional endgroups is complicated for long chain molecules because of low concentration. On the other hand, these methods do not yield information on how different monomer units or functional groups are distributed in the polymer molecule. Finally, these methods generally do not provide molar mass information.

With respect to methods sensitive to the size of the macromolecule, one can face other difficulties. SEC which is most frequently used to separate polymer molecules from each other according to their molecular size in solution, must be used very carefully when analyzing complex polymers. The molecular size distribution of macromolecules can generally be unambiguously correlated with MMD only within one heterogeneity type. For samples consisting of a mixture of molecules of different functionalities, the distribution obtained represents a sum of distributions of molecules having a different functionality and, therefore, cannot be attributed to a specific functionality type without additional assumptions.

For the analysis of copolymers by SEC either the chemical composition along the molar mass axis must be known or detectors must be used which, instead of providing a concentration information, can provide molar mass information. To this end, SEC has to be coupled to composition-sensitive or molar mass-sensitive detectors. Another option for the analysis of complex polymers is the separation with respect to chemical composition or functionality by means of interaction chromatography. In this case, functionally or chemically homogeneous fractions are obtained which then can be subjected to molar mass determination.

To summarize, for the complete analysis of complex polymers a minimum of two different characterization methods must be used. It is most desirable that each method is selective towards a specific type of heterogeneity. Maximum efficiency can be expected when, similar to the 2D distribution in properties, 2D analytical techniques are used. A possible approach in this respect is the coupling of different chromatographic modes in 2D chromatography or the coupling of a separation technique with selective detectors.

1.2 Liquid Chromatography of Polymers

Any chromatographic process relates to the selective distribution of an analyte between the mobile and the stationary phase of a given chromatographic system. In LC the separation process can be described by

$$V_e = V_i + V_p K_d, \tag{1.5}$$

where V_e is the retention volume of the solute, V_i is the interstitial volume of the column, V_p is the volume inside the pores of the packing, i.e. the "stationary" volume, and K_d is the distribution coefficient which is equal to the ratio of the analyte concentration in these fractions of the liquid phase inside the column. In other words, a molecule moves along the column as long it is in the interstitial volume, and it is retained as long as it is inside the pores.

It should be mentioned that the stationary volume can comprise a volume fraction of the mobile phase and the volume of the boundary layer at the surface of the stationary phase, which can have considerably different compositions. In typical reversed phase systems, there is a layer of bonded alkyl chains (which may be swollen or collapsed) as well as an adsorbed layer of almost pure organic component of the mobile phase [8–10], and there is no dividing interface between these layers and the bulk liquid. A similar situation is observed on hydrophilic interaction columns, which contain an aqueous layer close to the surface of the stationary phase [11].

As it is not possible to determine these fractions of the stationary volume, it does not make sense to consider another distribution coefficient for the partitioning of a solute between the bulk phase inside the pores and the boundary layer at the surface of the stationary phase. In both situations, the molecule will be retained. If there is an enthalpic interaction between the solute and the stationary phase, it will stay longer in the pore, which results in a larger distribution coefficient.

K_d is related to the change in Gibbs free energy ΔG related to the analyte partitioning between interstitial and pore volume [12].

$$\Delta G = \Delta H - T\Delta S = -RT \ \ln K_d. \tag{1.6}$$

In a logarithmic plot of the distribution coefficients as a function of $1/T$, one may determine the entropic and enthalpic contributions (van t'Hoff plot). As will be discussed later on:

$$\ln K_d = \frac{\Delta S}{R} - \frac{\Delta H}{RT}. \tag{1.7}$$

The change in Gibbs free energy may be due to different effects:

1. Inside the pore, which has limited dimensions, the macromolecule cannot occupy all possible conformations and, therefore, the conformational entropy ΔS decreases.
2. When penetrating the pores, the macromolecule may interact with the pore walls resulting in a change in enthalpy ΔH. Obviously, the interaction of a polymer chain with the stationary phase has also an entropic contribution: when the chain interacts with the surface, it will loose degrees of freedom. Instead of a random coil, there will be adsorbed trains, loops and free ends, see Table 1.2. Depending on the chromatographic system and the chemical structure of the macromolecule, there may be different entropic or enthalpic contributions.

In SEC separation is accomplished with respect to the hydrodynamic volume of the macromolecules. The stationary phase is a swollen gel with a characteristic pore size distribution, and depending on the size of the macromolecules a larger

Table 1.2 Model presentation of the adsorption of macromolecules

Model	Relation between adsorbed amount	Molar mass and surface thickness
Flat layer 	Independent of M	Independent of M
Coil 	Independent of M	$\sim M^{0.5}$
Collapsed coil 	$\sim M^{1/3}$	$\sim M^{1/3}$
Brush 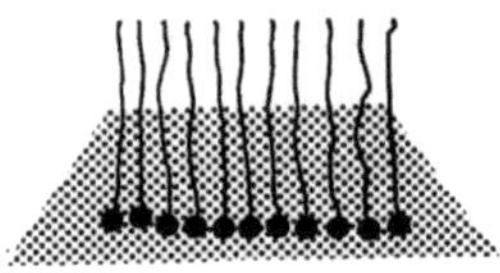	$\sim M$	$\sim M$
Loops and trains 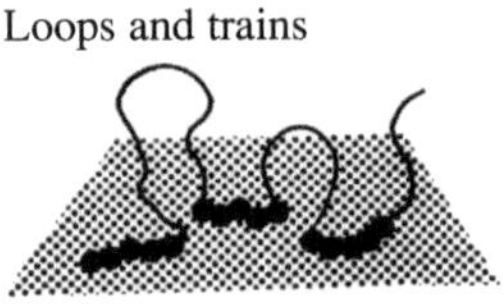	$\sim M^a$ $0 < a < 0.5$	Independent of M

Reprinted from [3] with permission of Springer Science + Business Media

or lesser fraction of the pores is accessible to the macromolecules. Very large molecules, which are excluded from the pores, will elute at the interstitial volume V_i, while small molecules, which have access to the entire pore volume, will elute at the void volume which is $V_0 = V_i + V_p$. Consequently, the separation range is $0 < K_{SEC} < 1$.

In *ideal* SEC, enthalpic interactions are absent, and the distribution coefficients are exclusively determined by the entropy change. In *real* SEC, this may not strictly be fulfilled. Especially with charged polymers it is often difficult to suppress enthalpic interactions completely. These interactions may be attractive or repulsive.

In liquid adsorption chromatography (LAC), where the separation is dominated by enthalpic interactions between the macromolecules and the stationary phase, an ideal and a real case may be defined as well. In *ideal LAC* (which may be observed with small molecules) conformational changes are assumed to be zero ($\Delta S = 0$) and the distribution coefficient is exclusively determined by enthalpic effects. In *real LAC* only a fraction of the pores of the packing is accessible for the polymer chains, which are more or less deformed, when they interact with the stationary phase. Therefore, entropic contributions must be assumed. Accordingly, the distribution coefficient is a function of ΔH and ΔS.

Real SEC and real LAC are often mixed-mode chromatographic methods with predominance of entropic or enthalpic interactions. With chemically heterogeneous polymers, effects are even more dramatic because exclusion and adsorption act differently on molecules with different compositions. In a more general sense, the size exclusion mode of LC relates to a separation regime where entropic interactions are predominant, i.e., $T\Delta S > \Delta H$. In the reverse case, $\Delta H > T\Delta S$, separation is mainly directed by enthalpic interactions. As both separation modes in the general case are affected by the size of the macromolecule and the pore size, a certain energy of interaction ε may be introduced, characterizing the specific interactions of the monomer unit of the macromolecule and the stationary phase. ε is a function of the chemical composition of the monomer unit, the composition of the mobile phase of the chromatographic system, the characteristics of the stationary phase and the temperature.

The theory of adsorption at porous adsorbents predicts the existence of a finite critical energy of adsorption ε_c, where the macromolecule starts to adsorb at the stationary phase. Thus, at $\varepsilon > \varepsilon_c$ the macromolecule is adsorbed, whereas at $\varepsilon < \varepsilon_c$ the macromolecule remains non-adsorbed. At $\varepsilon = \varepsilon_c$ the transition from the non-adsorbed to the adsorbed state takes place, corresponding to a transition from SEC to LAC. This transition is termed "critical point of adsorption" or "critical adsorption point" (CAP) and relates to a situation, where the adsorption forces are exactly compensated by entropy losses [13–15].

$$T\Delta S = \Delta H. \tag{1.8}$$

Accordingly, at the critical point of adsorption the Gibbs free energy is constant ($\Delta G = 0$) and the distribution coefficient is $K_d = 1$, irrespective of the molar mass of the macromolecules and the pore size of the stationary phase.

The critical point of adsorption relates to a very narrow range between the size exclusion and adsorption modes of LC, a region which is very sensitive towards temperature and mobile phase composition. The transition from one to another chromatographic separation mode by changing the temperature or the composition of the mobile phase was reported for the first time by Tennikov et al. [16] and Belenkii et al. [13, 17]. They showed that a sudden change in elution behaviour may occur by small variations in the solvent strength. Thus, just by simply gradually changing the eluent composition, a transition from the SEC to the LAC mode and vice versa may be achieved. The point of transition from SEC to LAC is the critical point of adsorption and chromatographic separations at this point are termed *liquid chromatography at the critical point of adsorption* or liquid chromatography at critical conditions (LCCC).

The retention behaviour of linear homopolymers in these separation modes is shown schematically in Fig. 1.2: with increasing molar mass, retention decreases in SEC, increases in LAC, while it is constant in LCCC. The separation modes

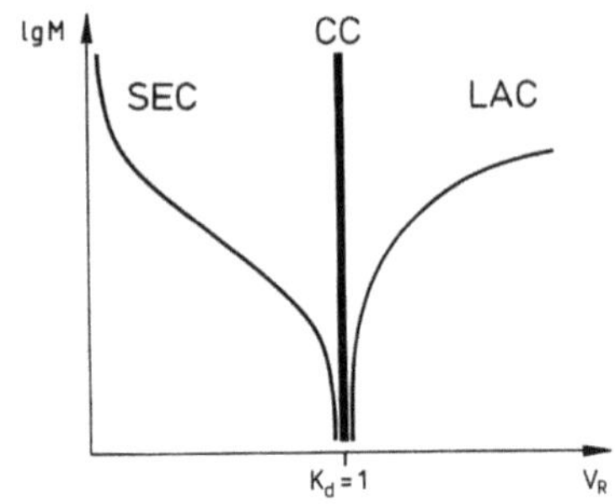

Fig. 1.2 Molar mass vs. retention volume behaviour in different chromatographic modes

described above can be combined in various ways in order to separate polymers according to the distributions of molar mass, chemical composition and functionality.

For SEC as the most important and well established separation method a number of approaches are in place to obtain chemical composition and molar mass information in the same chromatographic run [18]:

1. Multiple detection SEC systems: n independent detector signals (different responses by components) allow the composition calculation of n components in the sample.
2. Universal calibration: measurement of Mark-Houwink coefficients for copolymers with homogeneous and known composition will give copolymer molar masses.
3. SEC with on-line viscometric detection: on-line measurement of Mark-Houwink coefficients for copolymers of various architectures. Here, copolymer M_n measurement according to the Goldwasser approach [19] is an additional benefit.
4. SEC with light-scattering detection: direct measurement of copolymer molar masses for chemically homogeneous and segmented copolymers independent of their chemical structure.

The different approaches of copolymer analysis, their requirements, benefits, and limitations are summarized in Table 1.3.

Although these approaches are most useful, they are not based on a chemical composition separation as in the case of interaction chromatography. SEC is, therefore, intrinsically not able to provide a CCD. The chemical composition information that is obtained is an average value that relates to a given molar mass fraction.

1.3 Multidimensional Separation of Complex Polymers

Complex polymers are distributed in more than one direction of molecular heterogeneity. Copolymers are characterized by the molar mass distribution and the chemical heterogeneity, whereas functional homopolymers are distributed in molar mass and functionality. Hence, the experimental evaluation of the different distribution functions requires analysis in more than one direction. The molecular

Table 1.3 Chromatographic methods for copolymer and blend analysis

Method	Requirements	Preconditions	Advantages	Limitations
Multiple detection	Two or more detectors, proper calibrants	No segment-segment interactions, no neighboring group effects	Bulk composition and compositional distribution, broad applicability, no additional sample preparation	Statistical copolymers, densely grafted polymer chains
Universal calibration	Base calibration, [η]-M relationship	Validity of universal calibration, homogeneous sample composition	Simple and accurate MMD	Chemically homogeneous samples only, detailed information about sample required
Light-scattering detection	Light-scattering and concentration detectors	Known dn/dc as a function of elution volume	Direct MMD measurement, no calibration, independent of architecture	No CCD information
Viscometric detection	Viscometric and concentration detectors	Validity of universal calibration	Easy MMD calculation, independent of architecture, K and a for copolymers	No CCD information, no heterogeneous samples

Reprinted with permission from [18]. Copyright (1995) American Chemical Society

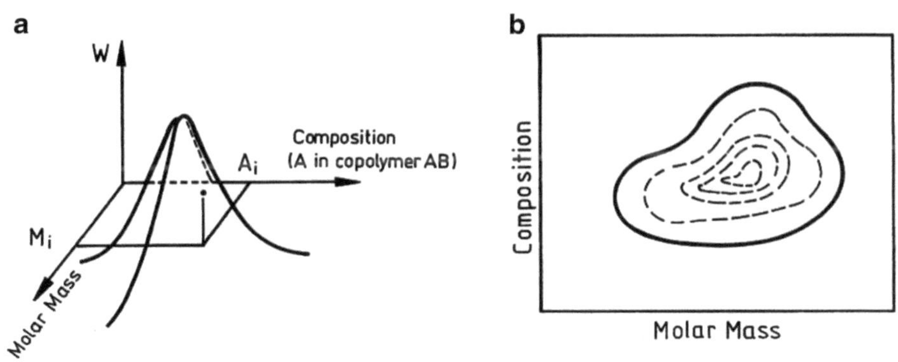

Fig. 1.3 Representation of the molecular heterogeneity of a random copolymer, (**a**) three-dimensional diagram, (**b**) contour diagram; A_i and M_i indicate the average composition and molar mass, respectively (reprinted from [3] with permission of Springer Science + Business Media)

heterogeneity of a random AB copolymer is presented in Fig. 1.3 showing the distributions in molar mass and chemical composition.

For a complete analysis both distributions must be determined. The classical approach is based upon the dependence of copolymer solubility on composition and chain length. A solvent/non-solvent combination fractionating solely by molar mass would be appropriate for the evaluation of MMD, another one separating with respect to chemical composition would be suited for determining CCD. Unfortunately, in most cases precipitation fractionation yields fractions which vary both in molar mass and chemical composition. Even high resolution fractionation would not improve the result and it is nearly impossible to obtain perfectly homogeneous fractions.

By the use of different modes of liquid chromatography it is possible to separate polymers selectively with respect to hydrodynamic volume (molar mass), chemical composition or functionality. Using these techniques and combining them with each other or with a selective detector, one can obtain two-dimensional information on different aspects of molecular heterogeneity. If, for example, two different chromatographic techniques are combined in a "cross-fractionation" mode, information on CCD and MMD can be obtained. Literally, the term "chromatographic cross-fractionation" refers to any combination of chromatographic methods capable of evaluating the distribution in size and composition of copolymers.

First attempts to make use of orthogonal chromatography were presented by Balke et al. [20] and Glöckner [12] in the 1980s. Balke et al. used the fact that macromolecules of the same chain length but different composition have different hydrodynamic volumes. Since SEC separates according to hydrodynamic volume, SEC in different eluents can separate a copolymer in two diverging directions. The authors coupled two SEC instruments together so that the eluent from the first one flowed through the injection valve of the second one. At any desired retention time the flow through SEC 1 could be stopped and an injection made into SEC 2. The first instrument was operated with THF as the eluent and polystyrene gel as the packing, whereas for SEC 2 polyether bonded-phase columns and THF-heptane were used. The schematic presentation of this system is given in Fig. 1.4. Both instruments utilized SEC columns. However, whereas the first SEC was operating so as to achieve conventional molecular size separation, the second SEC was used to fractionate by composition, utilizing a mixed solvent to encourage adsorption and partition effects in addition to size exclusion.

Much work on chromatographic cross-fractionation was carried out with respect to combination of SEC and gradient HPLC. In most cases SEC was used as the first separation step, followed by HPLC. In a number of early papers the cross-fractionation of model mixtures was discussed. Investigations of this kind demonstrated the efficiency of gradient HPLC for separation by chemical composition. Mixtures of statistical copolymers of styrene and acrylonitrile were separated by Glöckner et al. [21]. In the first dimension a SEC separation was carried out using THF as the eluent and polystyrene gel as the packing. In total, about 10 fractions were collected and subjected to the second dimension, which was gradient HPLC on a CN bonded-phase using isooctane-THF as the mobile phase. Model mixtures of statistical copolymers of styrene and 2-methoxyethyl methacrylate were separated in a similar way, the mobile phase of the HPLC mode being isooctane-methanol in this case [22]. Graft copolymers of methyl methacrylate onto EPDM rubber were analyzed by Augenstein and Stickler [23] whereas Mori reported on the fractionation of block copolymers of styrene and vinyl acetate [24].

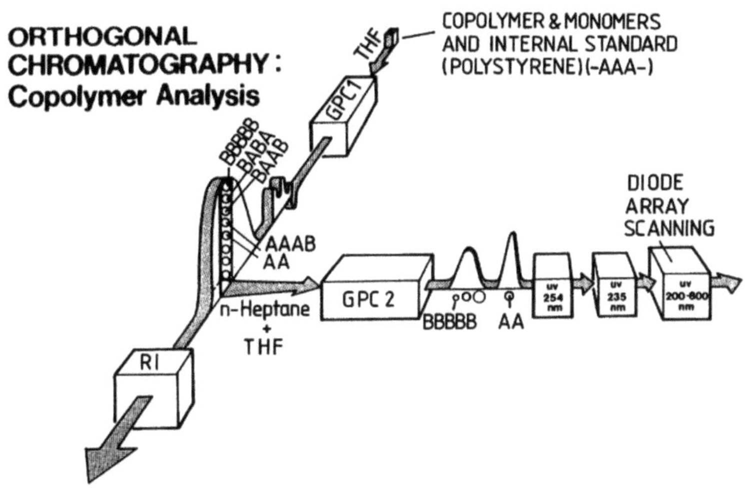

Fig. 1.4 Schematic representation of an orthogonal chromatographic system showing size fractionation of a linear copolymer by SEC 1 and the variety of molecules of the same molecular size within a chromatogram slice, A-styrene and B-butyl methacrylate units (reprinted from [3] with permission of Springer Science + Business Media)

A more feasible way of analyzing copolymers is the pre-fractionation through HPLC in the first dimension and subsequent analysis of the fractions by SEC [25, 26]. HPLC was found to be rather insensitive towards molar mass effects and yields very uniform fractions with respect to chemical composition. Principal considerations of LC couplings will be discussed more in detail in Chapter 6.

The major disadvantage of all early investigations on chromatographic cross-fractionation was related to the fact that both separation modes were combined to each other either off-line or in a stop-flow mode. Regardless of the separation order SEC vs. HPLC or HPLC vs. SEC, in the first separation step fractions were collected, isolated, and then subjected to the second separation step. This procedure was very time-consuming and the reliability of the results at least to a certain extent depended on the skills of the operator.

A fully automated two-dimensional chromatographic system was developed by Kilz et al. [27–29] in the 1990s. It consists of two chromatographs, one which separates by chemical composition or functionality and a SEC instrument for subsequent separation by size. Via a storage loop system, fractions from the first separation step are automatically transferred into the second separation system. The operation of the column switching device is automatically driven by the software, which at the same time organizes the data collection from the detector. The design of a typical system is presented schematically in Fig. 1.5.

Another option to address the molecular heterogeneity of complex polymers is the combination of selective fractionation methods with information-rich detectors, see Fig. 1.6.

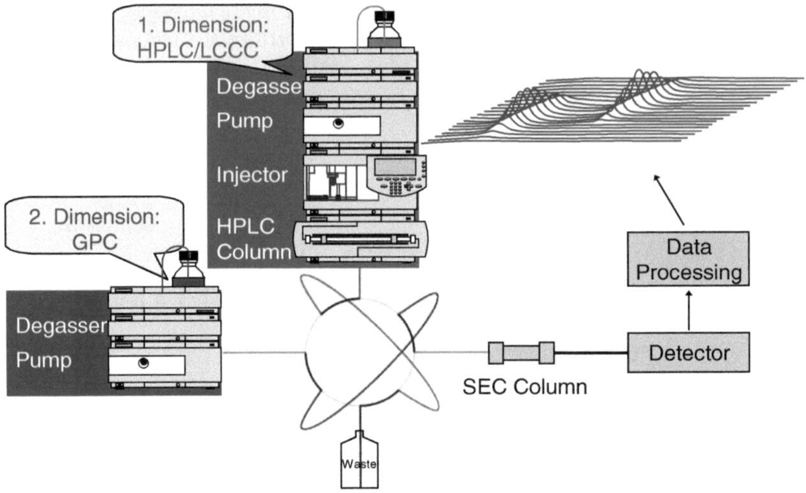

Fig. 1.5 Schematic representation of an automated two-dimensional chromatographic system (reprinted from [30] with permission of Elsevier)

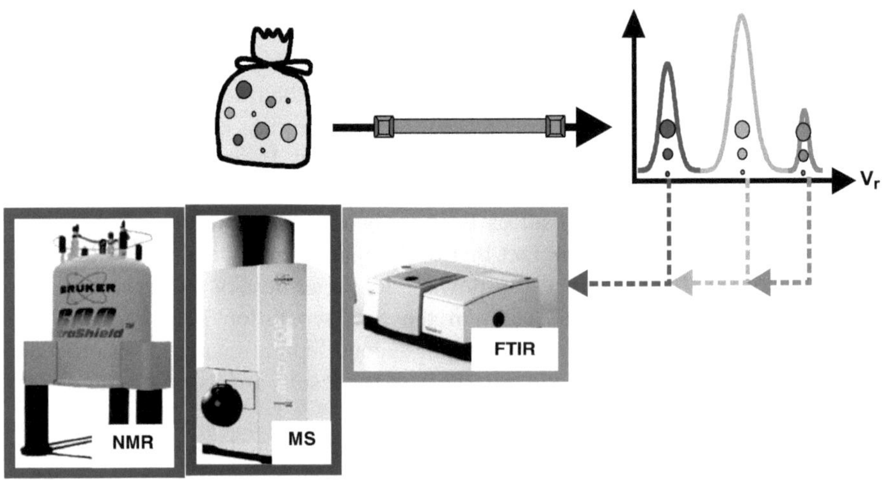

Fig. 1.6 Schematic representation of the hyphenation of a selective chromatographic separation and spectroscopic analysis for the analysis of a sample that is distributed regarding composition (different colours) and molar masses (different sizes)

During the last two decades a number of techniques have been introduced in organic chemistry and applied to polymer analysis, combining chromatographic separation with spectroscopic detection [31]. GC-MS has been used in polymer analysis, but, due to the low volatility of high molar mass compounds it is limited to the oligomer region. The combination of pyrolysis and GC-MS, however, is of great value for polymer characterization [32, 33]. It provides for the analysis of complex polymers with respect to chemical composition. Much more important are the different techniques of liquid chromatography. Using SEC, liquid adsorption chromatography (LAC), or liquid chromatography at the critical point of adsorption (LCCC) polymers can be fractionated with respect to different aspects of molecular heterogeneity, including molar mass, functionality, and chemical composition. As will be shown in the next chapters, liquid chromatography can be efficiently coupled to infrared spectroscopy [34–39], to mass spectrometry, and to nuclear magnetic resonance spectroscopy [40, 41]. Another most feasible approach is multidetector SEC where molar mass separation is hyphenated with molar mass sensitive detectors like on-line viscometry and on-line static light scattering [1, 42].

References

1. Striegel AM (2004) In: Striegel AM (ed) Multiple detection in size exclusion chromatography, vol 893, ACS Symposium Series. ACS, Washington, DC
2. Pasch H (2000) Adv Polym Sci 150:1
3. Pasch H, Trathnigg B (1998) HPLC of polymers. Springer, Berlin
4. Yau WW, Kirkland JJ, Bly DD (1979) Modern size exclusion chromatography. Practice of gel permeation and gel filtration chromatography. Wiley Interscience, New York
5. Janca J (1984) Steric exclusion liquid chromatography. Dekker, New York
6. Glöckner G (1987) Polymer characterization by liquid chromatography. Elsevier, Amsterdam
7. Mori S, Barth HG (1999) Size exclusion chromatography. Springer, Berlin
8. Rustamov I, Farcas T, Ahmed F, Chan F, LoBrutto R, McNair HM, Kazakevich YV (2001) J Chromatogr A 913:49
9. Kazakevich YV, LoBrutto R, Chan F, Patel T (2001) J Chromatogr A 913:75
10. Bocian S, Vajda P, Felinger A, Buszewski B (2008) J Chromatogr A 1204:35
11. Jandera P (2008) J Sep Sci 31:1421
12. Glöckner G (1991) Gradient HPLC and chromatographic cross-fractionation. Springer, Berlin
13. Belenkii BG, Gankina ES, Tennikov MB, Vilenchik LZ (1978) J Chromatogr 147:99
14. Entelis SG, Evreinov VV, Gorshkov AV (1986) Adv Polym Sci 76:129
15. Entelis SG, Evreinov VV, Kuzaev AI (1985) Reactive oligomers. Khimiya, Moscow
16. Tennikov MB, Nefedov PP, Lazareva MA, Frenkel SJ (1977) Vysokomol Soedin A19:657
17. Belenkii BG, Gankina ES, Tennikov MB, Vilenchik LZ (1976) Dokl Acad Nauk USSR 231:1147
18. Kilz P, Krüger RP, Much H, Schulz G (1995) In: Provder T, Urban MW, Barth HG (eds) Chromatographic characterization of polymers: hyphenated and multidimensional techniques, vol 247, Advances in Chemistry Series. American Chemical Society, Washington, DC
19. Goldwasser JM, Rudin A (1983) J Liq Chromatogr 6:2433
20. Balke ST, Patel RD (1980) J Polym Sci B Polym Lett 18:453
21. Glöckner G, van den Berg JHM, Meijerink NL, Scholte TG (1986) In: Kleintjens I, Lemstra P (eds) Integration of fundamental polymer science and technology. Elsevier Applied Science, Barking
22. Glöckner G, Stickler M, Wunderlich W (1989) J Appl Polym Sci 37:3147
23. Augenstein M, Stickler M (1990) Makromol Chem 191:415
24. Mori S (1990) J Chromatogr 503:411
25. Mori S (1988) Anal Chem 60:1125
26. Mori S (1981) Anal Chem 53:1813
27. Kilz P (1993) Labor Praxis 6:64
28. Kilz P, Krüger R-P, Much H, Schulz G (1995) ACS Adv Chem 247:223
29. Kilz P, Krüger R-P, Much H, Schulz G (1993) PMSE Preprints 69:114

30. Pasch H (2012) In: Matyjaszewski K, Moeller M (eds) Comprehensive polymer science, 2nd edn. Elsevier, Amsterdam
31. Cortes HJ (1992) J Chromatogr 626:3
32. Liebman S, Levy E (1985) Pyrolysis and GC in polymer analysis. Dekker, New York
33. Geißler M (1997) Kunststoffe 87:194
34. Sabo M (1985) Anal Chem 57:1822
35. Hellgeth JW, Taylor LT (1987) Anal Chem 59:295
36. Hellgeth JW, Taylor LT (1986) J Chromatogr Sci 24:519
37. Griffiths PR, Conroy CM (1986) Adv Chromatogr 25:105
38. Gagel JJ, Biemann K (1986) Anal Chem 58:2184
39. Wheeler LM, Willis J (1993) Appl Spectrosc 47:1128
40. Lindon JC (1996) Progr Nucl Magn Res Spectrosc 29:1
41. Albert K (1995) J Chromatogr A 703:123
42. Striegel AM, Yau WW, Kirkland JJ, Bly DD (2009) Modern size-exclusion liquid chromatography. Practice of gel permeation and gel filtration chromatography. Wiley, Hoboken, NJ

Theory of Polymer Chromatography

2

The retention of a species in liquid chromatography can be expressed in terms of the distribution coefficient relevant to the partition function of a polymer between the flowing and the stagnant liquid phase in the column, which correspond to the interstitial volume V_{i} (i.e. the volume of the solvent outside the particles of the packing) and the pore volume V_{p}.

$$K = \frac{V_{\mathrm{e}} - V_{\mathrm{i}}}{V_{\mathrm{p}}}. \tag{2.1}$$

The regimes in liquid chromatography, in which the distribution coefficient may assume different values, may be classified in terms of the adsorption interaction parameter c, which has been introduced by de Gennes [1]. It is measured in inverse length (nm^{-1}). The value of c strongly depends on the mobile phase composition (for a given polymer-sorbent system), and on temperature.

2.1 Size Exclusion Chromatography

Size exclusion chromatography (SEC) is governed by entropy. In SEC, the interaction parameter is strongly negative, and $0 < K < 1$. In ideal SEC, the elution volumes of linear macromolecules in wide slit-like pores [2, 3] are given by

$$V_{\mathrm{SEC}} = V_{\mathrm{i}} + V_{\mathrm{p}}\left(1 - \frac{4}{\sqrt{\pi}}\frac{R}{D}\right), \tag{2.2}$$

wherein D is the pore diameter and R is the radius of gyration of the analyte molecule, which can be expressed by the length a and number n of the repeat units:

$$R = a\sqrt{\frac{n}{6}}. \tag{2.3}$$

H. Pasch and B. Trathnigg, *Multidimensional HPLC of Polymers*, Springer Laboratory, DOI 10.1007/978-3-642-36080-0_2, © Springer-Verlag Berlin Heidelberg 2013

SEC separates polymers according to molecular dimensions, regardless of their functionality.

2.2 Liquid Adsorption Chromatography

Liquid adsorption chromatography (LAC) is based on the enthalpic interactions of the polymer with the stationary phase. In LAC the interaction parameter is positive, and K increases exponentially with the number of repeat units, hence $K \gg 1$.

In LAC, retention is often described by Martin's rule [4–8], an empirical equation between the retention factor $k = (V_e - V_0)/V_0$ and the number of repeat units n within a homologous series of oligomers [5–8].

$$\ln k = A + Bn. \tag{2.4}$$

The dimensionless retention factor k is given by the elution volume V_e of the solute and the hold-up volume of a column, i.e. the elution volume of an unretained compound. It must be mentioned, that the hold-up volume and the total volume of the mobile phase in the column (the void volume $V_0 = V_i + V_p$) need not be identical [9–11]. Moreover the extra-column volume is often ignored, which leads to erroneous retention factors [12–14]. There have been numerous papers concerning the question of the hold-up volume. An excellent review has been given by Rimmer et al. [10].

As has been shown [15–17], Martin's rule can be explained using the theory of polymer chromatography [18–22]. There is, however, a considerable difference between these equations. In Eq. (2.4) the retention factor k is calculated using the void volume instead of the accessible volume. Many chromatographers consider the elution volume of the solvent peak as the void volume, which leads to considerable deviations from linearity at low n.

The theory of chromatography of flexible homopolymers which can interact with the pore walls of a stationary phase (for Gaussian polymer chains and slit-like pores) has been developed by Gorbunov and Skvortsov [23, 24]. According to this theory, the elution volume of a non-functional chain in LAC is given by

$$V_e = V_i + V_p\left(1 - \frac{4}{\sqrt{\pi}}\frac{R}{D}\right) - \frac{2V_p}{cD} + \frac{2V_p}{cD}\exp{(cR)^2}[1 + \operatorname{erf}{(cR)}], \tag{2.5}$$

wherein R is the radius of gyration of the macromolecule, D is the pore diameter of the stationary phase, c is the interaction parameter, and erf(cR) is the error function. It must be mentioned, that the value of c does not depend on D, V_p, V_i or R (i.e. on the degree of polymerization n).

At sufficiently strong interaction (which is typical for LAC), the term erf(cR) approaches unity, hence one may write

$$V_e = V_i + V_p\left(1 - \frac{4}{\sqrt{\pi}}\frac{R}{D}\right) - \frac{2V_p}{cD} + \frac{4V_p}{cD}\exp{(cR)^2} \tag{2.6}$$

or

$$V_e = V_0^* + \frac{4V_p}{cD}\exp\left(n\frac{c^2a^2}{6}\right), \tag{2.7}$$

wherein V_0^* is the accessible volume, which can be entered by a polymer chain.

$$V_0^* = V_i + V_p\left(1 - \frac{4R}{D\sqrt{\pi}}\right) - \frac{2V_p}{cD} = V_{SEC} - \frac{2V_p}{cD}. \tag{2.8}$$

Evidently, the accessible volume V_0^* is always smaller than the void volume $V_0 = V_i + V_p$, and it is also smaller than the elution volume of the same polymer chain in ideal SEC.

As has been shown previously [16], the accessible volume in LAC can be easily obtained from a plot of the elution volumes V_n of oligomers with n repeat units versus the differences $\Delta V_n = V_n - V_{n-1}$ in the elution volumes of consecutive oligomers.

$$V_n = V_0^* + \gamma\Delta V_n. \tag{2.9}$$

The intercept in such a plot represents the accessible volume, and the interaction parameter c can be calculated from the slope $\gamma = e^B/(e^B - 1)$ (wherein $B = (c^2a^2)/6$ is the slope in Martin's rule).

$$c = \frac{1}{a}\sqrt{6\ln\frac{\gamma}{\gamma - 1}}. \tag{2.10}$$

It must be mentioned, that the determination of the accessible volume and the interaction parameter does not require the knowledge of n for the individual peaks (provided that they belong to the same series). This works well in the range of sufficiently strong interaction (as long as the approximation of $\mathrm{erf}(cR) = 1$ holds).

There is, however, another approach, which has been developed by Gorbunov et al. [25]. Using special software the interaction parameter can be determined in all modes of LC (from LAC to SEC). This approach requires a set of measurements with non-functional polymer standards, the molar mass of which is known.

Equation 2.7 may also be written in terms of the retention factor k^*:

$$k^* \frac{V_e - V_0^*}{V_0^*} = \frac{4V_p}{cDV_0^*}\exp\left(n\frac{c^2a^2}{6}\right). \tag{2.11}$$

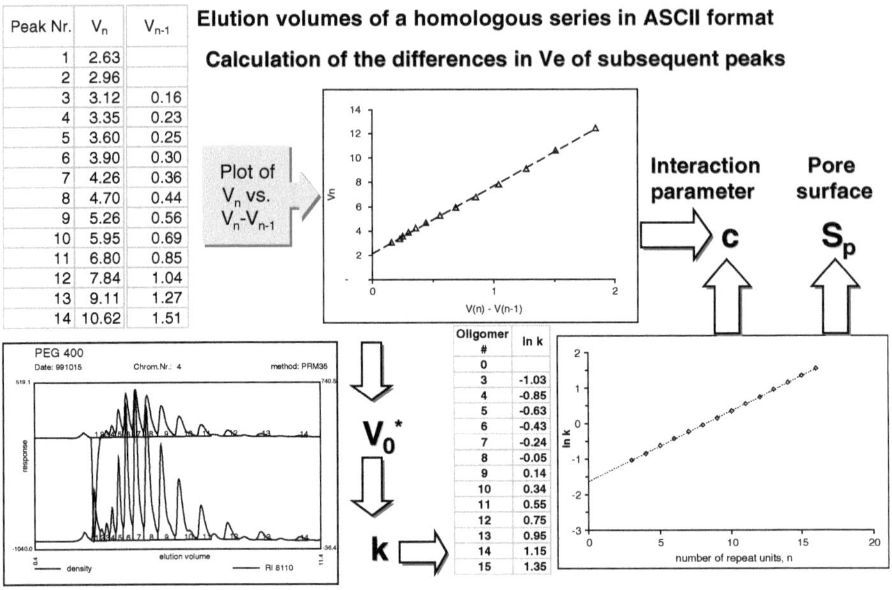

Fig. 2.1 Determination of the interaction parameter and the pore surface (schematically)

In logarithmic form this yields an equation, which corresponds to Martin's rule.

$$\ln k^* = \ln\left(\frac{4V_\mathrm{p}}{cDV_0^*}\right) + n\frac{c^2a^2}{6}. \tag{2.12}$$

In slit-like pores, the internal surface of the pores is given by the ratio of pore volume and pore diameter [17]:

$$S_\mathrm{p} = \frac{2V_\mathrm{p}}{D}. \tag{2.13}$$

For non-functional chains, Martin's rule can be written in the form

$$\ln k^* = \ln\frac{2S_\mathrm{p}}{cV_0^*} + n\frac{(ca)^2}{6}. \tag{2.14}$$

Using Eq. (2.9), the interaction parameter can be determined from a chromatogram of a polymer homologous series, in which a sufficient number of peaks is reasonably resolved. Once the interaction parameter is known, one may determine the pore surface from the intercept in Martin's plot. This requires, however, the identification of the peaks (i.e. the number of repeat units for each peak). The procedure is shown schematically in Fig. 2.1.

For monofunctional chains, an additional parameter q has to be taken into account [19, 26]. The specific endgroup parameter q measures the difference of the free energy of adsorption $\Delta\chi$ (in RT units) between endgroup and repeat unit (δ is the interaction distance between the specific group and the sorbent) [20, 21]. A simple procedure for the determination of q has been described previously by Trathnigg et al. [27].

In LAC, retention of monofunctionals with an adsorbing endgroup can be decribed by

$$\ln k_{\mathrm{m}}^{*} = \ln\left(\frac{4}{cD}\frac{V_p}{V_{0,\mathrm{m}}^{*}}(1+qc)\right) + n\frac{c^2a^2}{6}, \tag{2.15}$$

wherein $V_{0,\mathrm{m}}^{*}$ is the accessible volume for monofunctionals, which is somewhat smaller than the accessible volume for non-functionals:

$$V_{0,\mathrm{m}}^{*} = V_0^{*} - \frac{V_\mathrm{p}}{D}\frac{2}{\sqrt{\pi}}\frac{q}{Rc}. \tag{2.16}$$

In both cases, k increases exponentially with the number of repeat units. In a plot of ln k versus n straight lines with the same slope, but a different intercept are obtained. For samples with a narrow molar mass distribution (MMD) this may be done with isocratic elution, otherwise gradient elution has to be applied.

After all, LAC can be applied to determine the MMD of non-functionals and monofunctionals with a weakly adsorbing endgroup. If the interaction of the endgroup is too strong, no resolution of the individual oligomers can be achieved by LAC.

2.3 Liquid Chromatography at Critical Conditions

In liquid chromatography at critical conditions (or at the critical point of adsorption) (LCCC) [18, 19, 28–36] entropic and enthalpic terms compensate each other, and the interaction parameter equals zero. At the so-called critical adsorption point (CAP), which is defined by mobile phase composition and temperature, the distribution coefficient of non-functional molecules equals unity: $K_0 = 1$. Consequently, non-functional chains are eluted (regardless of their molar masses) at the void volume V_0.

For monofunctional molecules (with an adsorbing endgroup a) an additional term q_a appears, which describes the influence of the endgroup on retention. As has been discussed previously by Trathnigg and Gorbunov [20–23], the elution volume of monofunctionals with an adsorbing endgroup at the CAP is larger than the void volume, but still independent of molar mass.

$$V_a \approx V_\mathrm{i} + V_\mathrm{p}(1+q_a) \approx V_0 + q_a V_\mathrm{p}. \tag{2.17}$$

As the retention of monofunctionals at the CAP depends only on the magnitude of the endgroup parameter q_a, but not on their molar mass, they elute as narrow peaks. LCCC allows the separation of monofunctionals according to their functionality, regardless of their molar mass.

A very different behaviour is, however, observed with difunctional chains (with adsorbing groups at both ends) [21, 37–39]. The elution volume of symmetrical difunctionals (with the same adsorbing groups a at both ends) is given by

$$V_{aa} \approx V_{\mathrm{i}} + V_{\mathrm{p}}\left(1 + 2q_a + \frac{q_a^2}{\sqrt{\pi}} \cdot \frac{D}{R}\right), \tag{2.18}$$

which means, that they are separated according to the radius of gyration R of the critical block, which can be expressed by the length a and number n of the repeat units, see Eq. (2.3). A similar relation holds for asymmetrical difunctionals having different endgroups a and b:

$$V_{ab} \approx V_{\mathrm{i}} + V_{\mathrm{p}}\left(1 + q_a + q_b + \frac{q_a \cdot q_b}{\sqrt{\pi}} \frac{D}{R}\right). \tag{2.19}$$

Difunctionals elute much later than monofunctionals: the shift in their elution volumes is, however, not only the sum of the endgroup contributions, but contains an additional term, which depends on the ratio of the pore diameter D of the stationary phase and the radius of gyration R of the polymer molecules [21, 38]. Consequently, difunctionals with adsorbing endgroups elute in SEC order: smaller molecules elute later than larger ones.

A special situation occurs, if the endgroup is adsorbed, while the repeat units elute in SEC mode. This is utilized in a regime that could be named liquid exclusion-adsorption chromatography [22]. If the interaction parameter of the repeat unit c is negative and the one of the endgroup c_B positive, monofunctional chains elute in SEC order, but far behind the void volume. For small polymer chain lengths the elution volume V_{AB} of monofunctionals with the repeat unit A and the end group B is given by

$$V_{AB} \approx V_B\left(1 - \frac{\sqrt{\pi}}{2} c_B R_A\right) = V_B\left(1 - \tilde{C}\sqrt{n_A}\right) \tag{2.20}$$

Consequently, the elution volumes decrease with increasing radius of gyration (and hence with the square root of the number of repeat units). A molecule containing just the endgroups but no repeat units elutes at V_B. This is the method of choice in the separation of monofunctionals with a strongly adsorbing endgroup and an average degree of polymerization below 10–15. As it is performed under isocratic conditions, one may use refractive index (RI) detection, which allows an accurate quantification in the analysis of non-ionic surfactants [40].

2.4 Thermodynamics of Polymer Chromatography

As has already been briefly discussed, the distribution coefficient corresponds to the change in Gibbs free energy, which results from the changes in enthalpy and entropy, see Eq. (1.6). Consequently, the distribution coefficient results from an entropic and an enthalpic term, see Eq. (1.7). SEC is governed by the entropy change. LAC is generally considered as an enthalpic process, but in real LAC, entropic contributions must also be taken into account. In LCCC the changes in entropy and enthalpy compensate each other: at the critical adsorption point (CAP) is $\Delta G = 0$, as $\Delta H = T\Delta S$.

As follows from Eqs. (1.6) and (1.7), these mechanisms may show different temperature dependences. In ideal SEC, where enthalpic interactions should be absent, distribution coefficients should not depend on temperature. In LCCC, a temperature change will require a change in mobile phase composition to maintain critical conditions. In LAC, retention depends on both entropy and enthalpy changes. Consequently, the retention of a polymer in LAC on a given stationary phase depends on mobile phase composition and on temperature, but in a different way and to a different extent. The enthalpy change (which reflects the interaction of the polymer with the stationary phase) is generally assumed to be negative in LAC, and its absolute value should increase with decreasing eluent strength. Consequently, retention is expected to decrease with increasing temperature.

As has been shown previously [41–43], this needs not always be the case. On typical reversed phase columns, polyethylene glycols (PEG) show a completely different behaviour in different mobile phases: while the expected behaviour is observed in methanol–water, the opposite is found in acetonitrile–water, where retention increases with temperature. In acetone–water, retention is almost independent on temperature.

The changes in entropy and enthalpy can be determined from the intercept and slope in a plot of ln K vs. $1/T$ according to Eq. (1.7) (van't Hoff plot). In the literature, different approaches have been used in the determination of the thermodynamic parameters [41–44]. These approaches differ in the calculation of the distribution coefficient (or the retention factor, which is typically used instead of the distribution coefficient).

In many papers, a relation is applied, which is based on the assumption of a direct proportionality between the distribution coefficient K and the retention factor $k = (V_e - V_0)/V_0$:

$$\ln k = -\frac{\Delta H^{\circ}}{RT} + \frac{\Delta S^{\circ}}{R} + \ln \varphi. \tag{2.21}$$

In Eq. (2.21), the term φ is the so-called phase ratio (the volume ratio of the stationary to the mobile phase), which is generally unknown. In principle, these volumes should correspond to the pore volume and the interstitial volume. The slope and the intercept in a plot of ln k vs. $1/T$ represent the thermodynamic parameters $\Delta H^{\mathrm{o}}/R$ and $(\Delta S^{*}/R) = (\Delta S^{\mathrm{o}}/R) + \ln \varphi$. There is, however, an important difference between Eqs. (1.7) and (2.21), as becomes clear from the definitions of the retention factor k and the distribution coefficient K.

$$k = \frac{V_{\mathrm{e}} - V_0}{V_0}, \tag{2.22}$$

$$K = \frac{V_{\mathrm{e}} - V_{\mathrm{i}}}{V_{\mathrm{p}}} = \frac{V_{\mathrm{e}} - V_{\mathrm{i}}}{V_0 - V_{\mathrm{i}}}. \tag{2.23}$$

From Eqs. (2.22) and (2.23) follows Eq. (2.24):

$$k = (K - 1)\frac{V_{\mathrm{p}}}{V_0}. \tag{2.24}$$

Obviously, there is no direct proportionality between k and K, as is typically assumed. The assumption of $k = K \cdot \varphi$ is an approximation which holds only for $K \gg 1$. The exact relation contains the distribution coefficient K, but this requires the determination of the characteristic volumes V_{i}, V_{p} and V_0.

It must be mentioned, that the determination (and even the definition) of V_0 is not trivial. There have been numerous papers on the determination of the void volume, dead volume or hold-up volume of a column [9–11]. In the literature, different definitions are given, which may be relevant in different situations [7, 10, 45–54]. The void volume is the total amount of solvent in the column, which may be determined by gravimetry: $V_0 = V_{\mathrm{i}} + V_{\mathrm{p}}$. The hold-up volume is considered to be the elution volume of an unretained sample. Various methods for the determination of the hold-up volume have been proposed [9–11]. The interstitial volume V_{i} can be easily determined by inverse SEC (ISEC): V_{i} is the elution volume of a polymer, which is completely excluded from the pores (i.e. a polymer with a molar mass above the exclusion limit). This is, however, not as easy for the pore volume: as ISEC is typically performed in a good SEC solvent, large errors in V_{p} may result in other mobile phase compositions. As has been shown previously [11], the pore volume may be considerably smaller in mixed mobile phases, while the interstitial volume remains the same. Consequently, the correct pore volume in LAC will only be obtained, if ISEC is performed in the same mobile phase composition. This would require a set of polymer standards, which do not interact with the stationary phase under such conditions.

Polyethylene glycol (PEG) elutes in SEC mode on typical reversed phase columns in acetone–water mobile phases containing 50–80 wt% acetone [38]. In such mobile phases, polypropylene glycols or similar polymers elute in LAC mode. In methanol–water mobile phases this approach cannot be applied, as PEG does not elute in SEC mode at any composition.

The pore volumes of reversed phase columns in acetone–water may, however, also be assumed in methanol–water mobile phases [30, 41, 55]. This is, of course, an approximation, but still better than the usual approach. Obviously, the results obtained with Eqs. (1.7) and (2.21) will be different. The entropy change can only be determined using Eq. (1.7), as the other equations yield the apparent parameter ΔS^* [50, 56].

The correct relation is Eq. (1.7), as has already been discussed. The distribution coefficient characterizes the distribution between the flowing part of the mobile phase (the interstitial volume) and the stagnant volume, which comprises the bulk phase inside the pores and the volume of the adsorption layer at the internal surface of the pores. As the assumption of a dividing interface between the adsorption layer and the bulk phase is problematic [50, 56], there is also no defined volume of the stationary phase.

The interstitial volume can be easily determined, the crucial point is, however, the determination of the pore volume [11], which may be subject to various sources of error. In LAC, this problem can be solved by the following approach, which does not require the knowledge of the pore volume.

For two consecutive oligomers with n and $(n - 1)$ repeat units and the elution volumes V_n and V_{n-1} one may write

$$\ln K_n - \ln K_{n-1} = \ln \frac{V_n - V_\mathrm{i}}{V_{n-1} - V_\mathrm{i}} = -\frac{\Delta H_n - \Delta H_{n-1}}{RT} + \frac{\Delta S_n - \Delta S_{n-1}}{R}. \quad (2.25)$$

From the intercept and slope in a plot of $\ln (V_n - V_\mathrm{i})/(V_{n-1} - V_\mathrm{i})$ vs. $1/T$, the increments $(\Delta S)_R$ and $(\Delta H)_R$ in the entropy and enthalpy change are obtained, which represent the additional entropy or enthalpy change caused by addition of one more repeat unit to the chain. One may assume, that the total entropy or enthalpy change $(\Delta S)_n$ or $(\Delta H)_n$ of a chain with n repeat units results from the contributions of the endgroup(s) and the repeat units, $(\Delta S)_\mathrm{E}$ and $(\Delta H)_\mathrm{E}$, respectively. In this case, straight lines should be obtained in a plot of $(\Delta S)_n$ or $(\Delta H)_n$.

$$(\Delta S)_n = (\Delta S)_\mathrm{E} + n \cdot (\Delta S)_\mathrm{R}, \quad (2.26)$$

$$(\Delta H)_n = (\Delta H)_\mathrm{E} + n \cdot (\Delta H)_\mathrm{R}. \quad (2.27)$$

References

1. de Gennes PG (1969) Rep Prog Phys 32:187
2. Casassa EF (1967) J Polym Sci, Part B: Polym Phys 5:773
3. Casassa EF, Tagami Y (1969) Macromolecules 2:14
4. Martin AJP (1949) Biochem Soc Symp 3:4
5. Tchapla A, Heron S, Lesellier E, Colin H (1993) J Chromatogr A 656:81
6. Tchapla A, Colin H, Guiochon G (1984) Anal Chem 56:621
7. Krstulovic AM, Colin H, Guiochon G (1982) Anal Chem 54:2438
8. Colin H, Krustulovic AM, Gonnord MF, Guiochon G, Yun Z, Jandera P (1983) Chromatographia 17:9
9. Engelhardt H, Muller H, Dreyer B (1984) Chromatographia 19:240

10. Rimmer CA, Simmons CR, Dorsey JG (2002) J Chromatogr A 965:219
11. Trathnigg B, Veronik M, Gorbunov A (2006) J Chromatogr A 1104:238
12. García Domínguez JA, Díez-Masa JC, Davankov VA (2001) Pure Appl Chem 73:969
13. Escuder-Gilabert L, Bermudez-Saldana JM, Villanueva-Camanas RM, Medina-Hernandez MJ, Sagrado S (2004) J Chromatogr A 1033:247
14. Gritti F, Felinger A, Guiochon G (2006) J Chromatogr A 1136:57
15. Skvortsov A, Trathnigg B (2003) J Chromatogr A 1015:31
16. Trathnigg B, Skvortsov A (2006) J Chromatogr A 1127:117
17. Trathnigg B, Jamelnik O, Skvortsov A (2006) J Chromatogr A 1128:39
18. Skvortsov AM, Gorbunov AA (1990) J Chromatogr 507:487
19. Gorbunov AA, Vakhrushev AV (2004) Polymer 45:7303
20. Gorbunov AA, Skvortsov AM, Trathnigg B, Kollroser M, Parth M (1998) J Chromatogr A 798:187
21. Gorbunov AA, Trathnigg B (2002) J Chromatogr A 955:9
22. Trathnigg B, Gorbunov AA (2001) J Chromatogr A 910:207
23. Gorbunov AA, Skvortsov AM (1986) Vysokomol Soedin Ser A 28:2170
24. Skvortsov AM, Gorbunov AA (1986) J Chromatogr 358:77
25. Gorbunov AA, Vakhrushev AV, Trathnigg B (2009) J Chromatogr A 1216:8883
26. Skvortsov AM, Fleer GJ (2002) Macromolecules 35:8609
27. Nguyen Viet C, Trathnigg B (2010) J Sep Sci 33:464
28. Berek D (1996) Macromol Symp 110:33
29. Gorshkov AV, Much H, Becker H, Pasch H, Evreinov VV, Entelis SG (1990) J Chromatogr 523:91
30. Trathnigg B, Rappel C, Fraydl S, Gorbunov A (2005) J Chromatogr A 1085:253
31. Trathnigg B, Gorbunov A (2006) Macromol Symp 237:18
32. Gorshkov AV, Evreinov VV, Entelis SG (1983) Zh Fiz Khim 57:2665
33. Evreinov VV, Filatova NN, Gorshkov AV, Entelis SG (1995) Vysokomol Soedin 37: 2076
34. Pasch H, Rode K (1996) Macromol Chem Phys 197:2691
35. Radke W, Rode K, Gorshkov AV, Biela T (2005) Polymer 46:5456
36. Trathnigg B (2005) Polymer 46:9211
37. Trathnigg B, Rappel C, Raml R, Gorbunov A (2002) J Chromatogr A 953:89
38. Rappel C, Trathnigg B, Gorbunov A (2003) J Chromatogr A 984:29
39. Trathnigg B, Rappel C, Hödl R, Fraydl S (2003) Tenside Surfact Deterg 40:148
40. Trathnigg B (2001) J Chromatogr A 915:155
41. Trathnigg B, Veronik M (2005) J Chromatogr A 1091:110
42. Trathnigg B, Nguyen Viet C, Ahmed H (2009) J Sep Sci 32:2857
43. Cho D, Park S, Hong J, Chang T (2003) J Chromatogr A 986:191
44. Kim Y, Ahn S, Chang T (2009) Anal Chem 81:5902
45. Smith RJ, Nieass CS, Wainwright MS (1986) J Liq Chromatogr 9:1387
46. Kazakevich YV, McNair HM (1993) J Chromatogr Sci 31:317
47. Mockel HJ (1994) J Chromatogr A 675:13
48. Oumada FZ, Roses M, Bosch E (2000) Talanta 53:667
49. Kazakevich YV, McNair HM (2000) J Chromatogr A 872:49
50. Rustamov I, Farcas T, Ahmed F, Chan F, LoBrutto R, McNair HM, Kazakevich YV (2001) J Chromatogr A 913:49
51. Dominguez JAG, Diez-Masa JC, Davankov VA (2001) Pure Appl Chem 73:969
52. Gritti F, Guiochon G (2005) J Chromatogr A 1097:98
53. Gritti F, Kazakevich Y, Guiochon G (2007) J Chromatogr A 1161:157
54. Wang M, Mallette J, Parcher JF (2008) J Chromatogr A 1190:1
55. Trathnigg B, Fraydl S, Veronik M (2004) J Chromatogr A 1038:43
56. Kazakevich YV, LoBrutto R, Chan F, Patel T (2001) J Chromatogr A 913:75

3 Interactive Modes of Polymer Chromatography

Interaction chromatography is based on the retention of solute molecules by interaction with the surface of the stationary phase including the pore surface. This interaction can be due to adsorption, hydrophobic, polar or ionic interactions or dispersive forces. Intermittent capture and release of solute molecules by the stationary phase are controlled by two basically different mechanisms or some combinations thereof. With regard to adsorption-desorption phenomena, an abrupt process is the critical step leading to sorption or desorption. This process is typified by molecular desorption from surfaces where molecules can detach, and then do so suddenly, if they possess sufficient activation energy to cause the necessary rearrangement or rupture of chemical or physical bonding. Quite different in effect are the diffusion-controlled sorption-desorption kinetics where a change occurs only gradually as molecules diffuse in and out of localized regions [1].

The interactive modes of polymer chromatography are based on selective interactions with the surface of the stationary phase inside and outside the pores resulting in separations that are mainly based on the chemical composition and/or functionality of the analyte. In contrast, size exclusion chromatography (SEC) is based on the diffusion of macromolecules into pores of different sizes and as a result a size separation is obtained. Modes, experimental setups and applications of SEC have been extensively described. Here, the authors just refer to a number of standard publications [2–8].

The behaviour of polymer molecules in interactive modes of liquid chromatography (LC) deviates in several respects from the behaviour of low molar mass compounds. The differences are caused by the following properties of polymers [9]: small diffusion coefficients of macromolecules in solution, size of the macromolecules, which may be of the same magnitude as the pores of the packing, retention of polymers via "trains" of numerous repeat units, the flexibility of chain molecules, which enables conformational changes to occur, limited solubility of polymers [9, 10]. Very frequently in a first step, a part of the repeat units of the polymer chain is bound to the active groups of the stationary phase. If the adsorption energy per repeat unit is too low, the macromolecule is repulsed from the surface and will not adsorb. In this case most frequently a SEC separation mode is operating. When the adsorption energy

H. Pasch and B. Trathnigg, *Multidimensional HPLC of Polymers*, Springer Laboratory, DOI 10.1007/978-3-642-36080-0_3, © Springer-Verlag Berlin Heidelberg 2013

exceeds a certain limit, the macromolecule is adsorbed. If the change in adsorption enthalpy ΔH is higher than the entropy losses ΔS, related to the transformation from a three-dimensional coil to a two-dimensional adsorbed structure, the macromolecule changes its conformation and becomes fully adsorbed with all repeat units. The type of the final conformation of the macromolecule at the surface of the stationary phase depends on the flexibility of the polymer chain, its chemical composition, the molar mass and the number of adsorbed repeat units.

Different modes of interaction chromatography (IC) can be identified depending on the type of interactions with the stationary phase, the effect of enthalpic and entropic factors as well as the temperature. Under liquid adsorption and gradient chromatography all methods are summarized that are based on the adjustment of the interactions by changes in the mobile phase composition. The section on liquid chromatography at critical conditions (LCCC) summarizes all methods that are based on enthalpy-entropy compensation phenomena while the section on temperature gradient interaction chromatography (TGIC) describes the adjustment of interactions by temperature changes. The different modes of interaction chromatography have also been named "non-exclusion liquid chromatography" (NELC) by Mori to indicate that size exclusion effects in these cases are less important than in SEC [11].

3.1 Liquid Adsorption and Gradient Chromatography

Each type of interaction chromatography requires a mobile phase that comprises two or more solvents. In liquid adsorption chromatography (LAC) the initial and final mobile phases are good solvents for the polymer. In most cases silica gel is used as the stationary phase. The sample injected into the column adsorbs on the silica surface at the initial mobile phase. For very low molar mass samples the adsorption might be sufficiently weak and desorption in the same mobile phase composition may take place. In this case elution takes place under isocratic conditions and in some cases single mobile phases are used.

In most cases and, in particular, for higher molar mass samples adsorption is too strong. The strength of the mobile phase has a dramatic effect on retention. With a weak mobile phase, the retention time of a polymer usually far exceeds any reasonable period of experimental work. Increase of the elution strength will eventually lead to a sudden change to the opposite behaviour: the polymer is not retained at all and leaves the column at interstitial volume V_i. A small alteration of elution conditions causes transition from zero retention to infinity. This "on or off" behaviour can be understood as a consequence of multiple attachment [9]. Synthetic polymers consist of a large number of repeat units. In principle, all of them have the chance of becoming adsorbed. A polymer chain is retained in the stationary phase as long as at least one of its repeat units is adsorbed. A chain can migrate only if all constituting units are in the mobile phase. Assuming independent adsorption-desorption equilibrium for each unit the mobility condition of a macromolecule is a function of the corresponding probabilities of the repeat units and the chain length. For weak eluents and long polymer chains it can be assumed that there is always a repeat unit interacting with the packing material. Accordingly, the probability is very high that the macromolecule is retained for a very long time. Consequently, in most cases solvent gradients must be used to enhance the solvent strength of the mobile phase and to balance the adsorption-desorption behaviour of the macromolecules.

LAC has been applied widely for the separation of copolymers according to chemical composition; the first paper was published in 1979 by Teramachi et al. [12]. Mori and others developed separation methods for different styrene-acrylate copolymers while Mourey et al. focused on different polyacrylate combinations [13–17].

Another important aspect of adsorption chromatography of polymers is solubility. Solubility in general requires a negative change in Gibbs free energy on mixing. This is easily fulfilled with low-molar-mass solutes due to the large entropy contribution. In low-molecular mixing processes, the contribution $T\Delta S_m$ is so large that even positive values of ΔH_m do not prevent dissolution. In contrast to low molar mass compounds, macromolecules normally have low order in the solid state. In addition to this, the regular arrangement of the repeat units along the polymer chain remains on dissolution. Furthermore, in solutions of equal weight concentrations the number of solute molecules is much greater in low molar mass samples than in polymer systems. Thus, an entropy contribution to ΔG_m will be small and the condition $\Delta G_m < 0$ requires a negative change in enthalpy, $\Delta H_m < 0$. Accordingly, the solubility parameters of the solvent must be very close to that of the polymer. In addition to sufficient solubility of the macromolecules in the mobile phase, a certain solvent strength of the mobile phase is required. In view of the separation mechanism in adsorption chromatography, the solvent strength must be high enough to promote desorption of the macromolecules from the packing.

One type of liquid chromatography (LC) that makes use of the solubility of macromolecules of different chemical compositions is high performance precipitation liquid chromatography (HPPLC) [9]. In HPPLC, the initial mobile phase is a non-solvent for the sample so that the sample injected into the column will precipitate. Gradient elution is then performed by adding a good solvent to the sample and fractions start to re-dissolve and elute with the mobile phase. As solubility is a function of chemical composition, HPPLC can be used for chemical composition separation of copolymers and polymer blends. Examples of this kind of separation have been presented by Glöckner and others for styrene and acrylate copolymers [18–21].

Very frequently it is difficult to classify a separation technique as LAC or HPPLC since it is not completely clear if the sample fully precipitates in the initial mobile phase (different chemical compositions of the macromolecules). In such cases the separation can be classified as normal-phase (NP-LC) or reversed-phase chromatography (RP-LC). In both cases the initial and final mobile phases should be good solvents for the sample. In NP-LC a polar (hydrophilic) stationary phase is used. The initial mobile phase is less polar and the solvent strength is increased by adding a polar solvent. Typical examples are *n*-hexane/THF [22] and *n*-hexane/chloroform [23]. RP-LC uses non-polar (hydrophobic) stationary phases and a polar

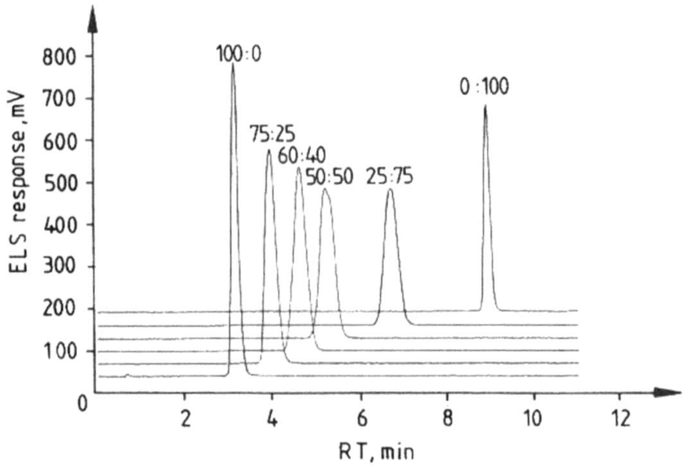

Fig. 3.1 Gradient HPLC chromatograms of random decyl methacrylate-methyl methacrylate copolymers, stationary phase: Nucleosil 5 CN, 200 × 4.6 mm i.d., mobile phase: isooctane-THF gradient, copolymer compositions are given as DMA/MMA in mol% (reprinted from [26] with permission of Springer Science + Business Media)

initial mobile phase that is modified in the course of the gradient with a less polar solvent. Typical examples for this case are ACN-THF [24] and ACN-methylene chloride [25].

A typical example for the separation of random copolymers by NP-LC is given in Fig. 3.1. Using a cyanopropyl-bonded silica gel as the stationary phase and a solvent gradient of isooctane-THF, copolymers of decyl methacrylate and methyl methacrylate were separated by composition [27].

In other examples, copolymers of styrene and methyl methacrylate were separated by composition in numerous eluents. Most of them represented normal-phase systems with gradients increasing in polarity and a polar stationary phase. Figure 3.2 shows the separation of a mixture of seven statistical poly(styrene-co-methyl methacrylate) samples on a silica column through a gradient of i-octane/(THF + 10% methanol) [19].

3.2 Liquid Chromatography at Critical Conditions

As has been pointed out in Sect. 3.1 both entropic and enthalpic interactions affect the chromatographic behaviour of macromolecules. They are adjusted to the required type of separation by selecting suitable stationary and mobile phases. In a specific mode of LC of polymers—LCCC—the adsorptive interactions are fully compensated by entropic interactions. $T\Delta S$ is equal ΔH and, therefore, ΔG becomes zero. K_d is 1 irrespective of molar mass and, consequently, homopolymer molecules of different molar masses co-elute in one chromatographic peak [19, 28–32].

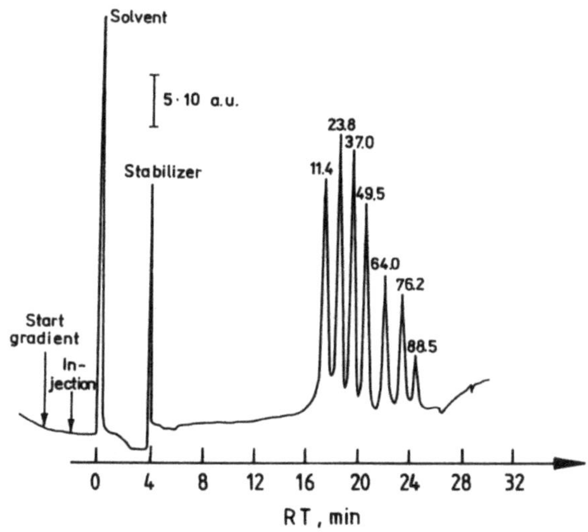

Fig. 3.2 Separation of a mixture of seven statistical poly(styrene-co-methyl methacrylate) samples by gradient HPLC, stationary phase: silica gel, mobile phase: i-octane-(THF + 10% MeOH), samples: mass% of MMA indicated (reprinted from [19] with permission of Elsevier)

The transition from one to another chromatographic separation mode can be accomplished by changing the temperature or the composition of the mobile phase. From an experimental point of view most frequently the composition of the mobile phase is adjusted rather than the temperature. However, as will be shown later it can be more advantageous to achieve a certain separation by adjusting the temperature.

To explain the experimental treatment of LCCC, the individual steps required to determine the critical point of polymethyl methacrylate are discussed. The critical point is usually obtained by investigating a number of samples of different molar masses, preferably calibration standards, in mobile phases of varying composition. The chromatographic system must be selected such that weak adsorptive interactions of the polymer chain with the stationary phase can be achieved. In the case of PMMA a polar stationary phase based on silica gel was selected.

Figure 3.3 shows the chromatographic behaviour of a number of PMMAs on a polar silica gel column using a mobile phase comprising mixtures of methyl ethyl ketone (MEK) and cyclohexane [33]. When the MEK concentration in the mobile phase was >73% by volume, the retention time decreased as the molar mass of the sample increased. Accordingly, retention corresponded to a size exclusion mode. In mobile phases, comprising a higher amount of cyclohexane, different retention behaviour was obtained. At a MEK concentration <73% by volume in the mobile phase, the retention time of the samples increased with increasing molar mass, indicating that the system was in the adsorption mode. At exactly 73% by volume of MEK in the mobile phase, the critical point of adsorption of PMMA was obtained. At this point all samples, regardless of their molar mass, eluted at the same retention time and a straight line parallel to the molar mass axis was obtained in the molar mass vs. retention time plot.

The fact that the molar mass effect on elution of a given homopolymer disappears has been named "chromatographic invisibility" to explain the LCCC behaviour of polymer blends and segmented copolymers. For diblock copolymers for example, chromatographic conditions can be found where one block elutes irrespective of

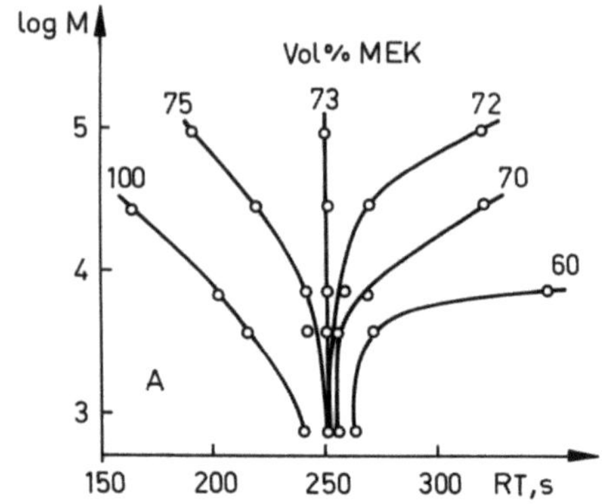

Fig. 3.3 Critical diagram molar mass vs. retention time of PMMA, stationary phase: Nucleosil Si-100, mobile phase: methyl ethyl ketone-cyclohexane (reprinted from [26] with permission of Springer Science + Business Media)

molar mass at LCCC conditions (the "invisible" block) while the other block elutes at SEC conditions (the "visible" block). Provided suitable detection and calibration procedures are used the molar mass of the "visible" block can be quantified in the diblock copolymer [19, 34–37]. Other applications of LCCC include the separation of homo- and copolymers according to functional endgroups, see the following chapters for more details.

A further refinement in the classification of enthalpy-entropy compensation phenomena has been proposed by Berek et al. Depending on the strength of the sample injection solvent relative to the mobile phase they proposed to use the terms "liquid chromatography at the critical adsorption point" (LC-CAP), "liquid chromatography under limiting conditions of adsorption" (LC-LCA), and "liquid chromatography under limiting conditions of desorption" (LC-LCD) [38–41].

3.3 Temperature Gradient Interaction Chromatography

As explained earlier, in addition to changing the mobile phase composition changes in the column temperature can be used to adjust chromatographic separations. It has been shown that the retention behaviour in interaction chromatography can be changed via temperature and even critical conditions of adsorption can be adjusted very delicately. As compared to changing the mobile phase composition the range of temperature variation is limited by the freezing and boiling points of the mobile phases [32].

A rough idea on the effect of temperature and mobile phase composition regarding the actual operating LC separation mechanism for a given stationary phase has been presented by Chang [32]. On a porous C_{18}-modified silica gel and dichloromethane-acetonitrile as the mobile phase all three modes of polymer LC were realized, see Fig. 3.4. It is clear from these experiments that for a given temperature (or mobile phase composition) SEC, LCCC and IC conditions can be obtained by changing the mobile phase composition (or temperature). This convincingly proves that all three modes are in agreement with a unified interaction mechanism of polymer liquid chromatography.

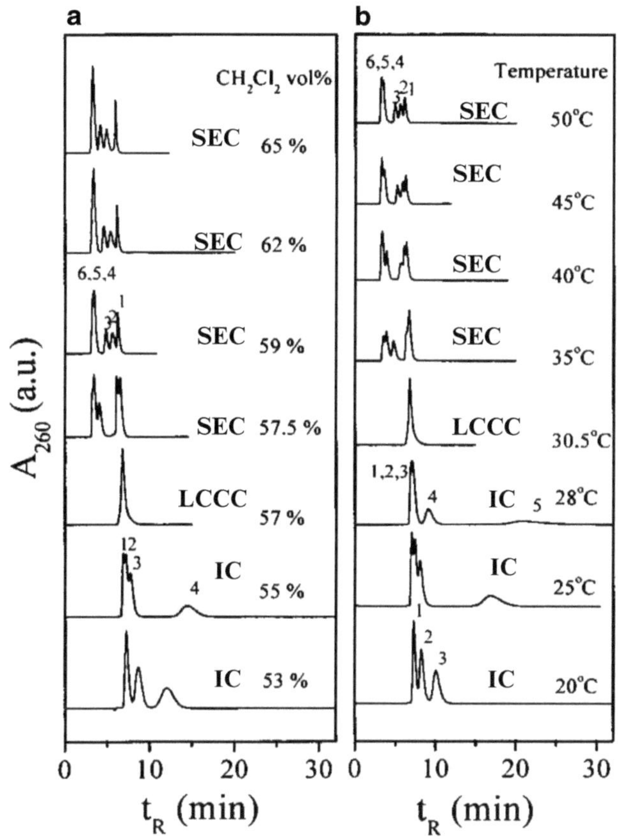

Fig. 3.4 Chromatograms of six polystyrene standards at different temperatures (**b**, dichloromethane-acetonitrile 57:43 v/v) and compositions of mobile phase (**a**, temperature 35 °C), operating modes: SEC, IC, LCCC, stationary phase: Nucleosil C18, 100 Å, mobile phase: dichloromethane-acetonitrile, sample molar masses: (1) 2.5, (2) 12, (3) 29, (4) 165, (5) 502, (6) 1,800 kg/mol (reprinted from [32] with permission of Springer Science + Business Media)

There are two major advantages of temperature gradient interaction chromatography (TGIC) over other comparable HPLC methods: as compared to SEC much higher molar mass resolution and lower band broadening can be achieved, as compared to solvent gradients higher resolution can be achieved with temperature gradients in selected applications.

The accurate analysis of molar mass distributions is a major topic in LC of polymers. Classically this is done by SEC, however, instrumental band broadening decreases the SEC resolution and is one of the disadvantages of MMD analysis. The application of TGIC for MMD analysis was first reported by Lee and Chang [42] and excellent separations were obtained for various polymers including polystyrene [43] and PMMA [44], see Fig. 3.5.

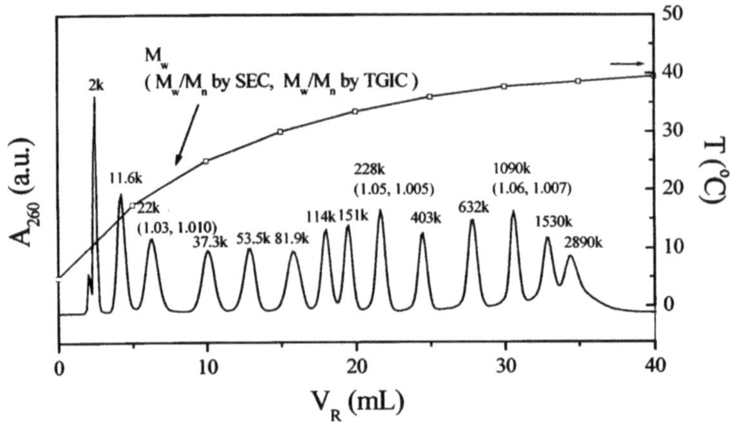

Fig. 3.5 TGIC separation of 14 PS standards, stationary phase: Nucleosil C18, mobile phase: methylene chloride—ACN, temperature program is shown (reprinted from [32] with permission of Springer Science + Business Media)

Using a similar approach for the MMD analysis of different polystyrenes and comparing these results with SEC data it was found that the TGIC data fit the calculated Poisson distributions much more closely due to the fact that in TGIC (different from SEC) instrumental band broadening was insignificant compared to MMD based peak dispersion [45]. Further results on the application of TGIC for the separation of segmented and branched copolymers and functional homopolymers have been published in a series of papers by Chang et al. [46–50]. They all indicate that TGIC is an interesting alternative to the conventional HPLC methods that are based on changes in the mobile phase composition.

References

1. Giddings JC (1965) Dynamics of chromatography. Marcel Dekker, New York, NY
2. Yau WW, Kirkland JJ, Bly DD (1979) Modern size exclusion chromatography. Practice of gel permeation and gel filtration chromatography. Wiley Interscience, New York, NY
3. Janca J (1984) Steric exclusion liquid chromatography. Dekker, New York, NY
4. Glöckner G (1987) Polymer characterization by liquid chromatography. Elsevier, Amsterdam
5. Mori S, Barth HG (1999) Size exclusion chromatography. Springer, Berlin
6. Striegel AM, Yau WW, Kirkland JJ, Bly DD (2009) Modern size-exclusion liquid chromatography. Practice of gel permeation and gel filtration chromatography. Wiley, Hoboken, NJ
7. Striegel AM (2000) In: Striegel AM (ed) Multiple detection in size exclusion chromatography. ACS Symposium Series 893, ACS, Washington, DC
8. Trathnigg B (2006) Size-exclusion chromatography of polymers. Encyclopedia of analytical chemistry, John Wiley & Sons, Chichester
9. Glöckner G (1991) Gradient HPLC and chromatographic cross-fractionation. Springer, Berlin
10. Glöckner G (1982) Polymercharakterisierung durch Fluessigchromatographie. Deutscher Verlag der Wissenschaften, Berlin
11. Mori S (1995) Size-exclusion chromatography and nonexclusion liquid chromatography for the characterization of styrene copolymers. In: Provder T, Barth HG, Urban MW (eds)

Chromatographic characterization of polymers. Hyphenated and multidimensional Techniques. Adv Chem Ser 247, ACS, Washington, DC
12. Teramachi S, Hasegawa A, Shima Y, Akatsuka M, Nakayama M (1979) Macromolecules 12:992
13. Mourey TH (1986) J Chromatogr 357:101
14. Mori S, Uno Y (1987) Anal Chem 59:90
15. Mori S, Mouri M (1989) Anal Chem 61:2171
16. Mori S (1991) J Chromatogr 541:375
17. Mori S (1990) Anal Chem 62:1902
18. Glöckner G, van den Berg JHM, Meijerink NLJ, Scholte TG, Koningsveld R (1984) Macromolecules 17:962
19. Glöckner G, van den Berg JHM (1986) J Chromatogr 352:511
20. Schulz R, Engelhardt H (1990) Chromatographia 29:325
21. Sparidans RW, Claessens HA, van Doremaele GHJ, van Herk AM (1990) J Chromatogr 508:319
22. Sato H, Takeuchi H, Tanaka Y (1986) Macromolecules 19:2613
23. Sato H, Takeuchi H, Suzuki S, Tanaka Y (1984) Macromol Chem Rapid Commun 5:719
24. Teramachi S, Hasegawa A, Motoyama K (1990) Polym J 22:480
25. Sato H, Mitsutani K, Shimizu I, Tanaka Y (1988) J Chromatogr 447:387
26. Pasch H, Trathnigg B (1998) HPLC of polymers. Springer, Berlin
27. Augenstein M, Mueller MA (1990) Makromol Chem 191:2151
28. Belenkii BG, Gankina ES, Tennikov MB, Vilenchik LZ (1978) J Chromatogr 147:99
29. Entelis SG, Evreinov VV, Gorshkov AV (1986) Adv Polym Sci 76:129
30. Entelis SG, Evreinov VV, Kuzaev AI (1985) Reactive oligomers. Khimiya, Moscow
31. Macko T, Hunkeler D (2003) Adv Polym Sci 163:61
32. Chang T (2003) Adv Polym Sci 163:1
33. Pasch H, Brinkmann C, Gallot Y (1993) Polymer 34:4100
34. Pasch H (1996) Macromol Symp 110:107
35. Pasch H, Esser E, Kloninger C, Hadjichristidis N (2001) Macromol Chem Phys 202:1424
36. Pasch H, Mequanint K, Adrian J (2002) e-Polymers No. 005
37. Mass V, Bellas V, Pasch H (2008) Macromol Chem Phys 209:2026
38. Hunkeler D, Janco M, Berek D (1996) In: Potschka M, Dubin PL (eds) Strategies in size exclusion chromatography. American Chemical Society, Washington, DC
39. Bartkowiak A, Hunkeler D (1999) In: Provder T (ed) Chromatography of polymers, hyphenated and multidimensional techniques. American Chemical Society, Washington, DC
40. Berek D (1998) Macromolecules 31:8517
41. Berek D (2001) Mater Res Innovat 4:365
42. Lee HC, Chang T (1996) Polymer 37:5747
43. Chang T, Lee W, Lee HC, Cho D, Park S (2002) Am Lab 34:39
44. Lee W, Lee HC, Chang T, Kim SB (1998) Macromolecules 31:344
45. Lee W, Lee H, Cha J, Chang T, Hanley KJ, Lodge TP (2000) Macromolecules 33:5111
46. Lee HC, Chang T, Harville S, Mays JW (1998) Macromolecules 31:690
47. Perny S, Allgaier J, Cho D, Lee W, Chang T (2001) Macromolecules 34:5408
48. Lee HC, Lee H, Lee W, Chang T, Roovers J (2000) Macromolecules 33:8119
49. Park S, Cho D, Ruy J, Kwon K, Lee W, Chang T (2002) Macromolecules 35:5974
50. Im K, Park S, Cho D, Chang T, Lee K, Choi N (2004) Anal Chem 76:2638

4 Equipment and Materials

A high performance liquid chromatography (HPLC) instrument must meet a number of challenging requirements to provide high-quality, reliable results. Typically, the design of a state-of-the-art HPLC instrument is such that all possible types of separations can be conducted. This requires a flexible pump to conduct isocratic and solvent gradient separations, a column oven for stable temperature conditions and a set of detectors that can provide quantitative information on a number of molecular parameters. Criteria that have to be met by any HPLC instrument are listed in Table 4.1.

4.1 Instrumental Setup

A general schematic of a SEC/HPLC system is presented in Fig. 4.1. The main components of any instrumental design are a solvent delivery system, a sample injection device, a set of detectors and a data acquisition and handling system. The core of any instrument is the column containing the stationary phase that is selected based on the required separation.

Very frequently solvent reservoirs are designed so that the mobile phase may be degassed automatically to prevent gas bubbles to form in the detector during separation. Elimination of oxygen by purging the mobile phase with inert gas is also required to prevent reactions with sensitive samples. Alternatively, mobile phases can be degassed by ultrasonication. In this case, however, re-dissolution of air may occur.

For solvent/mobile phase delivery the following types of pumps are typically used in HPLC:

1. *Syringe pumps* work like a large syringe, the plunger of which is actuated by a screw-feed drive (usually by a stepper motor), hence they deliver a completely pulseless flow.
2. *Reciprocating pumps* exist in various modifications:
   (a) *Single piston pumps* are cheap, but are generally not well suited for SEC/HPLC.
   (b) *Dual piston pumps* with parallel pistons deliver a smooth flow.

H. Pasch and B. Trathnigg, *Multidimensional HPLC of Polymers*, Springer Laboratory,
DOI 10.1007/978-3-642-36080-0_4,

Table 4.1 Important criteria for HPLC equipment (adapted from [1] with permission of J. Wiley & Sons)

Equipment design feature	Analytical accuracy	Retention reproducibility	Analysis speed	Separation versatility
Precise flow rate	X	X		
Temperature control	X	X		X
Precise sampling	X	X		
Stable detection	X			
High signal/noise ratio	X		X	X
Fast detection	X		X	
High-pressure pumping	X		X	X
Efficient columns	X		X	
Automatic data handling	X		X	
Low dead volume	X		X	
Flow rate sensing	X			
Range of column packings				X
Chemical resistance				X
Variety of detectors	X			X

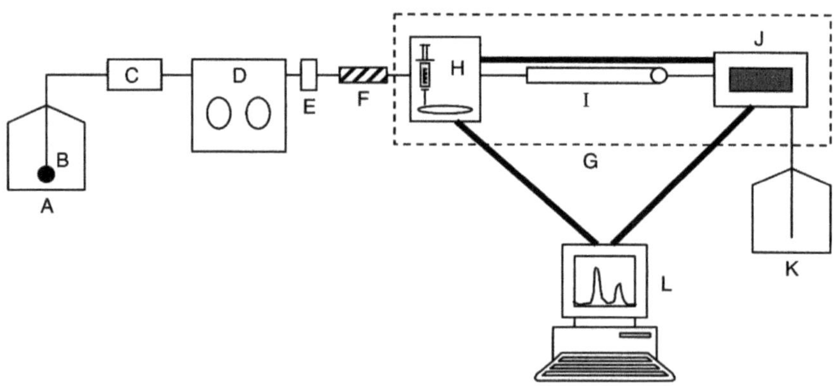

Fig. 4.1 Typical SEC/HPLC instrument equipped with inlet reservoir (A), inlet filter (B), degasser (C), pump (D), in-line filter (E), pulse dampener (F), column oven (G), sample injector and autosampler (H), chromatographic column (I), detector (J), waste reservoir (K), data system (L) (Reprinted from [1] with permission of J. Wiley & Sons)

(c) *Dual piston pumps* with pistons in series are easier to maintain, because they have only two check valves instead of four. The slightly higher pulsations, however, can be reduced by a pulse dampener to a level comparable to that of the parallel arrangement.

There are a number of general requirements for SEC/HPLC that include (1) a flow rate precision of better than 0.2%, (2) a pressure output of at least 6,000 psi, (3) a pulse-free or pulse-dampened output with pressure pulsations of less than 1% at

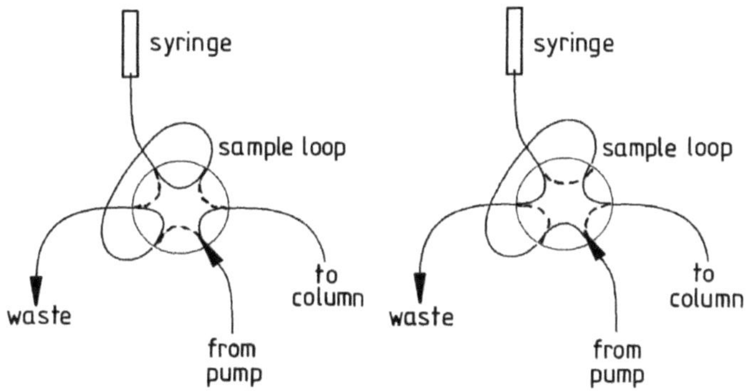

Fig. 4.2 Function of LC injection systems, load (*left*), inject (*right*) (reprinted from [2] with permission of Springer Science + Business Media)

1 mL/min, (4) flow rates in the range of 0.01–10 mL/min in either 0.1 or 0.01 mL/min increments, (5) chemical resistance to a wide range of mobile phases, (6) small hold-up volume for rapid solvent changes [1].

When selecting the pump for a given separation, one has to decide according to the separation technique: syringe pumps offer considerable advantages in SEC (flow stability) and LCCC (no evaporation of solvent, and thus absolutely constant composition of the mobile phase), but cause problems, when the mobile phase has to be changed. In gradient elution, only high-pressure mixing is possible with this type of pump, which makes the whole system more expensive, because it requires two pumps. On the other hand, reciprocating pumps allow gradient elution also with low-pressure mixing (i.e. with just one pump).

Typically, two-position six-port valves are used for sample injection, which may be operated manually or automatically (such as in an autosampler). The principle is shown schematically in Fig. 4.2.

For high precision, the loop should always be filled completely with the sample, and a sufficient amount of the sample solution (at least three volumes of the loop) should be used to rinse it thoroughly. The optimum size of the sample loop is determined by the column dimensions, the sensitivity of the detector(s), and the nature of the separation. In SEC of high molar mass samples, it is recommended to use a larger loop (50–100 μL) rather than a higher concentration of the sample solution. In LAC and LCCC, higher concentrations may be injected, and a smaller loop (10–50 μL) is preferred.

4.2 Stationary Phases

Depending on the mechanism of the separation, different types of stationary phases and dimensions of columns are used in liquid chromatography of polymers. As stationary phases for SEC are presented in detail in Refs. [1–3] and elsewhere, this

Table 4.2 Classification of HPLC columns with respect to size

Type	Column diameter (mm)	Column length (mm)	Particle size (μm)
Microbore	1–2	5	3
Narrow bore	2–3	5–10	3
Analytical	4–5	10–25	3–5
Semipreparative	<10	25	5–10
Preparative	>10	>25	>10

section will focus on columns for adsorption chromatography and liquid chromatography under critical conditions.

Unlike in SEC, where high separation efficiency can only be achieved with long columns (large volumes of the stationary phase), there is a trend towards the use of smaller columns in HPLC. As retention is determined by the distribution coefficients between stationary phase and mobile phase, the composition of the latter is the parameter governing the separation. Smaller columns mean faster analyses as well as lower solvent consumption. In analytical applications the diameter as well as the length of columns may be reduced. In practice, there are, however, limitations of miniaturization:

- The quality of the column packing (the plate height) determines the length of the column that is required to achieve the desired plate number and resolution.
- High efficiency (a lower plate height) can only be achieved with smaller particles, which can be packed more densely. Hence a higher back pressure will result for such a column, the magnitude of which is determined by the flow rate, the viscosity of the mobile phase, and the diameter of the column.
- The length and diameter of the connecting capillaries and the internal volume of the detector cell have also to be small to maintain the overall efficiency of the system.

A classification according to column dimensions is shown in Table 4.2. Microbore LC requires special injection systems, detectors and capillaries, while narrow bore systems can be operated frequently with normal equipment.

Basically, HPLC columns can be packed with different stationary phases, the nature of which is determined by the separation problem. Porous particles are used in SEC where the separation occurs exclusively by limited accessibility of the pores (see Chap. 2), and LAC because of their higher surface area. In LCCC, where both effects (exclusion and adsorption) compensate each other, the pore size distribution is also highly important. In the last years, nonporous phases with a particle diameter of 1.5 μm have been introduced, which may reduce analysis time and solvent consumption in LAC dramatically. These packings are available in column sizes of 3.3–5 cm length and a diameter of 4.6 mm.

From a chemical point of view, most stationary phases are based on silica or on cross-linked organic polymers. For special applications, alumina can also be used as a matrix. A common classification refers to the polarity of the stationary and mobile phases: in normal-phase separations the stationary phase is more polar than the mobile phase, and the opposite is true for reversed-phase LC. Some typical

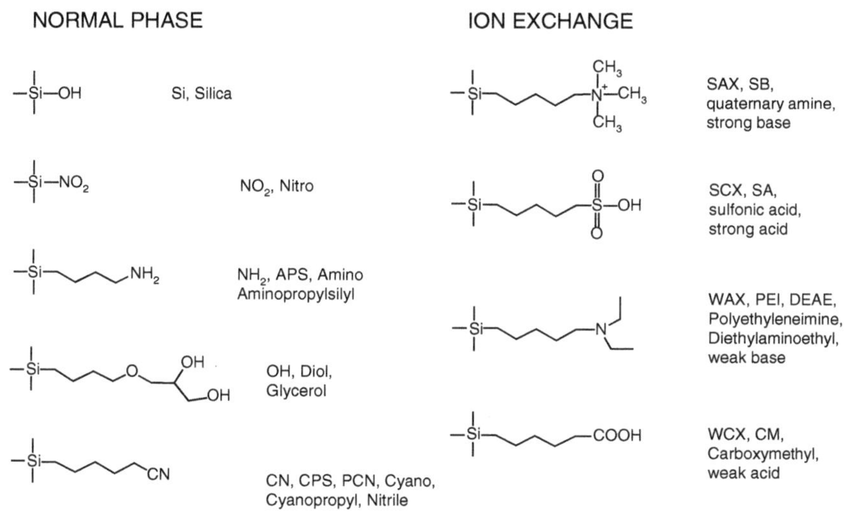

Fig. 4.3 Typical polar stationary phases for liquid chromatography (reprinted from [2] with permission of Springer Science + Business Media)

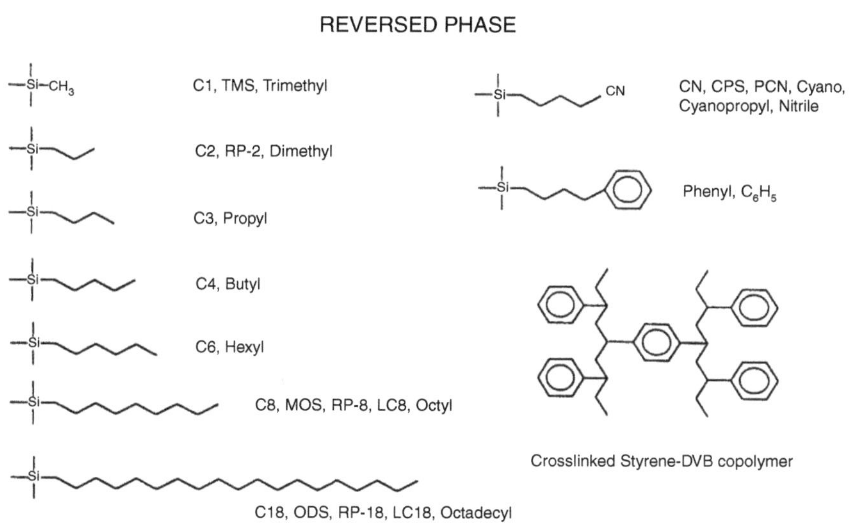

Fig. 4.4 Typical nonpolar stationary phases for liquid chromatography (reprinted from [2] with permission of Springer Science + Business Media)

polar stationary phases, as they are used in normal-phase or in ion-exchange LC, are shown in Fig. 4.3. Some typical nonpolar stationary phases are given in Fig. 4.4.

In reversed phase LC, stationary phases based on modified silica are most frequently used, which are typically obtained by reacting the silica with alkylsilanes. As this modification is seldom complete, the residual silanol groups may affect the separation, especially in the analysis of basic compounds, such as amines. Most producers offer packings with high carbon load and a high degree of end-capping, which should minimize these effects. The safest way, however, is the use of polymer-based packings (mainly crosslinked styrene-divinylbenzene copolymers). Note that the nitrile phase can be used either in normal or in reversed phase chromatography, depending on the polarity of the mobile phase.

4.3 Mobile Phases

The mobile phase is chosen primarily based on sample solubility and complex interactions between the sample, the stationary phase and the solvent. With macromolecules, diffusion is slow, and the need to maintain high column efficiency by using small particles and low-viscosity solvents is another important consideration.

Solvents can be classified by their chemical nature and by their polarity. The latter is the most important criterion in interaction chromatography (LAC, LCCC). With regard to polarity, solvents can be classified in terms of 'eluotropic series'. There are varying values reported for solvent polarity by different sources, but the order is in most cases roughly the same. Some typical HPLC solvents/mobile phases are listed in Table 4.3. It is obvious, that the polarity may vary considerably within each class of solvents even for chemically quite similar solvents.

Another limitation in mobile phase selection concerns the mode of detection. If a UV detector is used, one has to take into account the absorption of the solvent (UV-cutoff). When a chromatographic separation is to be coupled to FTIR or NMR spectroscopy, the spectroscopic behaviour of the mobile phase must also be considered. Typical solvents used in LC of polymers are presented in Table 4.3 together with their polarity indices and the UV-cutoff.

Another measure for polymer solubility is the comparison of the solubility parameters δ of the polymer with those of different solvents. When the values are close to each other then there is a good chance that the polymer is soluble in the given solvent. A refinement of this approach is the consideration of different types of polymer-solvent interactions such as dispersive (δ_d), polar (δ_p) and hydrogen-bonding (δ_h) interactions. It is helpful to compare these different types of interactions to evaluate polymer solubility [4–7]. Tables 4.4 and 4.5 give an overview of different solvents and polymers.

4.4 Detectors

When the separated polymer molecules leave the column, they have to be detected by one or more detectors, the signal of which must represent the concentration of the polymer with good accuracy. In the analysis of polymers by SEC/HPLC most of the detectors that are used for low molar mass analytes can be used. However, specific requirements and approaches must be considered that result from the

Table 4.3 Typical solvents used in liquid chromatography of polymers

Class	Solvent	Polarity index	UV cutoff
Alkanes	Hexane, heptane	0.0	200
	Cyclohexane	0.2	200
Aromatics	Benzene	2.7	280
	Toluene	2.4	285
	Xylene	2.5	290
Ethers	Diisopropyl ether	2.2	220
	Methyl-t-butyl ether	2.5	210
	Tetrahydrofuran	4.0	215
	Dioxane	4.8	215
Alkyl halides	Tetrachloromethane	1.6	263
	Dichloromethane	2.5	235
	Dichloroethane	4.0	225
	Chloroform	4.8	245
Esters	Butyl acetate	4.0	254
	Ethyl acetate	4.4	260
Ketones	Methyl ethyl ketone	4.7	329
	Acetone	5.1	330
Alcohols	*n*-Butanol	3.9	215
	i-Propanol	3.9	210
	n-Propanol	4.0	210
	Methanol	5.1	205
Nitriles	Acetonitrile	5.8	190
Amides	Dimethyl formamide	6.4	268
Carboxylic acids	Acetic acid	6.2	230
Water		9.0	200

Table 4.4 Solubility parameters of solvents [in $(J/cm^3)^{1/2}$]

Solvent	δ_s	δ_d	δ_p	δ_h
Acetone	20.0	15.5	10.4	7.0
Benzene	21.3	18.7	8.6	5.3
Chloroform	18.8	17.7	3.1	5.7
Cyclohexane	16.7	16.7	0	0
DMSO	26.5	18.4	16.4	10.2
Dioxane	20.5	19.0	1.8	7.4
Ethanol	26.4	15.8	8.8	19.4
Formamide	36.2	17.2	26.2	19.0
Pyridine	21.7	18.9	8.8	5.9
Water	48.1	12.3	31.3	34.2

Table 4.5 Solubility parameters of polymers [in $(J/cm^3)^{1/2}$]

Polymer	δ_m	δ_d	δ_p	δ_h
Polyethylene	15.8–17.1			
Polypropylene	16.8–18.8			
Polyisobutylene	16.0–16.6	16.0	2.0	7.2
Polystyrene	17.4–19.0	17.6	6.1	4.1
Poly(vinyl chloride)	19.2–22.1	19.2	9.2	7.2
Poly(vinyl acetate)	22.5–24.1	19.0	10.2	8.2
Poly(methyl methacrylate)	22.0–25.0	18.8	10.2	8.6
Poly(propylene oxide)	15.4–20.3			

specific nature of very large molecules. Basically, one has to distinguish between the following groups of detectors:

Concentration sensitive detectors		
Selective detectors	Universal detectors	Molar mass sensitive detectors
UV detector	RI detector	Single capillary viscometer
IR detector	Conductivity detector	Differential viscometer
Fluorescence detector	Density detector	Light scattering detectors: LALLS, MALLS, RALLS
Electrochemical detector	Evaporative light scattering detector	

In addition, there are spectroscopic detectors that provide direct chemical composition information.

4.4.1 Concentration Sensitive Detectors

The signal (response) of a concentration sensitive detector is determined by the concentration of the solute in the mobile phase. Among the concentration sensitive detectors, one has to distinguish detectors measuring a property of the solute and detectors measuring a (bulk) property of the mobile phase. The first group can thus be regarded as selective, the second one as universal (even though this is not a general rule). In the analysis of copolymers, two detectors may be required to monitor the concentrations of the two comonomers.

In HPLC, various *selective detectors* are available, but not all of them can be applied to polymers. Among the photometric detectors, the UV detector is the most frequently used instrument. IR detectors are very useful, but limited to certain mobile phases, which do not absorb at the detection wavelength. A good alternative is off-line coupling of FTIR with HPLC using an evaporative interface, which provides valuable qualitative information. In such a system, the eluate is sprayed onto a Germanium disk, which can be transferred to any FTIR spectrometer to yield the full spectral information over any peak of the chromatogram. It must, however, be mentioned, that such a device should be combined with an additional concentration detector, because quantification may be problematic sometimes. Fluorescence detection cannot be applied to most polymers, and the same is true for electrochemical detection. In the case of oligomers, these detectors may be used after derivatization of the end groups, which is, however, typically not feasible in chromatography of high polymers.

4.4.1.1 UV Detector

The most familiar solute property detector is the UV absorption detector, which exists in different modifications and is available from most producers of HPLC instruments. This detector measures the absorption of light of a selected wavelength and can be applied to polymers containing chromophoric groups. Typical detection wavelengths are in the range of 180–350 nm, however, UV can be utilized only in solvents with a sufficiently low absorbance, see Table 4.3. Many typical solvents for polymers (aromatics, esters, ketones, DMF, $CHCl_3$ etc.) do not allow UV detection at low wavelengths (<250 nm). In ethers the UV-cutoff may be dramatically influenced by contaminations, which are formed by oxidation, such as peroxides in THF etc., or by stabilizers.

Basically, one has to distinguish between three types of UV detectors, which differ in the way monochromatic light is obtained: fixed wavelength detectors, variable wavelength detectors, diode-array detectors (DAD). In most fixed wavelength detectors, a low-pressure mercury lamp emitting at 254 nm wavelength is used as the light source. In variable wavelength detectors, monochromatic light is obtained by means of a holographic grating. In some instruments, wavelength programming is also provided. DADs allow simultaneous measurement of an entire UV spectrum at any point of the chromatogram. The main difference between classical variable wavelength detectors and the DAD is the arrangement of monochromator and sample cell: in the DAD, the monochromator is placed in the light beam behind the sample cell ("inverse optics").

There are four important types of *universal detectors*, which are applied to HPLC: refractive index (RI) detector, conductivity detector, density detector, evaporative detectors. Even though universal detectors are in general less sensitive than selective detectors, they are applied to the analysis of many polymer samples, such as polyolefins, aliphatic polyethers or the like, which cannot be detected by a selective detector. The most familiar instrument in this group is the RI detector, which exists in various modifications. The conductivity detector is not a common option in polymer chromatography. The density detector (operating according to the mechanical oscillator principle) is very useful in polymer analysis, especially in combination with other detectors. Evaporative detectors vaporize the mobile phase and the nonvolatile components of the sample can be detected by measuring the scattering of a transversal light beam, as is the case in the evaporative light scattering detector (ELSD). It is also possible to use such an evaporation device as an interface to a mass spectrometer or an FTIR spectrometer. In LC of polymers, only the ELSD is used as a routine instrument.

4.4.1.2 RI Detector

There are basically three types of RI detectors, the most common one being the deflection refractometer. Deflection refractometers have a better linear range than Fresnel refractometers but have larger measuring cells. Interferometric refractometers

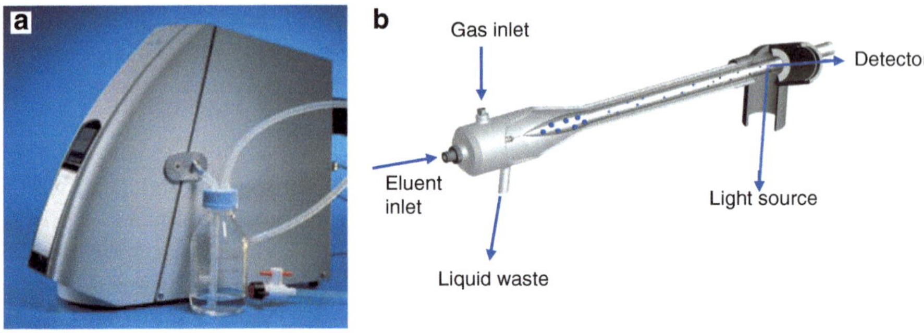

Fig. 4.5 PL-ELS 1000 detector with the solvent condensation device allowing recovery of material from the nebulizer drain and the solvent vapor from the exhaust (**a**) and scheme of the evaporator tube (**b**) (Courtesy of Varian/Polymer Labs)

are more sensitive (by 1 order of magnitude) than the other RI detectors. It must be mentioned, that response factors of the RI detector depend on molar mass and on chemical composition, as is also the case for the density detector. While molar mass dependence can be compensated rather easily, the use of a second concentration detector is inevitable in the analysis of copolymers or polymer blends. Moreover, preferential solvation of the polymer coils by one component of the mobile phase may affect detector signals [8, 9], which can also require a second concentration detector.

4.4.1.3 Density Detector

The density detector provides additional information in isocratic elution experiments, when combined with a UV or RI detector. This instrument utilizes the mechanical oscillator principle [10–12]. Its measuring cell is an oscillating U-shaped capillary the period of which depends on the density of its content. Period measurement is performed by counting the periods of an oven-controlled 10 MHz quartz during a predetermined number of periods of the measuring cell. The signal of such a detector is thus inherently digital, which means that no A/D-converter is required in data acquisition, and its response is integrated over each measuring interval.

4.4.1.4 Evaporative Light Scattering Detector

Although the ELSD measures a property of the solute, it can be regarded as universal detector, because it detects any non-volatile components of a sample [13–15]. While there are lots of UV detectors in the market, only a few producers offer this type of detector. In such an instrument, the eluate is nebulized and the solvent evaporated from the droplets. Each droplet containing non-volatile material will form a particle. When the aerosol thus obtained crosses a light beam, the light will be scattered. This effect can be utilized for detection, even though there are some problems in quantification of the signal due to the non-linearity of the detector response (Fig. 4.5).

Even though such an instrument is very useful in interaction chromatography because it allows solvent gradient elution (provided that no non-volatile buffers have to be used), quantification is a challenge. The dependence of the detector response on molar mass and chemical composition of the sample as well as mobile phase composition has to be addressed. Moreover, the size of the droplets formed in the nebulizer depends on operation conditions of the instrument. Hence some additional parameters influence the sensitivity, such as oven temperature, flow rate of carrier gas and eluate, viscosity and surface tension of the latter (which will of course change when a surface active substance is eluted, and also in gradient elution) etc. This means, that an ELSD must be calibrated very carefully in order to yield reliable results, as has been shown by several authors [16–18].

4.4.2 Molar Mass Sensitive Detectors

Molar mass sensitive detectors are particularly useful in SEC because they provide the molar mass of each fraction of a polymer peak. Since the response of such a detector depends on both concentration and molar mass of the fraction, it has to be combined with a concentration sensitive detector. For experiments in interaction chromatography these detectors are used to a much lesser extent because (A) the separation occurs mainly regarding chemical composition and (B) frequently solvent gradients are used that make it difficult to use these detectors.

The following types of molar mass sensitive detectors are commonly used [19–22]: differential viscosimeters, low angle laser light scattering detectors (LALLS), right angle laser light scattering detectors (RALLS), multi-angle laser light scattering detectors (MALLS). Since viscosity and light scattering yield different information, it makes sense to combine both of them [23]. From light scattering detection, the absolute molar mass distribution (MMD) can be determined directly. Using more than one angle, one may also obtain the radius of gyration. On the other hand, SEC with viscosity detection yields the intrinsic viscosity distribution (IVD). In this case, the MMD is, however, determined indirectly and is thus subject to retention errors. Hence, a combination of a concentration detector with both a light scattering detector and a viscosity detector provides the highest reliability of results. Moreover, information on branching can be obtained in this way [24, 25].

4.4.2.1 Viscosity Detector [1, 26–32]

The viscosity of a polymer solution is closely related to the molar mass (and architecture) of the polymer molecules. The viscosity of a dilute polymer solution depends on the nature of polymer and solvent, the concentration of the polymer, its average molar mass and molar mass distribution, the temperature and the rate of deformation. In the following discussion it is assumed that the rate of deformation is so low, that its influence can be neglected.

The ratio of the viscosity of a solution η to the viscosity of the pure solvent η_0 is called the viscosity ratio or the relative viscosity η_{rel}.

$$\eta_{rel} = \eta/\eta_0. \tag{4.1}$$

The relative increase of the viscosity of a polymer solution to that of the solvent is called the specific viscosity η_{sp}.

$$\eta_{sp} = (\eta - \eta_0)/\eta_0 = \eta_{rel} - 1. \tag{4.2}$$

This quantity is divided by the concentration c to obtain the reduced viscosity η_{sp}/c which expresses the average contribution of the solute molecules at concentration c to the viscosity. The limiting reduced viscosity or the Staudinger index $[\eta]$ which is also called the *intrinsic viscosity* is the value of the reduced viscosity at infinite dilution, i.e.

$$[\eta] = \lim \eta_{sp}/c = \lim[(\eta - \eta_0)/\eta_0\ c]. \tag{4.3}$$

The viscosity of polymer solutions, especially high molar mass polymers, is often appreciably dependent on the rate of shear in the range of measurement. The intrinsic viscosity should, therefore, be given as the limiting value of η_{sp}/c not only at infinite dilution but also at a shear rate of zero, or the value of the rate of shear should be specified.

The intrinsic viscosity is a measure of the capacity of a polymer molecule to enhance viscosity, which depends on the size and the shape of the polymer molecule. Within a given series of polymer homologues, $[\eta]$ increases with M; hence it is a measure of M. The empirical Mark-Houwink-Sakurada equation describes the viscosity-molar mass relationship for Gaussian coils:

$$[\eta] = K\ M^a \tag{4.4}$$

where K and a are coefficients that are characteristic of the polymer, solvent, and temperature.

Viscosity measurements as an online tool in SEC can be performed by measuring the pressure drop P across a capillary, which is proportional to the viscosity η of the flowing liquid (the viscosity of the pure mobile phase is denoted as η_0). The relevant parameter is, however, the intrinsic viscosity $[\eta]$, see Eq. (4.3). When a polymer passes the capillary, the pressure drop is increased by ΔP. In viscosity detection, one has to determine the viscosity η of the sample solution as well as the viscosity η_0 of the pure mobile phase. The specific viscosity $\eta_{sp} = \Delta\eta/\eta$ is obtained from $\Delta P/P$. As the concentrations in SEC are typically very low, $[\eta]$ can be approximated by η_{sp}/c.

The design of various viscometric detectors is presented in Fig. 4.6. The simple approach using one capillary (see Fig. 4.6a) and one differential pressure transducer will not work very well, because the viscosity changes $\Delta\eta = \eta - \eta_0$ will typically be very small compared to η_0, which means, that one has to measure a very small change of a large signal. Moreover, flow-rate fluctuations due to pulsations of a

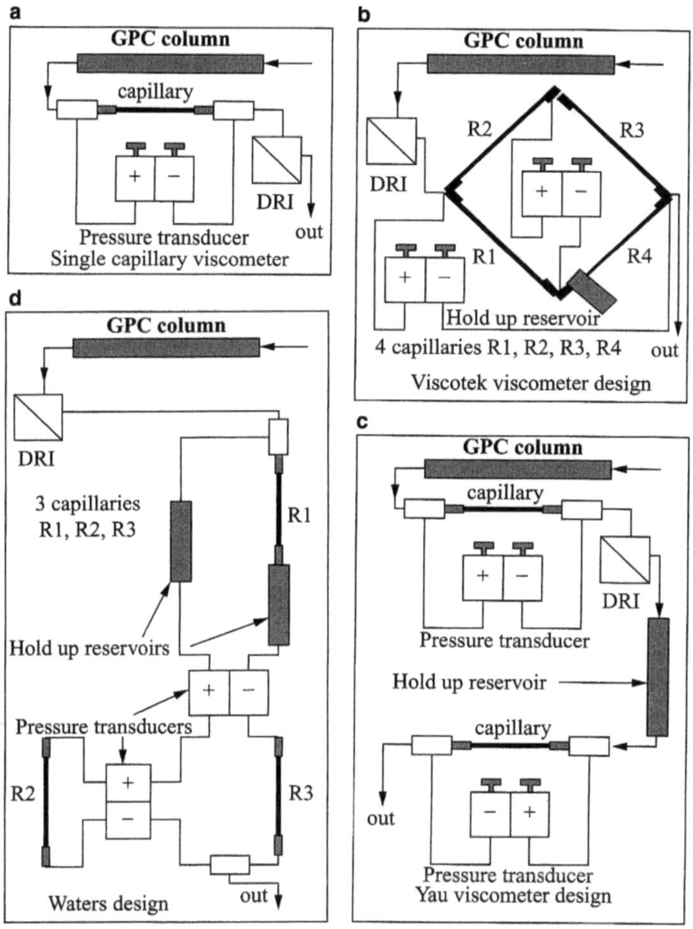

Fig. 4.6 Design of various viscometric detectors (reprinted from [33] with permission of Marcel Dekker Inc)

reciprocating pump will lead to much greater pressure differences than the change in viscosity due to the eluted polymer. Instruments of this type can only be used with a syringe pump.

A significantly better approach is the use of two capillaries in series, each of which is connected to a differential pressure transducer and a sufficiently large holdup reservoir in between, as is shown in Fig. 4.6c. With this approach, one measures the sample viscosity η from the pressure drop across the first capillary, and the solvent viscosity η_0 from the pressure drop across the second capillary. Pulsations are eliminated in this setup, because they appear in both transducers simultaneously. Another design is the differential viscometer in which four capillaries are arranged similar to a Wheatstone bridge, see Fig. 4.6b. In this 'bridge' design, a holdup reservoir in front of the reference capillary (R4) ensures that only pure mobile phase flows through the reference capillary, when the peak passes the sample capillary (R3). This design offers considerable advantages: the detector measures actually the pressure difference ΔP at the differential pressure transducer (DP) between the inlets of the sample capillary and the reference capillary, which have a common outlet and the overall pressure P at the inlet of the bridge. The specific viscosity $\eta_{sp} = \Delta\eta/\eta_0$ is thus obtained from $\Delta P/P$. One concern with this type of detector is that the flow must be divided 1:1 between both arms of the bridge. This is achieved by capillaries R1 and R2, which must have a sufficiently high back pressure. Nevertheless, when a peak passes through the sample capillary, a slight deviation of the 1:1 ratio might be observed.

4.4.2.2 Light Scattering Detector

In a light scattering detector, the scattered light of a laser beam passing through the cell is measured at angles different from zero. The (excess) intensity $R(\theta)$ of the scattered light at the angle θ is related to the weight-average of molar mass M_w:

$$\frac{K^*c}{R(\theta)} = \frac{1}{M_w P(\theta)} + 2A_2c, \tag{4.5}$$

wherein c is the concentration of the polymer, A_2 is the second virial coefficient, and $P(\theta)$ describes the scattered light angular dependence. K^* is an optical constant containing Avogadro's number N_A, the wavelength λ_0, the refractive index n of the solvent, and the refractive index increment dn/dc of the sample:

$$K^* = 4\pi^2(\mathrm{d}n/\mathrm{d}c)^2/(\lambda^4 N_A). \tag{4.6}$$

In a plot of $K^*c/R(\theta)$ versus $\sin^2(\theta/2)$, M_w can be obtained from the intercept and the radius of gyration from the slope. A multi-angle measurement provides additional information. The accuracy of the results strongly depends on the method of data treatment [34].

A typical experimental SEC-MALLS setup is shown in Fig. 5.4. The design of a multi-angle laser light scattering detector [35] and the experimental result from a SEC-MALLS measurement are presented in Fig. 4.7. The arrangement of photodiodes at different observation angles can be seen clearly. The MALLS detector is coupled in series to a concentration detector. Therefore, the delay volumes between the detectors must be determined with very high precision.

Product specifications of various commercially available light scattering (LS) detectors are summarized in Ref. [1].

The analysis of a polymer sample by SEC-MALLS-RI is presented in Fig. 4.8. It shows the RI and LS (angle 90°) detector readings and the angular dependence of the MALLS detector reading. The extrapolation to an angle of 0° provides the weight-average molar mass M_w and the slope of the extrapolation curve provides the radius of gyration R_g.

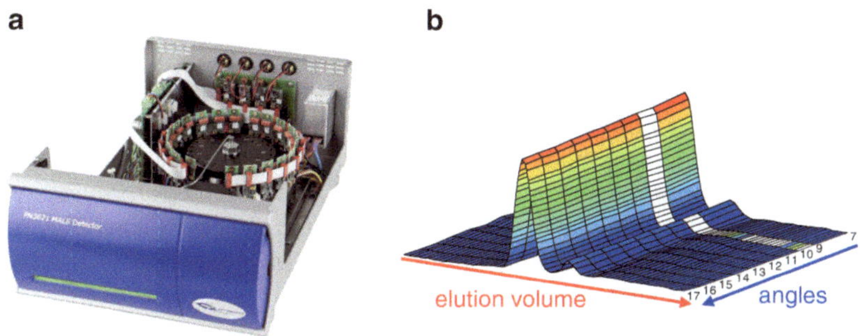

Fig. 4.7 Design of a MALLS detector (**a**) and typical experimental result from a multi-angle measurement showing the MALLS detector reading at different angles (**b**)

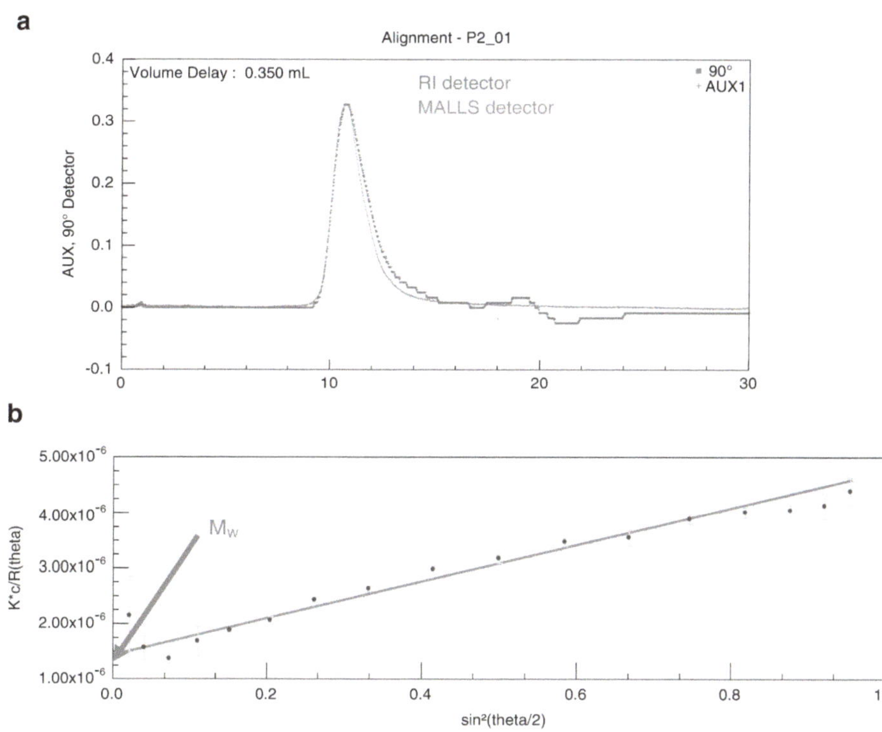

Fig. 4.8 SEC-MALLS-RI analysis of a polymer (**a**) and the angular dependence plot from the MALLS detector reading (**b**)

An important point relevant to all SEC-MALLS experiments is the fact that the refractive index increment dn/dc must be known. For a number of polymers these values are tabulated for different polymer-solvent combinations. If dn/dc is not known it must be determined experimentally. This is typically done with a DRI detector [1]. Selected dn/dc values for some of the common polymers are given in Table 5.3.

4.4.3 Spectroscopic Detectors

Accurate molar mass analysis can only be conducted when the chemical composition of the analyte is known. For homopolymers this information can be obtained easily by spectroscopic methods such as FTIR and NMR. For chemically heterogeneous polymers (copolymers, polymer blends) these methods provide average information with no details about the chemical heterogeneity. Generally, it is not possible by FTIR or NMR to define a sample as being a copolymer, or a blend of homopolymers or a blend of homo- and copolymers.

For the analysis of the chemical heterogeneity of a complex polymer, a separation step is strictly required that is mainly based on chemical composition. As has been discussed before there is a variety of interactive modes of polymer chromatography in place that can be used for chemical composition, functionality or microstructure fractionation.

The separation according to chemical composition is preferably monitored not only by a concentration detector but also by a spectroscopic detector that can provide direct chemical composition information on chromatographic fractions. The most common spectroscopic methods that are used as chemistry-selective detectors are FTIR, NMR and mass spectrometry. The fundamentals and experimental approaches of these detector couplings are discussed in detail in Chap. 7.

References

1. Striegel AM, Yau WW, Kirkland JJ, Bly DD (2009) Modern size-exclusion liquid chromatography. Practice of gel permeation and gel filtration chromatography. Wiley, Hoboken, NJ
2. Pasch H, Trathnigg B (1998) HPLC of polymers. Springer, Berlin
3. Mori S, Barth HG (1999) Size exclusion chromatography. Springer, Berlin
4. Giddings JC (1991) Unified separation science. Wiley-Interscience, New York, NY
5. Hildebrand J, Scott R (1949) The solubility of non-electrolytes. Reinhold, New York, NY
6. Van Krevelen DW (1990) Properties of polymers. Elsevier, Amsterdam
7. Gnamm H, Fuchs O (1980) Loesungsmittel und Weichmachungsmittel. Wiss. Verlagsgesellschaft, Stuttgart
8. Trathnigg B, Yan X (1993) J Chromatogr A 653:199
9. Trathnigg B, Thamer D, Yan X, Maier B, Holzbauer HR, Much H (1994) J Chromatogr A 665:47
10. Trathnigg B, Jorde C (1987) J Chromatogr 385:17
11. Trathnigg B (1990) J Liq Chromatogr 13:1731
12. Trathnigg B (1991) J Chromatogr 552:507
13. Lafosse M, Elfakir L, Morin-Allory L, Dreux M (1992) J High Res Chromatogr 15:312
14. Rissler K, Fuchslueger U, Grether HJ (1994) J Liq Chromatogr 17:3109
15. Brossard S, Lafosse M, Dreux M (1992) J Chromatogr 591:149
16. Dreux M, Lafosse M, Morin-Allory L (1996) LC-GC Int 9:148
17. Vandermeeren P, Vandermeeren J, Baert L (1992) Anal Chem 64:1056
18. Hopia AI, Ollilainen VM (1993) J Liq Chromatogr 624:69
19. Podzimek S (1994) J Appl Polym Sci 54:91
20. Dayal U, Mehta SK (1994) J Liq Chromatogr 17:303
21. Dayal D (1994) J Appl Polym Sci 53:1557

22. Balke S, Rao B, Thitiratsakul R, Mourey TH, Schunk TC (1996) In: Prepr 9th Int Symp Polym Anal Char (ISPAC 9), Oxford, July 1–3
23. Jackson C, Yau WW (1993) J Chromatogr 645:209
24. Jackson J (1994) J Chromatogr A 662:1
25. Pang S, Rudin A (1993) In: Provder T (ed) Chromatography of polymers. ACS Symposium Series 521. American Chemical Society, New York, NY, p 254
26. Yau WW, Abbott SD, Smith GA, Keating MY (1987) In: Provder T (ed) Detection and data analysis in size exclusion chromatograph. ACS Symposium Series 352. American Chemical Society, Washington, DC, p 80
27. Styring MG, Armonas JE, Hamielec AE (1987) In: Provder T (ed) Detection and data analysis in size exclusion chromatography. ACS Symposium Series 352. American Chemical Society, Washington, DC, p 104
28. http://www.malvern.com/labeng/products/viscotek/viscotek_gpc_sec/viscotek_tdamax/viscotek_tdamax.htm
29. Haney MA, Armonas JE, Rosen L (1987) In: Provder T (ed) Detection and data analysis in size exclusion chromatography. ACS Symposium Series 352. American Chemical Society, Washington, DC, p 119
30. Brower L, Trowbridge D, Kim D, Mukherjee P, Seeger R, McIntyre D (1987) In: Provder T (ed) Detection and data analysis in size exclusion chromatography. ACS Symposium Series 352. American Chemical Society, Washington, DC, p 155
31. Mori S (1993) J Chromatogr 637:129
32. Cheung P, Lew R, Balke ST, Mourey TH (1993) J Appl Polym Sci 47:1701
33. Lesec J (2005) In: Cazes J (ed) Encyclopedia of chromatography, 2nd edn. Marcel Dekker, New York, NY
34. Jeng L, Balke ST, Snyder LR (1993) J Appl Polym Sci 49:1359
35. http://www.postnova.com/pn3621-mals-detector.html

Multidetector Size Exclusion Chromatography

5

Size exclusion chromatography (SEC) is the premier polymer characterization method for determining molar masses. As a polymer fractionation method it provides not only average molar masses (like viscometry or light scattering) but also molar mass distributions (MMD). In SEC, the separation mechanism is based on molecular hydrodynamic volume. For linear homopolymers, condensation polymers and strictly alternating copolymers, there is a direct relationship between elution volume and molar mass. Thus, chemically similar polymer standards of known molar mass can be used for calibration. However, for SEC of random and block copolymers and branched polymers, no simple correspondence exists between elution volume and molar mass because of possible compositional heterogeneity of these materials. As a result, molar mass calibration with polymer standards can introduce a considerable amount of error. To address this problem, selective detection techniques in addition to universal concentration detectors were introduced to be combined with SEC separation. In this chapter the focus is on using these powerful detectors for the SEC analysis of complex polymers.

The experimental setup of SEC is quite simple and the components to be used are similar to the components that are used in interactive modes of liquid chromatography (LC). Figure 5.1 shows the first part of a SEC experiment being the fractionation of the sample and the generation of a chromatogram.

The sample is injected into the SEC column filled with a porous stationary phase as a dilute solution in the mobile phase. The sample molecules are transported by the stream of the mobile phase through the column. Based on entropic interactions with the pores of the stationary phase the molecules are separated according to their hydrodynamic volumes, the largest molecules eluting first followed by smaller molecules. The concentration of the eluting molecules is monitored by one or more detectors. As a result a chromatogram is generated where the detector response corresponding to the concentration profile of the eluting species is plotted against the elution time (or elution volume). The elution time t_e is the time that elapses between the injection of the sample into the column and the detection of the sample by the detector. It is related to the elution volume V_e (equivalent to retention volume V_r) through the flow rate ($=V_e/t_e$). Typical stationary phases for SEC are

H. Pasch and B. Trathnigg, *Multidimensional HPLC of Polymers*, Springer Laboratory, 55
DOI 10.1007/978-3-642-36080-0_5, © Springer-Verlag Berlin Heidelberg 2013

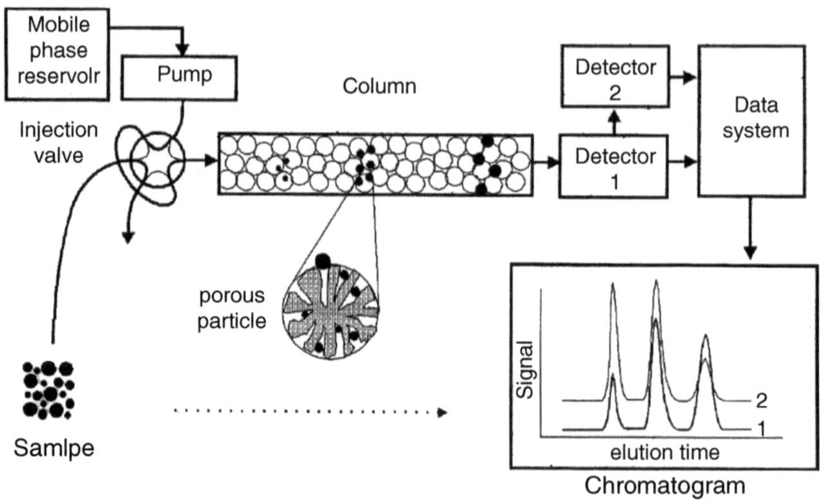

Fig. 5.1 Schematic presentation of a SEC experiment (reprinted from [1] with permission of Springer Science + Business Media)

cross-linked polymers, e.g. cross-linked polystyrene, or porous silica gel. Typical mobile phases are thermodynamically good solvents for the polymer. Extensive overviews on the experimental conditions of SEC are given in Refs. [1–6]. Further important sources are the multi-author volumes published in the ACS Symposia Series and edited by Provder and Barth [7–11]. Updated reviews on SEC have been published by Barth and Boyes in Analytical Chemistry [12]. Most recently, Striegel et al. summarized the state-of-the-art in multidetector SEC [13, 14].

As has been discussed previously, SEC is a relative method and needs to be calibrated. In a typical case calibration is done with well characterized narrow dispersed polymers, e.g. polystyrene, polymethyl methacrylate, polyethylene oxide or others. The data processing requires four basic steps, as schematically presented in Fig. 5.2.

Following the conversion of the elution time to elution volume, the elution volume must be converted to molar mass using the calibration curve. Alternatively, molar mass sensitive detectors can be used as will be discussed later. In the next step the detector signal must be converted to concentration. Depending on the detector used, a detector calibration might be necessary. Finally, polymer concentration is converted into weight fraction through a normalization procedure. Typically, the refractive index detector is assumed to be a universal concentration detector. Other options for concentration-sensitive detectors are UV, RI and ELS (evaporative light scattering) detectors [14].

When analyzing complex systems such as copolymers and polymer blends by SEC, one must take into account that the dimensional distribution of macromolecules can in general be unambiguously correlated with MMD only within one heterogeneity type. For samples consisting of molecules of different chemical composition, the

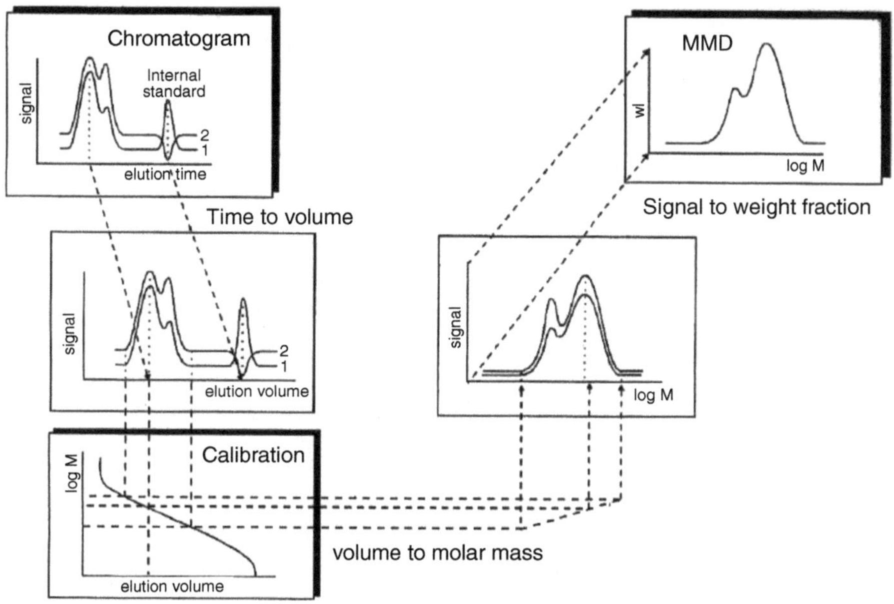

Fig. 5.2 Basic steps in obtaining a molar mass distribution from a chromatogram (reprinted from [1] with permission of Springer Science + Business Media)

distribution obtained represents an average of dimensional distributions of molecules having different compositions. As a result, in the same SEC volume fraction molecules elute that are different in molar mass, chemical composition and branching. The inadequacy of using SEC without further precaution for the determination of molar mass distribution (MMD) of polymer blends or copolymers can be explained with reference to Fig. 5.3 [16].

For a linear homopolymer distributed only in molar mass, fractionation by SEC results in one molar mass being present in each elution volume. The polymer fraction at each elution volume is monodisperse. If a blend of two linear homopolymers is fractionated then two different molar masses can be present in one retention volume. If now a copolymer is analysed then a multitude of different combinations of molar mass, composition, and sequence length can be combined to give the same hydrodynamic volume. In this case, fractionation with respect to molecular size is completely ineffective in assisting the analysis of composition or MMD.

To overcome the problems related to SEC of complex polymers, multidetector systems have been developed over the years. One approach is the use of a combination of SEC with multiple concentration detectors. If the response factors of the detectors for the components of the polymer are sufficiently different, the chemical composition of each slice of the elution curve can be determined from the detector signals. Typically, a combination of ultraviolet (UV) and refractive index (RI) detection is used; another possibility is the use of a diode-array detector. In the case of non-UV absorbing polymers a combination of RI and density detection yields information on

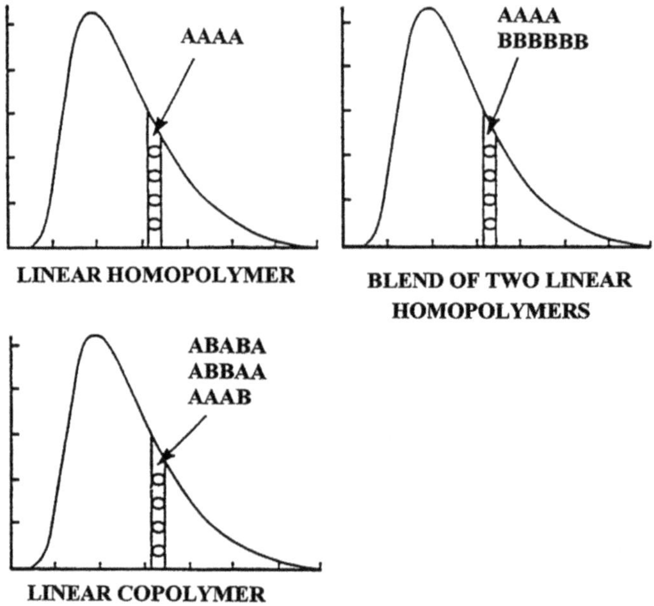

Fig. 5.3 SEC fractionation showing composition of fraction at a given elution volume (reprinted from [15] with permission of Springer Science + Business Media)

chemical composition [13, 17–22]. Similar information can be obtained by coupling SEC with spectroscopic detectors like FTIR, NMR or mass spectrometry. This approach will be addressed in Chap. 7. Such detector combinations, however, are normally not able to differentiate between copolymers and polymer blends. In this case it might be more suitable to carry out a separation according to chemical composition in a first step followed by a molar mass analysis (see for more details Chap. 6).

The most useful approach for the molar mass analysis of complex polymers is the coupling of molar mass-sensitive detectors to the SEC. Since the response of such detectors depends on both concentration and molar mass, they have to be combined with a concentration-sensitive detector, see Fig. 5.4. The following types of molar mass-sensitive detectors are used frequently [13, 23–28]: differential viscometer, low-angle laser light scattering detector (LALLS), multi-angle laser light scattering detector (MALLS). The details of these detectors are outlined in Chap. 4.

Frequently a triple-detector SEC technology is used, where three on-line detectors are used together in a single SEC system. In addition to the concentration detector, an on-line viscometer and a MALLS instrument are coupled to the SEC (TriSEC). With TriSEC, absolute molar mass determination is possible for polymers that are very different in chemical composition and molecular conformation. The usefulness of the TriSEC approach has been demonstrated in a number of applications [29–33].

The combination of SEC and molar mass-sensitive detectors is a powerful tool for the analysis of complex polymers. However, it is important to distinguish between claimed vs. actual capabilities and between potential expectations and demonstrated

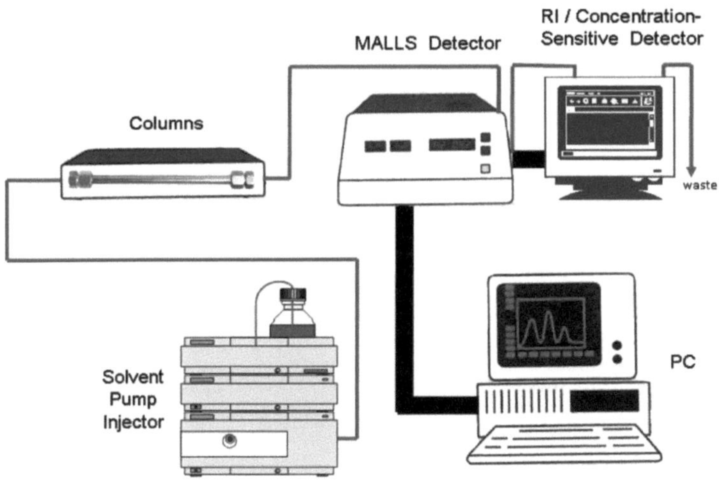

Fig. 5.4 Schematic representation of an SEC-MALLS setup

Table 5.1 SEC analysis using molar mass-sensitive detectors

	Information content	
Method	Primary	Secondary
Regular SEC		MMD
SEC-LALLS	MMD	
SEC-MALLS	MMD	RGD
SEC-VIS	IVD	MMD, RGD, copolymer M_n
SEC-VIS-LS	IVD, MMD, RGD	Copolymer M_n

MMD molar mass distribution, *IVD* intrinsic viscosity distribution, *RGD* radius of gyration distribution

performances [34]. The following two Tables 5.1 and 5.2, taken from a critical review of different techniques summarize the information content and additional details of SEC-light scattering and SEC-viscometry coupling [35]. The information content is classified into two categories: "primary" information is of high precision and accuracy, insensitive to SEC operation variables, and not requiring molar mass or universal calibration; "secondary" information is less precise and requires calibration.

5.1 Analysis of Copolymers by SEC with Dual Concentration Detection

In the molar mass analysis of binary copolymers or polymer blends two different concentration detectors can be used. Typically, a combination of UV and RI detection is used, but other detector combinations have also been described. If the

Table 5.2 Generalization of molar mass-sensitive detectors

Intended measurements	LALLS/MALLS	Viscometer
MMD	Requires precise n and dn/dc, not affected by non-exclusion effects	Requires universal calibration and K, a-parameters
IVD		Directly from experiment, not affected by non-exclusion effects
RGD	MALLS only	Calculated from $[\eta]M$
Chain conformation and branching	R_g vs. M plot, MALLS only	$[\eta]$ vs. M plot, R_g vs. M plot
Chemically heterogeneous polymer analysis	Limited	Better
Noise, particulates, bubbles	Strongly affected	Less affected

components of the copolymer have different UV spectra, a diode array detector will be the instrument of choice. One has, however, to keep in mind that non-linear detector response may also occur with UV detection.

The principle of dual detection is rather simple: when a mass m_i of a copolymer, which contains the weight fractions w_A and w_B (=1 − w_A) of the monomers A and B, is eluted in the slice i (with the volume ΔV) of the peak, the area $x_{i,j}$ of slice i obtained from detector j depends on the mass m_i (or the concentration $c_i = m_i/\Delta V$) of polymer in the slice, its composition (w_A), and the corresponding response factors $f_{j,A}$ and $f_{j,B}$, wherein j denotes the individual detector.

$$x_{i,j} = m_i(w_A f_{j,A} + w_B f_{j,B}). \tag{5.1}$$

The weight fractions w_A and w_B of the monomers can be calculated using

$$1/w_A = 1 - \{[(x_1/x_2)f_{2,A} - f_{1,A}]/[(x_1/x_2)f_{2,B} - f_{1,B}]\}. \tag{5.2}$$

Once the weight fractions of the monomers are known, the correct mass of polymer in the slice can be calculated using

$$m_i = x_i/\ [w_A(f_{1,A} - f_{1,B}) + f_{1,B}] \tag{5.3}$$

and the molar mass M_C of the copolymer is obtained by interpolation between the calibration lines of the homopolymers [36].

$$\ln M_C = \ln M_B + w_A(\ln M_A - \ln M_B), \tag{5.4}$$

wherein M_A and M_B are the molar masses of the homopolymers, which would elute in this slice of the peak (at the corresponding elution volume V_e).

It is clear, that the interpolation between the calibration lines cannot be applied to mixtures of polymers (polymer blends): if the calibration lines are different, different molar masses of the homopolymers will elute at the same volume.

The universal calibration is not capable of eliminating these errors either, which originate from the simultaneous elution of two polymer fractions with the same hydrodynamic volume but different composition and molar mass. Ogawa [37] has shown by a simulation technique, that the molar masses of polymers eluting at the elution volume V_e are given by the corresponding coefficients K and a in the Mark-Houwink equation.

In SEC of a polymer blend, molar masses of the homopolymers eluting in the same interval can be calculated using

$$\ln M = [AV_e/(1+a)] + [(B - \ln K)/(1+a)]. \quad (5.5)$$

The architecture of a copolymer (random, block, graft) has also to be taken into account, as Revillon [38] has shown by SEC with RI, UV, and viscosity detection. Intrinsic viscosity varies largely with molar mass according to the type of polymer, its composition, and the nature of its components. Tung [39] found, that for block copolymers in good SEC solvents the simpler first approach [Eq. (5.4)] is more precise.

The common combination of the RI with the UV detector can only be applied if at least one of the monomers of the complex polymer absorbs a suitable wavelength, and if the UV spectra of both components are sufficiently different. Successful applications of this set-up are the analysis of mixtures of polystyrene with PMMA, polybutadiene, polyvinyl chloride or polybutyl methacrylate. The RI detector provides the total elution profile, whereas the UV detector yields the elution profile of polystyrene. Substracting the latter from the former, the elution profile of the non-absorbing component can be generated.

For polymer systems without UV activity the combination of the RI detector with a density (D) detector can be used. The working principle of the density detector is based on the mechanical oscillator method. Since this detector yields a signal for every polymer, provided that its density is different from the density of the mobile phase, this detector can be regarded as universal [20, 21, 40]. The separation of mixtures of polystyrene and polybutadiene by SEC with dual density-RI detection is presented in Fig. 5.5. In a first set of experiments, the response factors of both polymers in both detectors have to be determined. Then from the intensity of each slice of the elution curves in both detectors, the mass distribution of both polymers across the elution volume axis can be calculated. As can be seen in Fig. 5.5a, a separation into the component peaks is obtained due to the fact that the molar masses of PS and PB are sufficiently different. For both components the individual elution profiles can be determined and using corresponding calibration curves for PS and PB the individual MMDs can be calculated. Similar information can be extracted from an experiment, in which the molar masses of the components are similar and SEC separation does not work, see Fig. 5.5b. Again the individual mass distributions are obtained and the MMDs for PS and PB can be determined.

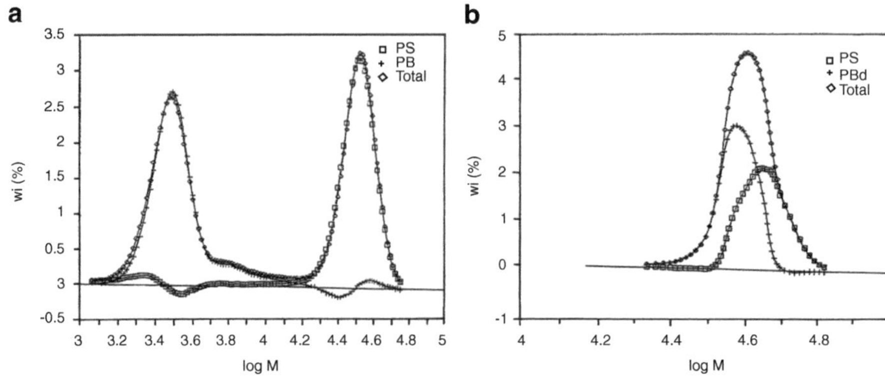

Fig. 5.5 Mass distribution and separated distributions of the components of a mixture of PS 50000 and PB 3000 (**a**) and PS 50000 and PB 31400 (**b**) from SEC with D-RI detection, stationary phase: styragel, eluent: chloroform (reprinted from [22] with permission of Springer Science + Business Media)

Besides obtaining information on the overall composition of blends and random copolymers, one may obtain valuable information in the case of block copolymers. If a block copolymer contains a homopolymer fraction, a shoulder or second maximum in the MMD curve can be obtained that should have a different chemical composition.

Further information on quantitative aspects of SEC with dual detection is provided by Trathnigg et al. [41]. Different applications of dual detection SEC in the analysis of segmented copolymers [42], block copolymers [43, 44], star polymers [45], and polymer blends [46] are also available. The limitation of SEC with dual detection is that only binary combinations of monomers can be investigated successfully. In the case of ternary combinations, more than two detectors must be used or one of the detectors must be able to detect two components simultaneously.

5.1.1 Analysis of PMMA-b-PDMA Block Copolymers by SEC with D-RI Detection

The analysis of copolymers by SEC is a challenge due to the fact that the samples exhibit a chemical composition distribution in addition to the molar mass distribution. In the general case it must be assumed that different molar mass fractions have different chemical compositions, see Fig. 5.3. A correct molar mass analysis by SEC with only one concentration detector is nearly impossible because the detector signal will be affected by the chemical composition. For example, in refractive index detection the refractive index increment dn/dc is a function of chemical composition and therefore the signal intensity depends not only on the concentration but also on the chemical composition of the respective molar mass fraction. One possible way to overcome this is to use two detectors that exhibit different sensitivities towards the different monomers in the macromolecules to obtain information about chemical composition at any point of the MMD. A very useful detector setup is the combination of a RI detector with a density detector. Both are universal detectors that detect different polymer species with different sensitivities. The advantage over the classical RI/UV detector combination is that in all cases both monomers give a detector signal.

Aim

The characterization of PMMA-b-PDMA block copolymers requires the analysis of MMD and chemical composition. With SEC a separation according to hydrodynamic volume is obtained which can be correlated to molar mass when the chemical composition of the molar mass fractions is known. The chemical composition along the polymer peak will be determined by dual detector SEC. This technique yields not only the chemical composition as a function of molar mass. It may provide additional information on the presence of homopolymers if their molar masses are different from that of the copolymer, even though it is not fully equivalent to a two-dimensional separation.

Materials

- **Calibration Standards**. Narrow-disperse PMMA (Polymer Laboratories, Church Stretton, UK), polydecyl methacrylate (PDMA) (laboratory products of Röhm GmbH, Darmstadt, Germany).
- **Sample**. PMMA-b-PDMA block copolymer (laboratory product of Röhm GmbH, Darmstadt, Germany).

Equipment

- **Chromatographic System**. Standard HPLC instrument, e.g. Agilent 1200 or similar.
- **Columns**. Phenogel M, average particle size 5 μm, column size 600 × 7.6 mm.
- **Mobile Phase**. 2-Butanone HPLC grade (Mallinckrodt, USA).
- **Detectors**. Density detection system DDS 70 (Chromtech, Graz, Austria) coupled with an ERC 7512 RI detector.
- **Column Temperature**. 25 °C.
- **Sample Concentration**. 1.0–4.0 g/L in the mobile phase.
- **Injection Volume**. 100 μL.

Preparatory Investigations

Before analyzing copolymers, one must analyze the homopolymers. In a first step the response factors for both homopolymers and for both detectors must be determined. If they are reliable, one should find the compositions 0 and 100%, respectively, in both cases (as shown in Fig. 5.6). In a second step the calibration curves for PMMA and PDMA must be obtained. The difference between the calibration

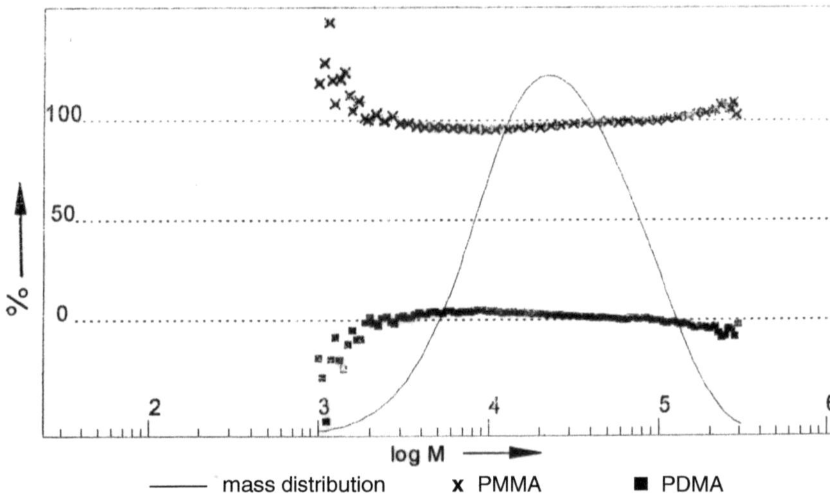

Fig. 5.6 Molar mass distribution and chemical composition of PMMA, stationary phase: Phenogel M, mobile phase: 2-butanone, detection: D-RI (reprinted from [1] with permission of Springer Science + Business Media)

curves is quite small: all the standards seem to fall on one line. In the case of PDMA, however, the order of the fit is a crucial point. Because of the limited number of available standards, too high an order produces erroneous results, as can be seen from Fig. 5.7.

Measurement

An analysis of copolymers by SEC with D-RI detection will require the following steps:

1. Determination of SEC calibration functions for both homopolymers, preferably with narrow MMD standards, if these are available.
2. Determination of the response factors of both detectors for the homopolymers. In the case of low molecular samples, the dependence of response factors on molar mass has also to be considered.
3. Determination of interdetector volume.
4. Analysis of the corresponding homopolymers. From the results the accuracy of steps 1–3 can be evaluated.
5. Analysis of the copolymer sample.

A PMMA-b-PDMA block copolymer which had been synthesized by anionic polymerization with sequential addition of monomers was separated by SEC. The chemical composition of each slice of the polymer peak was obtained from coupled density and RI detection as shown in Fig. 5.8.

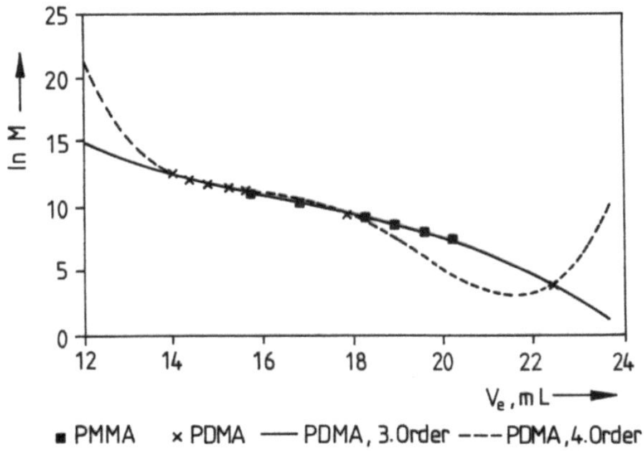

Fig. 5.7 Calibration functions for PMMA and PDMA, stationary phase: Phenogel M, mobile phase: 2-butanone, detection: D-RI (reprinted from [1] with permission of Springer Science + Business Media)

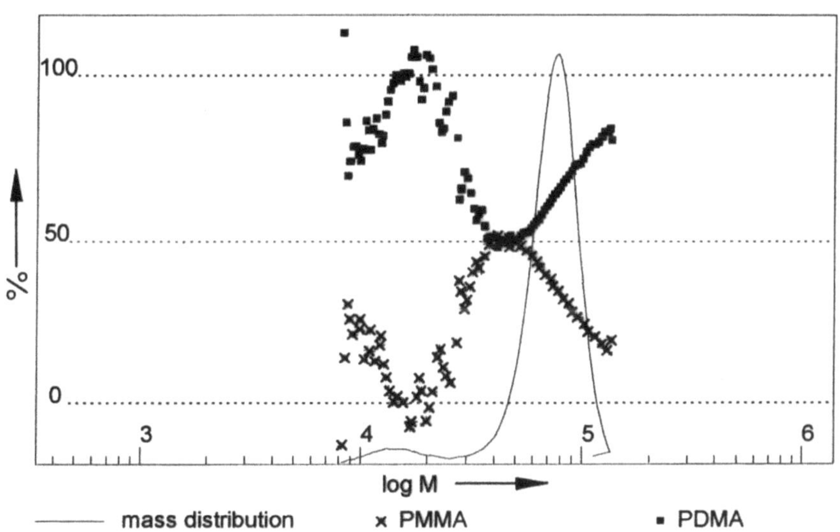

Fig. 5.8 MMD and chemical composition of a block copolymer PMMA-b-PDMA determined by dual detection SEC, stationary phase: Phenogel M, mobile phase: 2-butanone, detection: D-RI (reprinted from [1] with permission of Springer Science + Business Media)

Evaluation

The analysis of homopolymer standards prior to copolymer analysis is very important, particularly because it provides information on

- the quality of the calibration functions for both homopolymers.
- the accuracy of response factors (the suitability of the mobile phase).

– the accuracy of the interdetector volume.
– the extent of peak dispersion between the detectors.

Once the quality of the calibration functions has been evaluated as described above, the response factors for the homopolymers should be sufficiently different. If this is not the case, a change of the mobile phase may bring about a considerable improvement. One should, however, avoid mixed mobile phases, because in this case preferential solvation of the polymer may affect the accuracy of results. It is evident that erroneous response factors will yield overall compositions different from 0 and 100%, respectively, for the homopolymers.

As chemical composition is obtained in dual detection SEC from the ratio of the detector signals (x_1/x_2) along the chromatogram, two other sources of error affect narrow peaks (such as monodisperse oligomers) much stronger than broad MMD polymers. Obviously, an ***incorrect interdetector volume*** and ***peak dispersion*** will affect the results, but in a different way. An incorrect interdetector volume will shift the second detector tracing against the first one. An insufficiently compensated delay will lead to a low value of x_1/x_2 at the beginning of the peak and a high value at the end. This will result in an apparent variation of chemical composition along the peak. Peak dispersion between the detectors, however, leads to a high ratio (x_1/x_2) on both sides of the peak. Especially at the very ends of narrow peaks, where both x_1 and x_2 approach zero, this will cause considerable errors in the chemical composition. In the first case, the composition shows a slope, but a linear shape, while a U-shaped curve will be found in the second case. It must, however, be mentioned that similar deviations from the expected shape (horizontal, straight line) may be caused by other reasons, such as molar mass dependence of response factors, or simply an incorrect baseline. Once these sources of error have been eliminated, the composition of copolymers can be determined with good accuracy.

Besides the overall composition of the copolymer, one may obtain valuable information on possible by-products. If the sample contains homopolymer(s) fractions, a shoulder (or a second maximum) in the MMD may be found, which should have a different composition. Figure 5.8 shows the MMD of the block copolymer in which a second (lower molecular) maximum can be identified as PDMA. The increase of the PDMA content with molar mass in the main fraction is quite reasonable and corresponds well with what should be expected from the synthesis. These findings can be supported by liquid chromatography under critical conditions (LCCC) for both of the blocks [47].

Obviously, a combination of LCCC and dual detector SEC will give a much better understanding of the composition of the copolymer than each of these techniques, when applied alone. SEC yields the composition as a function of molar mass, but does not give information on whether a peak is a copolymer or a mixture of co-eluting homopolymers. LCCC yields the length of the blocks, but does not provide information on whether they are linked together.

5.1.2 Analysis of PEO-b-PPO Block Copolymers by SEC with D-RI Detection

Aim

Block copolymers of ethylene oxide (EO) and propylene oxide (PO) are useful for many applications: their low temperature viscosity behaviour makes them attractive as lubricants (because of their low toxicity even in the food industry). As amphiphilic molecules they may form micelles, which can be utilized in pharmacology and many other fields.

Reliable characterization of these materials is important, but also complicated. The problems in the characterization of EO-PO diblock copolymers are related to the fact that they may contain poly(ethylene glycol) (PEG) and poly(propylene glycol) (PPG) homopolymers. Even more interesting materials are the three-block copolymers, which may contain not only the PEG and PPG but also two-block copolymers. These materials are typically prepared either by ethoxylation of PPG or by propoxylation of PEG, mostly with basic catalysts. Under these conditions chain-transfer reactions may occur, which lead to the formation of homopolymer fractions besides the copolymer. Since these by-products have considerably lower molar masses than the desired copolymer, they can be separated from the main fraction by SEC and identified by dual detection.

Materials

- **Calibration Standards**. Narrow-dispersed PEG (Polymer Laboratories, Church Stretton, UK), polydisperse PEG and PPG (Fluka/Sigma-Aldrich, Buchs, Switzerland).
- **Oligomers**. Oligoethylene glycols ($n = 2$–6) and oligopropylene glycols ($n = 2$–3) (Fluka/Sigma-Aldrich, Buchs, Switzerland), block copolymers were laboratory products (H.-R. Holzbauer, Institute for Applied Chemistry Adlershof, Berlin, Germany).

Equipment

- **Chromatographic System**. Standard HPLC instrument, e.g. Agilent 1200 or similar.
- **Columns**. Phenogel M, average particle size 5 μm, column size 600 × 7.6 mm.
- **Mobile Phase**. Chloroform (stabilized with 2-methyl-butene) HPLC grade (Mallinckrodt, Hazelwood, USA).
- **Detectors**. Density detection system DDS 70 (Chromtech, Graz, Austria) coupled with an ERC 7512 RI detector.
- **Column Temperature**. 25 °C.
- **Sample Concentration**. 3.0–10.0 g/L.
- **Injection Volume**. 100 μL.

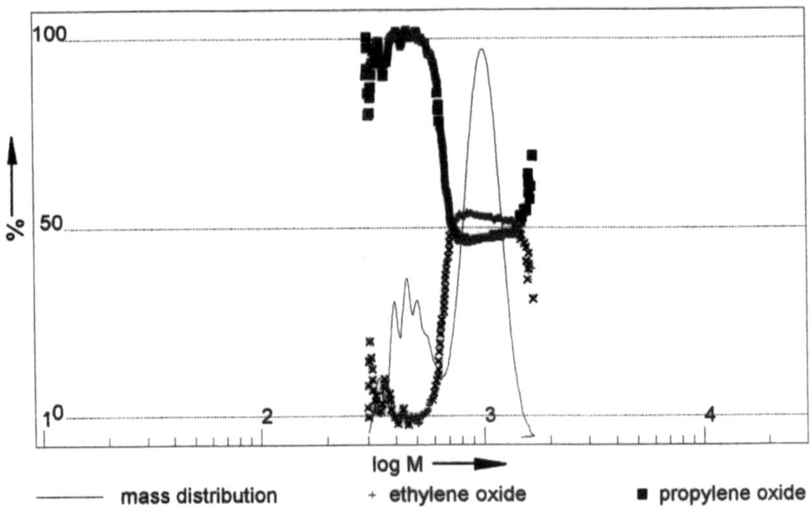

Fig. 5.9 MMD and chemical composition of an EO-PO block copolymer, as obtained from SEC with coupled density and RI detection, stationary phase: Phenogel M, mobile phase: chloroform, detection: D-RI (reprinted from [1] with permission of Springer Science + Business Media)

Preparatory Investigations

As the first step, the response factors of PEG, PPG, and the available monodisperse oligomers were determined for both detectors. With the same standards, SEC calibrations for PEG and PPG were established, which were found to be considerably different. This problem can be solved by using the dual detector setup. With dual detection the chemical composition can be calculated for each SEC slice. Once the composition is known, one may extrapolate between the two calibration curves to obtain a 'copolymer' calibration curve.

Separations

Once the necessary calibrations had been performed, a block copolymer was analyzed as described above. In Fig. 5.9 the MMD of the block copolymer is shown, which was obtained by propoxylation of PEG 600. This sample contains a fraction with lower molar mass which was assumed to be PPG formed by chain transfer to water in the reaction mixture. From the signals of the density and the RI detectors the chemical composition was determined at any point of the MMD, proving that the low molar mass fraction was indeed pure PPG.

Evaluation

As can be seen in Fig. 5.9 there is sufficient separation between the block copolymer and the homopolymer to be able to determine their molar masses and chemical compositions. The fact that the lower molar mass component indicates an EO content of 0% clearly indicates that the dual detector setup works with high accuracy. This is sufficient proof for the assumption that this component is residual PPG that did not take part in the copolymerization. Using the extrapolated 'copolymer' calibration curve the molar masses of both the PPG and the copolymer can be analyzed. Knowing these and assuming that the PPO block size in the copolymer is the same as in the residual PPG, the molar mass of the PEO block can be calculated. Here, however, it must be assumed that only a diblock copolymer was formed.

5.2 Analysis of Polymers by SEC with Molar Mass Sensitive Detection

As has been pointed out, for SEC of complex polymers no simple correspondence exists between elution volume and molar mass. It is, therefore, useful to determine the molar mass not via a calibration curve but directly from the SEC effluent. This can be done by using molar mass-sensitive detectors based on Rayleigh light scattering or intrinsic viscosity measurements [24]. Molar mass sensitive detectors are useful because they yield the molar mass of each fraction of a polymer peak directly from the detector reading. Since the response of such a detector depends on both the concentration and the molar mass of the fraction, it has to be combined with a concentration-sensitive detector. The following types of molar mass sensitive detectors are most common [23, 48, 49]:

- Differential viscosimeters
- Low angle laser light scattering detectors (LALLS)
- Right angle laser light scattering detectors (RALLS)
- Multi-angle laser light scattering detectors (MALLS)

Since viscosity and LS yield different types of information, it is sensible to combine both of them in a triple detector setup (TriSEC) [50]. From light scattering detection, the absolute MMD can be determined directly. Using more than one angle, one may also obtain the radius of gyration. On the other hand, SEC with viscosity detection yields the intrinsic viscosity distribution (IVD). In this case, the MMD is, however, determined indirectly and is thus subject to retention errors. Hence, a combination of a concentration detector with both a light scattering detector and a viscosity detector provides the highest reliability of results. Moreover, information on branching can be obtained in this way [51, 52].

5.2.1 RI-MALLS Detection

In a LS detector, the scattered light of a laser beam passing through the cell is measured at angles different from zero. The (excess) intensity $R(\theta)$ of the scattered light at the angle θ is related to the weight-average of molar mass M_w:

$$K^*c/R(\Theta) = [1/M_w P(\Theta)] + 2A_2 c, \tag{5.6}$$

wherein c is the concentration of the polymer, A_2 is the second virial coefficient, and $P(\theta)$ describes the scattered light angular dependence. K^* is an optical constant containing Avogadro's number N_A, the wavelength λ_0, the refractive index n_0 of the solvent, and the refractive index increment dn/dc of the sample:

$$K^* = 4\pi^2 n_0{}^2 (\mathrm{d}n/\mathrm{d}c)^2 / \left(\lambda_0{}^4 N_A\right). \tag{5.7}$$

In a plot of $K^* c/R(\theta)$ versus $\sin^2(\theta/2)$, M_w can be obtained from the intercept and the radius of gyration from the slope. A multi-angle measurement provides additional information.

In most cases the injected concentration is small and A_2 can be neglected. Thus, if the optical properties (n_0 and dn/dc) of the polymer solution are known, the molar mass at each elution volume increment can be determined.

$$M_{w,i} = R(\Theta)_i / K^* P(\Theta)_i c_i. \tag{5.8}$$

If a low-angle LS instrument is used, $P(\Theta)$ is close to unity and $M_{w,i}$ can be calculated directly. For a multi-angle LS instrument, the mean-square radius of gyration $< R_g{}^2 >$ at each elution volume can also be obtained from $P(\Theta)$:

$$\begin{aligned} 1/P(\Theta)_i &= 1 + q^2 < R_g{}^2 >_i /3, \\ q &= (4\pi/\lambda_0)\sin(\Theta/2). \end{aligned} \tag{5.9}$$

In practice, however, the radius of gyration can only be determined for molecules larger than 20 nm in diameter. By measuring the radius of gyration as a function of M_w, insight into the molecular conformation of the polymer can be obtained [14].

The molar mass determination requires the knowledge of the specific RI increment dn/dc which in the case of complex polymers depends on chemical composition. Copolymer RI increments $(\mathrm{d}n/\mathrm{d}c)_{copo}$ can accurately be calculated for chemically monodisperse fractions, if comonomer weight fractions w_i and homopolymer values are known:

$$(\mathrm{d}n/\mathrm{d}c)_{copo} = \Sigma w_i \, (\mathrm{d}n/\mathrm{d}c)_i. \tag{5.10}$$

However, in some cases additional effects on $(\mathrm{d}n/\mathrm{d}c)_{copo}$ must be considered. Due to cooperative interactions between the monomer units in the polymer chain, copolymer RI increments may deviate from the summation scheme. As a result of different sequence length distributions, different $(\mathrm{d}n/\mathrm{d}c)_{copo}$ can be obtained for the same gross composition. Copolymer $(\mathrm{d}n/\mathrm{d}c)_{copo}$ values can be obtained by multiple detection SEC providing the chemical composition at each slice of the elution curve.

Table 5.3 Selected dn/dc values at room temperature at a detector wavelength of 633–690 nm

Polymer	Solvent	dn/dc (mL/g)
Polystyrene	THF	0.185
Polystyrene	Toluene	0.105
Polystyrene	TCB at 135 °C	0.047
Poly(methyl acrylate)	THF	0.068
Poly(methyl methacrylate)	THF	0.084
Poly(butyl methacrylate)	THF	0.076
Polyisoprene	THF	0.127
Polybutadiene	THF	0.130
Poly(vinyl acetate)	THF	0.059
Polyethylene	TCB at 135 °C	−0.104

Selected values for dn/dc for a number of polymers are shown in Table 5.3. The refractive index increment depends on both the polymer and the solvent. In cases where dn/dc for a specific polymer-solvent combination (e.g. PS in toluene) cannot be found, it can be estimated from the values of a known polymer-solvent combination (e.g. PS in THF) and the pure solvent (e.g. toluene) using the following simple equation: $\mathrm{d}n/\mathrm{d}c^{\mathrm{PS-toluene}} = \mathrm{d}n/\mathrm{d}c^{\mathrm{PS-THF}} - (n_{\mathrm{toluene}} - n_{\mathrm{THF}})$. The calculated value in this case is 0.098 mL/g which is close to the value in Table 5.3 [53].

Unfortunately, LS investigations of copolymers are even further complicated by the fact that SEC does not separate into chemically monodisperse fractions. Accordingly, due to compositional heterogeneity the RI increment of a particular scattering center may be different from the total dn/dc of the corresponding SEC slice. Therefore, in general only apparent molar masses for copolymers can be measured [54]. Another influencing factor is the RI of the solvent. As has been shown by Kratochvil, the solvent RI should be significantly different from the values of the copolymer fractions and the corresponding homopolymers [55].

The evaluation of light scattering detectors for SEC was conducted by Jeng et al. with respect to precision and accuracy [56] and the proper selection of the light scattering equation [57]. The results obtained for PS and polyethylene were compared for low-angle and multi-angle light scattering instruments. The application of SEC-light scattering has been discussed in numerous papers. In addition to determining M_w values, the formation of microgels has been studied by Pille and Solomon [58]. Mourey and Coll investigated high molar mass PS and branched polyesters, and discussed the problems encountered in molar mass and radius of gyration determination [59, 60]. Grubisic-Gallot et al. proved that SEC-light scattering is useful for analyzing micellar systems with regard to determining molar masses, qualitative evaluation of the dynamics of unimer micelles re-equilibration, and revealing the mode of micelle formation [61, 62].

5.2.1.1 Analysis of Polyamides-11 and -12 by SEC with RI-MALLS Detection [63]

Aim

Polyamides are thermoplastics that exhibit high strength, abrasion resistance, stiffness and perpetuation of their physical and mechanical properties over a wide range of temperatures. The semicrystalline morphology and the intermolecular hydrogen bonds of the amide groups are responsible for the advantageous properties of the polyamides. On the other hand, morphology and hydrogen bonding influence the dissolution behaviour of polyamides and, hence, the determination of the molar mass and its distribution. Poor solubility in common organic solvents and strong adsorptive interactions are a challenge for the molar mass analysis by SEC. Today typically hexafluoro isopropanol (HFiP) is used as solvent for the molecular mass characterization of polyamides. As has been shown in a number of studies, salts have to be added to the solvent/mobile phase to prevent aggregate formation and to obtain exact molar masses [64–70]. The aim here is to develop a suitable SEC procedure for polyamide-11 and polyamide-12 using HFiP as the solvent and mobile phase. This is an important subject since PA-11 is becoming a fast developing biobased polymer.

Materials

- **Calibration Standards**. Narrow-dispersed PMMA (PSS GmbH, Mainz, Germany).
- **Samples**. Commercial samples of PA-11 and PA-12 as well as laboratory products produced by Arkema, Collombes, France.

Equipment

- **Chromatographic System**. Standard HPLC instrument, e.g. Agilent 1200 or similar.
- **Columns**. PSS PFG 100 Å, PSS PFG 300 Å, PSS PFG 1,000 Å, two columns PSS PFG linear all 8 × 300 mm i.d., 7 μm average particle size (PSS GmbH, Mainz, Germany), two PL HFiP Gel columns, 7.5 × 300 mm i.d. (Polymer Laboratories, Shropshire, UK).
- **Mobile Phase**. HFiP (1,1,1,3,3,3-hexafluoro-2-propanol) (Fluorochem Ltd., Derbyshire, UK), used as received; KTFAc (potassium trifluoroacetate) with a purity of >99.9 % (Fluka/Sigma-Aldrich, Buchs, Switzerland).
- **Detectors**. RI (G1362A) detector (Agilent Technologies, Santa Clara, USA), MALLS detector Dawn Eos, wavelength of 690 nm (Wyatt Technologies, Santa Barbara, USA).
- **Column Temperature**. 35 °C.
- **Sample Concentration**. 2.0 g/L.
- **Injection Volume**. 50 μL.

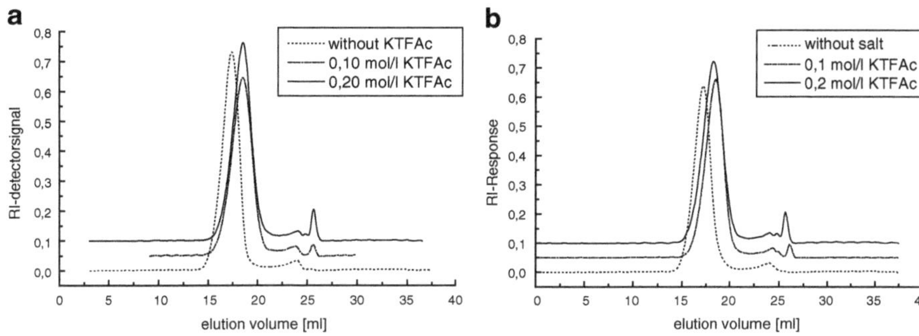

Fig. 5.10 Elugrams of (**a**) PA-11 (sample B1) and (**b**) PA-12 (sample A1) at different mobile phase compositions; stationary phase: PSS PFG 100 Å and PSS PFG 1,000 Å; mobile phase: HFiP + KTFAc (reprinted from [63] with permission of Elsevier)

Preparatory Investigations

For suitable SEC conditions, the composition of the mobile phase and the type of stationary phase must be optimized. In agreement with previous findings HFiP was used as the mobile phase and potassium trifluoroacetate was added to prevent aggregate formation. Using different concentrations of KTFAc in HFiP the minimum salt concentration was found to be 0.05 mol/L KTFAc for PA-11 and PA-12. The elution curves of a polyamide-11 (PA-11) in Fig. 5.10 show typical SEC profiles. When no salt was added to the mobile phase, early elution is observed that could indicate some aggregate formation. Salt addition increases the elution volume of the sample and at salt concentrations of 0.1 and 0.2 mol/L identical elution behaviour is observed. Further increase of the salt concentration does not change the elution behaviour. As was explained before, the early elution of the polyamides in HFiP without salt can be caused by aggregate formation, polyelectrolyte effects or repulsive interactions with the stationary phase. The addition of salt suppresses these effects and at salt concentrations >0.05 mol/L stable conditions are obtained. Multiple measurements show that a minimum salt concentration of 0.05 mol/L is required. For stable operating conditions a salt concentration of 0.1 mol/L is used throughout.

For the optimization of the stationary phase two different types of materials were tested, (a) a non-polar crosslinked styrene-divinylbenzene copolymer (Polymer Labs), and (b) a polar functionalized perfluoro silicagel (PSS). In order to evaluate the resolution and the linearity of the calibration function, different column sets were tested with PMMA calibration standards. The calibration curves for two typical column sets using a mobile phase of HFiP + 0.1 mol/L KTFAc are presented in Fig. 5.11. They indicate that good separation is obtained over more than three decades of molar masses (10^3–10^6 g/mol) being fully sufficient for the polyamides under investigation.

One of the major problems of molar mass analysis of polyamides by SEC is the lack of suitable calibration standards. Narrow dispersed polyamides are not

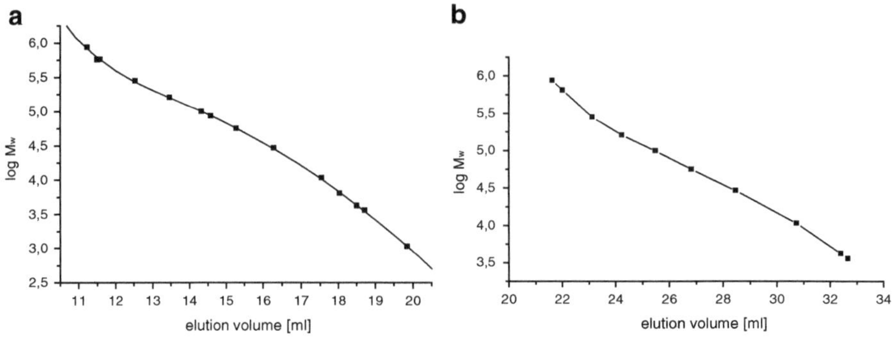

Fig. 5.11 PMMA calibration curves for different stationary phases, stationary phase: (**a**) 2× PL HFiP Gel, (**b**) PSS PFG 100 Å + 300 Å + 1,000 Å; mobile phase: HFiP + 0.1 mol/L KTFAc (reprinted from [63] with permission of Elsevier)

commercially available and the suitability of a calibration with PMMA or polyethylene oxide (PEO) is questionable. A suitable alternative is the molar mass analysis by SEC coupled to a multiangle light scattering detector. To use a MALLS detector, the d*n*/d*c* values for the polyamides must be known. They were determined by measuring sets of samples of different molar masses, giving averages of 0.211 mL/g for PA-11 and 0.219 mL/g for PA-12.

Separations

A number of samples having different molar masses were analysed by SEC-MALLS. A representative elugram and the corresponding calibration curve generated from the laser LS measurement are shown in Fig. 5.12. For molar masses determined by SEC-MALLS one has to bear in mind that the relative sensitivity of the LS detector at low molar masses is rather low. Measurements at the high molar mass end of the elution curve are much more precise. Therefore, very frequently the calculated M_n values are gradually too high.

The calculated molar masses and the refractive index increments are summarized in Table 5.4.

Another feasible way to accurately analyze molar masses of PA-11 and PA-12 is to use well characterized broadly distributed polyamides. The measurements of these polyamides by SEC-MALLS produce calibration curves that can be used in simple SEC-RI experimental setups. The SEC-MALLS calibration curves for PA-11 and PA-12 produced from samples "B1 PA11" and "A2 PA12" are presented in Fig. 5.13.

The calibration curves for PA-11 and PA-12 in Fig. 5.13 are absolutely identical and prove similar chromatographic behaviour of the two polyamides. They show that the accuracy of LS measurements at high molar masses (low elution volume of the peak) is very good. At low molar masses (high elution volumes) the accuracy is rather low due to the strong scattering of the LS signal. This is due to low absolute LS signal intensity at low molar masses.

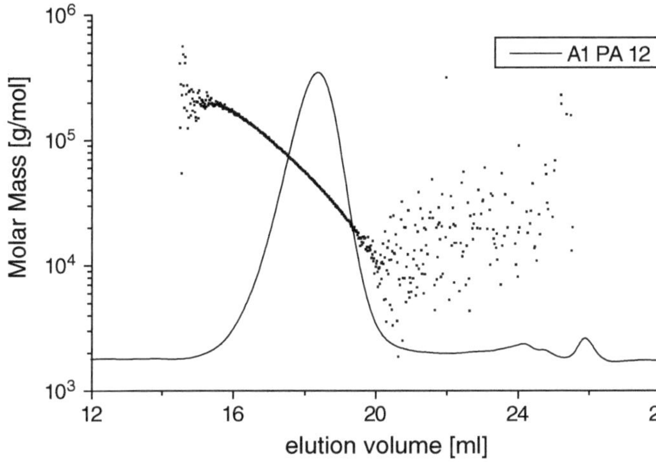

Fig. 5.12 SEC-MALLS analysis of PA-12 (sample A1); stationary phase: 2× PL HFiP Gel, mobile phase: HFiP + 0.1 mol/L KTFAc (reprinted from [63] with permission of Elsevier)

Table 5.4 Polyamide molar masses and dn/dc determined by SEC-MALLS

Sample	M_w (g/mol)	dn/dc (mL/g)
A1, PA 12	63,500	0.210
A2, PA 12	60,250	0.205
A3, PA 12	39,500	0.210
A4, PA 12	36,750	0.215
A5, PA 12	27,750	0.215
B1, PA 11	49,750	0.224
B2, PA 11	43,000	0.220
B3, PA 11	30,750	0.215
B4, PA 11	27,250	0.217

The molar mass data obtained by SEC-RI using the calibration curves presented in Fig. 5.13 are summarized in Table 5.5 together with a comparison to molar masses obtained by SEC-MALLS.

Evaluation

The data show clearly that both the PA-11 and the PA-12 calibration curve are equally suitable for molar mass analysis. The agreement with the SEC-MALLS data is very good in all cases and an excellent repeatability is obtained. This means that depending on the availability of a SEC-MALLS system there are two routes, (1) direct molar mass analysis by SEC-MALLS or (2) 'external' measurement of one broadly distributed sample by SEC-MALLS and use of the obtained specific calibration curve for the measurement of other samples by SEC with a concentration detector.

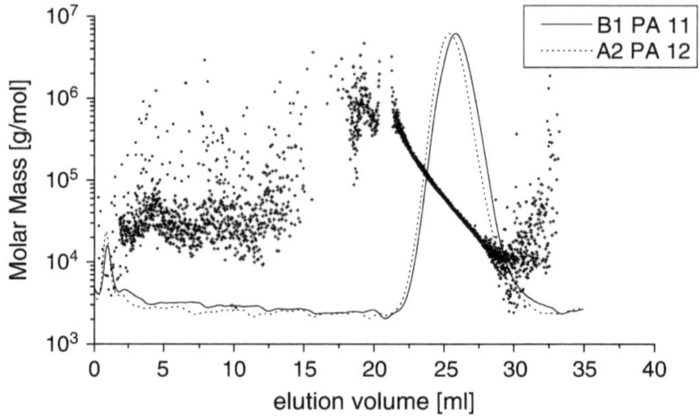

Fig. 5.13 Elution profiles and calibration curves from SEC-MALLS; stationary phase: PSS PFG 100 Å + 300 Å + 1,000 Å, mobile phase: HFiP + 0.1 mol/L KTFAc (reprinted from [63] with permission of Elsevier)

Table 5.5 Molar masses calculated using broadly distributed polyamides

Sample	M_w (LS) [g/mol]	M_w 1 [g/mol]	M_w 2 [g/mol]
A1 PA12	63,500	60,200	60,000
A2 PA12	60,250	59,900	60,300
A3 PA12	39,500	39,200	39,100
A4 PA12	36,750	37,500	37,800
A5 PA12	27,750	28,300	27,400
B1 PA11	49,750	51,100	51,500
B2 PA11	43,000	46,000	46,700
B3 PA11	30,750	31,600	31,900
B4 PA11	27,250	28,400	28,500

M_w 1: Based on PA-11 calibration, M_w 2: Based on PA-12 calibration

5.2.1.2 Analysis of Hydrophilic Copolymers by SEC with RI-MALLS Detection [71]

Aim

Hydrophilic copolymers are extremely important precursors for surfactants, dispergating agents and drug carriers in the chemical and pharmaceutical industries. In particular, they are used as binder and coating materials. Very frequently these materials are polymers containing hydrophilic and hydrophobic segments. In addition to the molar mass distribution they frequently exhibit a chemical composition distribution. Anionic methacrylic acid-(meth)acrylic ester copolymers belong to a group of pharmaceutical excipients that are primarily used as controlled release film coating agents in oral capsule and tablet formulations [72]. Their methacrylic acid content is up to 50 wt% making them water soluble in some cases and water insoluble in other cases.

Table 5.6 Chemical compositions and molar mass ranges of the EUDRAGIT® copolymers; molar mass estimates are based on solution viscometry measurements

Sample	Expected M_w (kg/mol)	Methacrylic acid (wt%)	(Meth)acrylic esters[a] (wt%)
A	50–100	30	70
B	100–200	50	50
C	100–300	50	50
D	500–1,000	0	100
E	100–500	10	90

[a]Esters are methyl methacrylate, methyl acrylate and ethyl acrylate

It is known that the molar mass separation of hydrophilic copolymers is not straightforward. Experimental difficulties arise from the presence of strongly polar functional groups at the polymer chains. Such very polar macromolecules tend to form aggregates with themselves and with solvent molecules. With regard to SEC, intermolecular electrostatic interactions (ion exchange, ion exclusion, ion inclusion) between the macromolecules and the stationary phase may occur. Other non-size exclusion effects encountered in aqueous SEC are intramolecular electrostatic interactions and adsorption due to hydrogen bonding and hydrophobic interactions between the polymer and the stationary phase [73–77]. For the suppression of non-size exclusion effects, the following measures have been proposed: (1) variation of the pH of the mobile phase, (2) addition of an electrolyte to the mobile phase, (3) addition of organic modifiers to the mobile phase [6, 78–82]. The aim of the present application is to optimize these experimental conditions and to use SEC-MALLS for absolute molar mass analysis.

Materials

- **Calibration Standards**. Narrow dispersed PMMA (PSS GmbH, Mainz, Germany).
- **Samples**. Commercial samples (EUDRAGIT®) as well as laboratory products produced by Roehm GmbH, Darmstadt, Germany, see Table 5.6.

Equipment

- **Chromatographic System**. Standard HPLC instrument, e.g. Agilent 1200 or similar.
- **Columns**. GRAM 3,000 Å, GRAM 100 Å and GRAM linear XL columns (PSS GmbH, Mainz, Germany), all of 300 × 8 mm i.d. and 10 μm average particle size. To condition new off-the-shelf GRAM linear XL columns, they were rinsed for 5 h with pure DMAC with a flow rate of 0.2 mL/min, with pure DMAC, then for 2 h with the mobile phase. Subsequently the system was heated to 60 °C and the columns were conditioned overnight with a flow rate of 0.1 mL/min. In all cases a GRAM precolumn was used.

Table 5.7 Molar masses of the methylated copolymers, calculated as PMMA equivalents

Sample	Expected M_w (kg/mol)	M_w (kg/mol)	M_n (kg/mol)
A	50–100	95.0	55.0
B	100–200	93.0	52.0
C	100–300	236.0	78.0
D	500–1,000	730.0	252.0
E	100–500	233.0	67.0

- **Mobile Phase**. N,N-dimethylacetamide (DMAC) HPLC grade (Fluka & Schopp, Karlsruhe, Germany), LiBr 99+ % (Acros, Belgium), AcOH >99.8 % (Acros, Belgium).
- **Detectors**. RI detector 1100 series (Agilent Technologies, Santa Clara, USA), MALLS detector Dawn Eos, wavelength of 690 nm (Wyatt Technologies, Santa Barbara, USA).
- **Column Temperature**. 60 °C.
- **Sample Concentration**. 3.0 g/L.
- **Injection Volume**. 100 μL.

Preparatory Investigations

For first information on the molar mass range of the samples and for optimizing the experimental conditions, the carboxylic groups of the samples were esterified by methylation with diazomethane. This yielded acrylate/methacrylate ester copolymers that can be analysed by standard SEC in THF using styrene-divinylbenzene copolymer (SDV) as the stationary phase. The molar masses that were obtained based on PMMA calibration are summarized in Table 5.7. These molar masses serve as reference data for the SEC analyses of the original samples.

In the second step the solubility of the original samples was tested. The most suitable solvent was DMAC that was modified with different concentrations of LiBr to screen the carboxylic groups. A number of stationary phases including SDV and cross-linked hydroxyethyl methacrylate (HEMA) were tested. These stationary phases produced multimodal elution profiles indicating non-ideal SEC behaviour. The test with a polar polyester-based material (GRAM), however, gave typical SEC profiles. Another problem is the adjustment of the mobile phase composition. In Fig. 5.14 elution profiles are summarized that were obtained in DMAC with different salt and acid additions.

As can be seen, the addition of acid is necessary to obtain proper elution profiles. Unfortunately, DMAC degrades very rapidly when high amounts of acetic acid and LiBr are added and 80 °C is used as column temperature. After only a few days the mobile phase becomes yellow and darkens further with time. In order to improve the stability of the mobile phase, the column temperature was kept at 60 °C and the concentration of additives was as low as possible. In another set of experiments different column combinations were tested in order to achieve optimum performance. Instead of only one column of 3,000 Å, a second column of 100 Å was

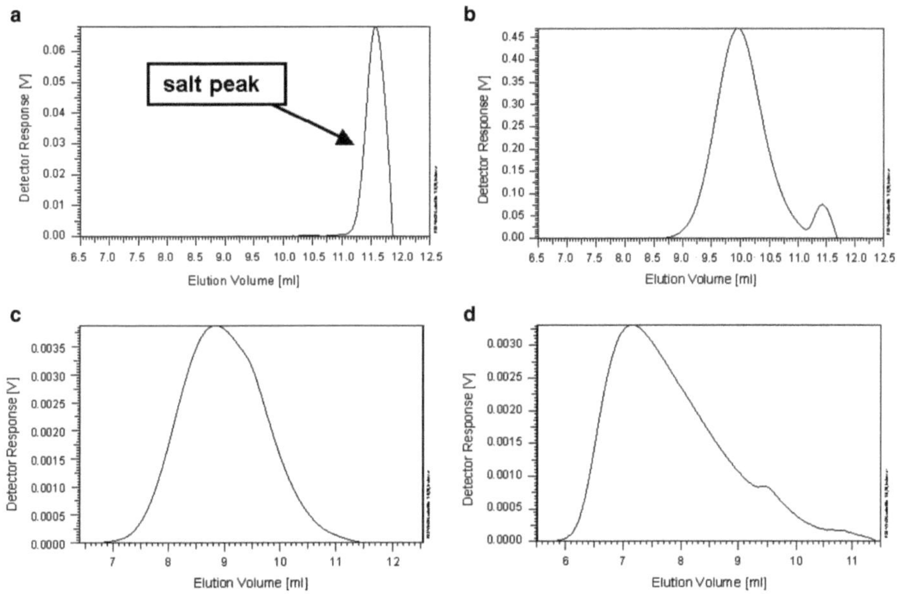

Fig. 5.14 SEC of copolymers B (**a–c**) and C (**d**) in different mobile phases; stationary phase: GRAM 3,000 Å, mobile phase: (**a**) DMAC + 5 g/L LiBr, (**b**) DMAC + 5 g/L LiBr + 90 mM AcOH, (**c**, **d**) DMAC + 150 mM AcOH, detector: RI (reprinted from [71] with permission of European Polymer Federation)

added. The better separation in the lower molar mass range enabled to separate the polymer peak from the salt peak. Subsequently, these columns were replaced by one and two linear columns GRAM XL, respectively, with a significantly broader separation range.

Separations

With respect to chromatographic behaviour of the samples and stability of the mobile phase, optimum performance was obtained for a mobile phase of DMAC with 6 g/L acetic acid and 3 g/L LiBr. The elution profiles indicate monomodal molar mass distributions for all samples as was expected from the behaviour of the methylated derivatives. The chromatograms of the copolymer samples are summarized in Fig. 5.15. As can be seen, not only the expected monomodality is obtained but also the salt peak is completely separated from the polymer peaks and does not disturb quantification.

To check the validity of the results, the absolute weight-average molar masses were determined by SEC coupled to laser light scattering. The samples A to E were investigated in order to see if there are deviations of the results as a function of the methacrylic acid content. Figure 5.16 shows a representative elution profile obtained from the RI detector and the MALLS detector at an angle of 90° for copolymer sample B. Similar to the methylated samples and the RI traces, the light

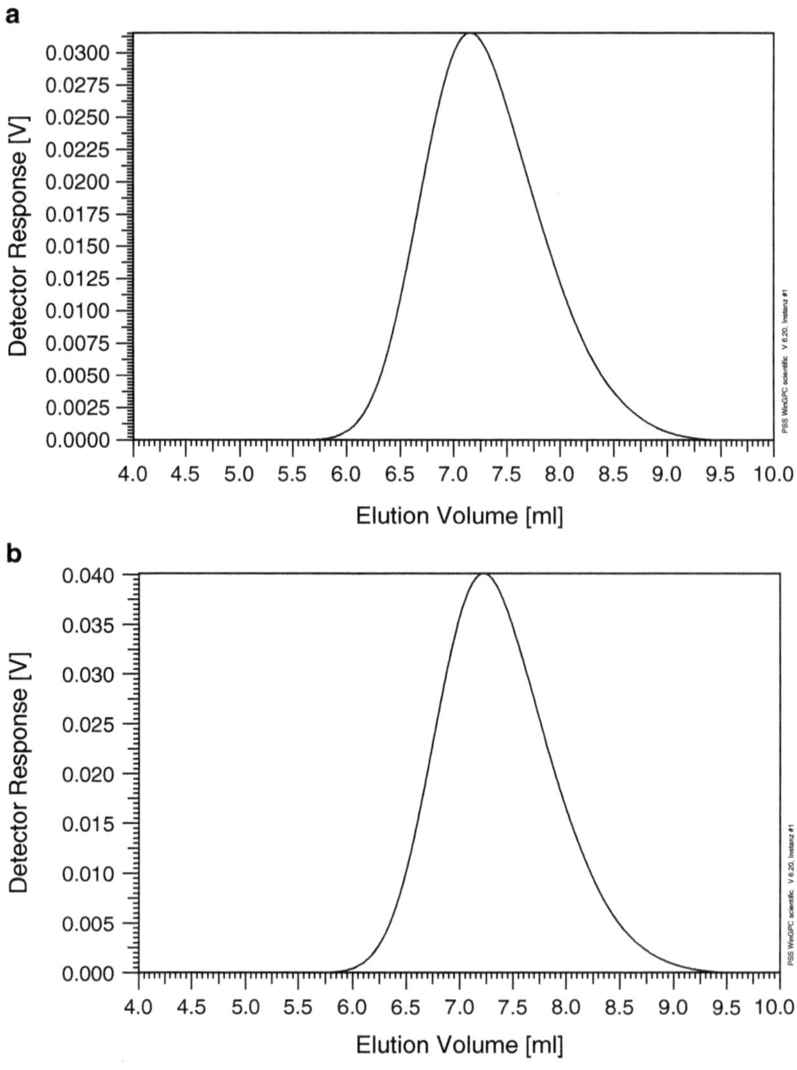

Fig. 5.15 SEC of the copolymers A (**a**) and B (**b**) under optimum conditions; stationary phase: GRAM XL, mobile phase: DMAC + 6 g/L (100 mM) AcOH + 3 g/L LiBr, detector: RI (reprinted from [71] with permission of European Polymer Federation)

scattering (LS) traces of the samples are monomodal. This is a strong indication that qualitatively the separation behaviour of the methylated and the original samples is similar.

From the combined RI and LS traces, the molar mass distributions of the samples can be calculated without a molar mass calibration. To do the calculations, two essential pieces of information are required—the injected sample amount and the refractive index increment dn/dc. The latter has to be determined after the SEC

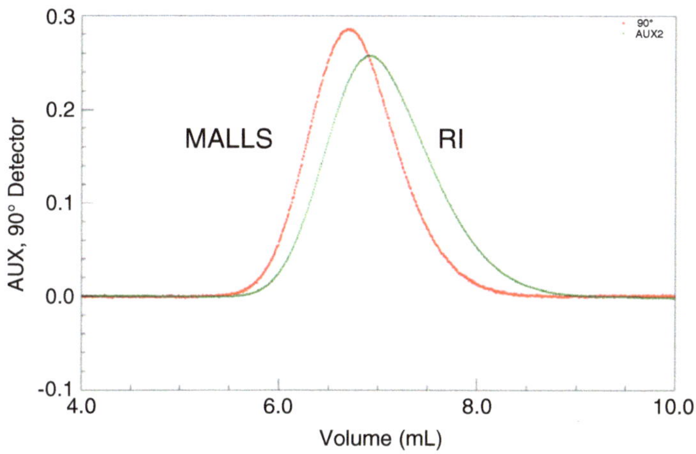

Fig. 5.16 SEC of copolymer B; stationary phase: GRAM XL, mobile phase: DMAC + 6 g/L AcOH + 3 g/L LiBr, detector: RI and MALLS (measuring angle 90°) (reprinted from Ref. [73] with permission of European Polymer Federation)

separation to ensure that the Donnan equilibrium is reached, meaning that the salt concentration inside the polymer coils and in the mobile phase are equal. The dn/dc was calculated by assuming 100% mass recovery of the sample. As for the exact determination of the injected sample amount, one can encounter problems related to the solubility of the samples. Since the samples have to be filtered prior to SEC, a good filterability is a precondition for full recovery after filtration. This is indeed the case for samples A–C, while for samples D and E a recovery of 100% cannot be assumed due to poor filterability and so the estimated dn/dc may be incorrect. Since $1/M_w$ is proportional to $(dn/dc)^2$ the calculation can give too high M_w values.

Figure 5.17 shows the molar mass distributions of two representative samples. As can be seen, the function $\log M = f(V_e)$ in all cases indicates typical SEC behaviour. This is additional proof for the finding that at the present experimental conditions the samples elute in the typical SEC mode. There are no indications for adsorptive or other non-SEC effects.

A comparison of the molar masses determined by SEC-MALLS and the molar masses of the methylated samples is given in Table 5.8. The largest deviation between both sets of data is observed for the samples with the smallest dn/dc. At the same time these samples are difficult to filter. The differences of the M_w values for the methylated samples C, D and E are due to the different hydrodynamic behaviours of these acrylate copolymers in THF and DMAC compared to pure PMMA. Only the methylations of A and B give pure PMMA and, therefore, these samples show the same molar masses in THF and in DMAC. The M_w data indicate that the calculated molar masses of the original samples taken from SEC-MALLS differ to a certain extent from the molar masses of the methylated samples taken from RI detection and PMMA calibration. This is reasonable since, in addition to

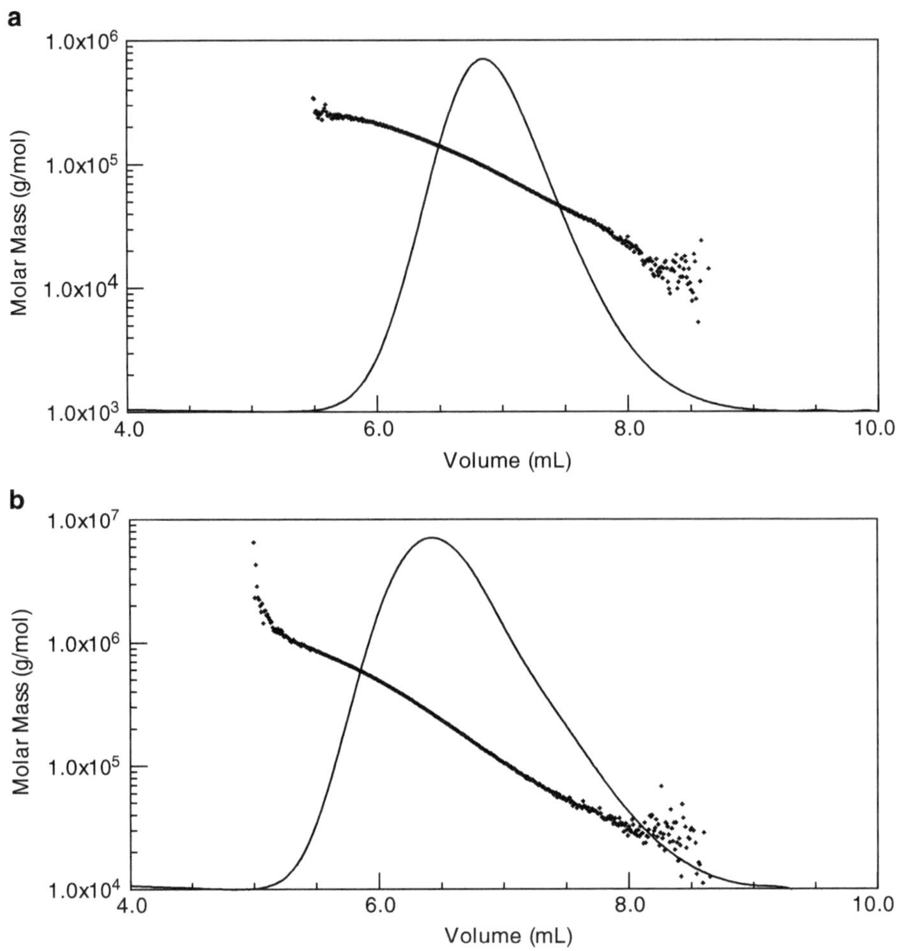

Fig. 5.17 Molar mass vs. elution volume functions of the copolymers A (**a**) and C (**b**), stationary phase: GRAM XL, mobile phase: DMAC + 6 g/L AcOH + 3 g/L LiBr, detector: RI and MALLS (reprinted from [71] with permission of European Polymer Federation)

Table 5.8 Molar masses of the original (non-methylated) samples determined by SEC-MALLS and of the methylated samples, methylated samples are calculated as PMMA equivalents

Sample	M_w (kg/mol) methylated	M_w (kg/mol) SEC-MALLS	d*n*/d*c* (mL/g)
A	95.0	100.0	0.071
B	90.0	90.0	0.081
C	270.0	244.0	0.066
D	808.0	732.0	0.039
E	271.0	306.0	0.048

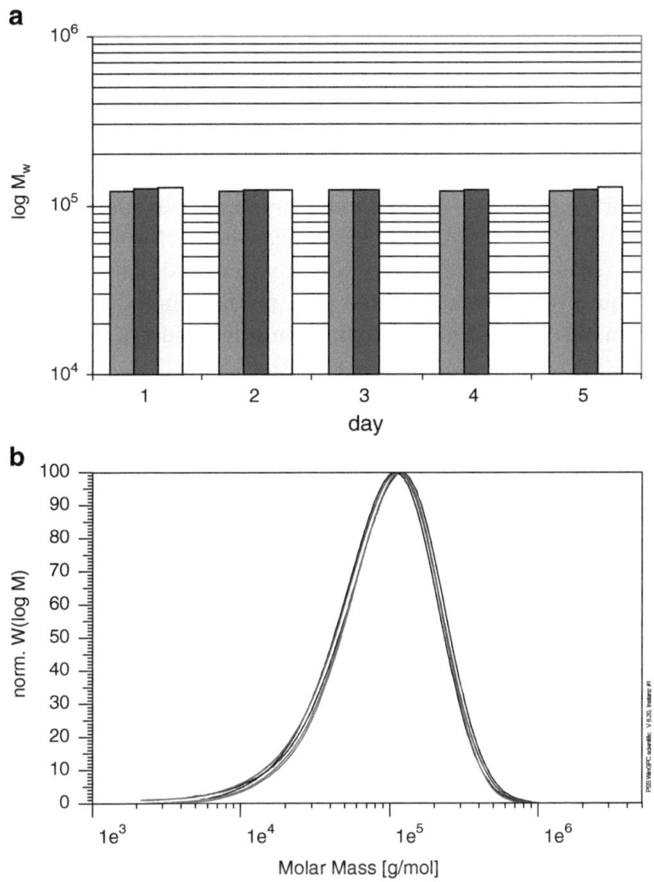

Fig. 5.18 Molar mass analyses of sample A at three different laboratories, (**a**) comparison of M_w data, (**b**) overlay of molar mass distributions from different laboratories, stationary phase: GRAM linear XL, mobile phase: DMAC + 6 g/L AcOH + 3 g/L LiBr, detector: RI (reprinted from [71] with permission of European Polymer Federation)

possible sensitivity and response problems of the different detectors, the hydrodynamic behaviour of the carboxy containing copolymers is different from that of the neutral methylated copolymers.

Evaluation

The final step in this application was the investigation of the robustness and the reproducibility of the SEC method. For a reproducibility test among three different laboratories the one-column set was used. The same samples were measured at five consecutive days on different equipment and with manual and automated injection. Selected results are presented as diagrams in Fig. 5.18 together with overlays of molar mass distributions measured at the three laboratories. These results indicate excellent reproducibility of the method.

5.2.2 RI-Viscometer Detection

Another very useful approach to obtaining molar mass information of complex polymers is the coupling of SEC to a viscosity detector [83–88]. The viscosity of a polymer solution is closely related to the molar mass (and architecture) of the polymer molecules. The product of polymer intrinsic viscosity $[\eta]$ times molar mass is proportional to the size of the polymer molecule (the hydrodynamic volume). Viscosity measurements in SEC can be performed by measuring the pressure drop ΔP across a capillary, which is proportional to the viscosity η of the flowing liquid (the viscosity of the pure mobile phase is denoted as η_0). The relevant parameter $[\eta]$ is defined as the limiting value of the ratio of specific viscosity $\left(\eta_{sp} = (\eta - \eta_0)/\eta_0\right)$ and concentration C for $C \to 0$:

$$[\eta] = \lim(\eta - \eta_0)/\eta_0 C = \lim \ \eta_{sp}/C \qquad \text{for } C \to 0. \tag{5.11}$$

The viscosity of a polymer solution as compared to the viscosity of the pure solvent is measured by the pressure drop ΔP across an analytical capillary-transducer system. The specific viscosity is obtained from $\Delta P/P$, where P is the inlet pressure of the system. As the concentrations in SEC are usually very low, $[\eta]$ can be approximated by η_{sp}/C.

Once it is possible to determine $[\eta]$ as a function of elution volume, one can now compare the hydrodynamic volumes V_h for different polymers. The hydrodynamic volume is, through Einstein's viscosity law, related to intrinsic viscosity and molar mass by $V_h = [\eta]M/2.5$. Einstein's law is, strictly speaking, valid only for impenetrable spheres at infinitely low volume fraction of the solute (equivalent to concentration at very low values). However, it can be extended to particles of other shapes, defining the particle radius then as the radius of a hydrodynamically equivalent sphere. In this case V_h is defined as the molar volume of impenetrable spheres which would have the same frictional properties or enhance viscosity to the same degree as the actual polymer in solution.

Assuming the validity of this approach and in agreement with the SEC mechanism, similar elution volumes correspond to similar hydrodynamic volumes.

$$V_{e,1} = V_{e,2} \quad \to \quad M_1 \ [\eta]_1 = M_2 \ [\eta]_2. \tag{5.12}$$

In a plot of log $(M[\eta])$ versus V_e identical calibration lines should be found for the two polymers 1 and 2, irrespective of their chemical composition. This "universal calibration" approach has been predicted and experimentally proved by Benoit et al. [89]. As a consequence, using the universal calibration curve established with known calibration standards (for example polystyrene), one can obtain the SEC-molar mass calibration for an unknown polymer sample.

The intrinsic viscosity is a function of molar mass via the Mark-Houwink relationship, wherein K and a are coefficients for a given polymer in a given solvent at a given temperature.

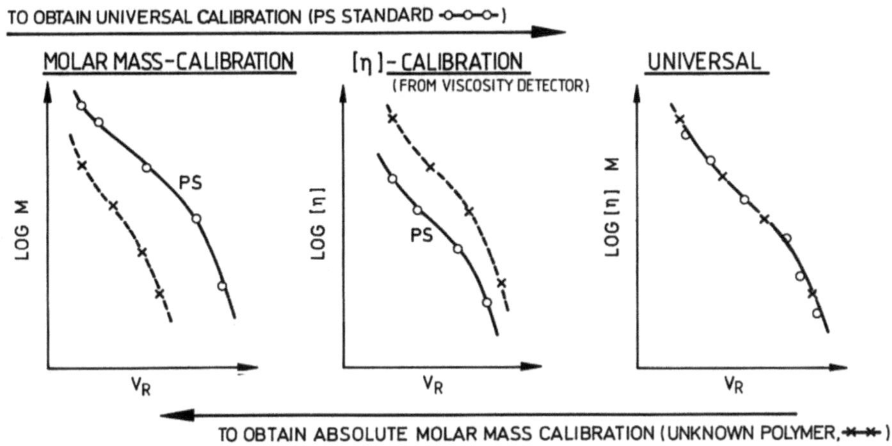

Fig. 5.19 Determination of absolute molar masses via universal SEC calibration (reprinted from [15] with permission of Springer Science + Business Media)

$$[\eta] = K\ M^a. \tag{5.13}$$

Combination of Eqs. (5.12) and (5.13) yields

$$K_1 {M_1}^{a(1)+1} = K_2 {M_2}^{a(2)+1}. \tag{5.14}$$

If a column has been calibrated with polymer 1 (e.g. polystyrene), the calibration line for polymer 2 can be calculated, provided that the coefficients K and a are known for both polymers with sufficient accuracy:

$$\ln M_2 = (1/1 + a_2)\ln(K_1/K_2) + (1 + a_1/1 + a_2)\ln M_1. \tag{5.15}$$

Thus, the concept of universal calibration provides an appropriate calibration also for polymers for which no calibration standards exist. The limiting factor of this approach is the accuracy of determining K and a. There are very high variations in the values reported in literature [2, 90]. The values of even common polymers such as PS and PMMA may differ considerably.

If the Mark-Houwink coefficients are not available, a universal calibration curve is established using polystyrene calibration standards and the SEC-viscometer combination. The basic steps involved in the MMD analysis are summarized in Fig. 5.19. First, the universal calibration curve of the SEC separation system has to be established by using narrow molar mass standards as indicated by the top arrow pointing to the right. Once the universal calibration curve is established, one can then reverse the procedure, by going from right to left following the bottom arrow, to obtain the molar mass calibration curve of any unknown polymer. The calibration curve is obtained literally by substracting the $[\eta]$ calibration curve of the unknown sample from the universal calibration curve. The $[\eta]$ calibration curve for the unknown sample is obtained from the on-line viscometer [35].

The application of RI and differential viscometer detection in SEC has been discussed by a number of authors [91–93]. Lew et al. presented the quantitative analysis of polyolefins by high-temperature SEC and dual refractive index-viscosity detection [94]. They applied a systematic approach for multidetector operation, assessed the effect of branching on the SEC calibration curve, and used a signal averaging procedure to better define intrinsic viscosity as a function of retention volume. The combination of SEC with RI, UV, and viscosity detectors was used to determine molar mass and functionality of polytetrahydrofurane simultaneously [95]. Long chain branching in EPDM copolymers by SEC-viscometry was analysed by Chiantore et al. [96].

One of the difficult problems in characterizing copolymers and polymer blends by SEC-viscometry is the accurate determination of the polymer concentration across the SEC elution curve. The concentration detector signal is a function of the chemical drift of the sample under investigation. To overcome this problem, Goldwasser proposed a method, where no concentration detector is required for obtaining M_n data [97]. In the usual SEC-viscometry experiment, the determination of the intrinsic viscosity at each slice of the elution curve requires a viscosity and a concentration signal:

$$[\eta]_i = (\ln\ \eta_{\mathrm{rel}}/C)_i, \tag{5.16}$$

where $\ln \eta_{\mathrm{rel}}$ is the direct detector response of the viscometer. From Eq. (5.16) one calculates the molar mass averages by

$$M_{\mathrm{n}} = \Sigma\ C_i/\Sigma\ \left[C_i/(V_{\mathrm{h,x}}/[\eta])_i\right], \tag{5.17}$$

$$M_{\mathrm{w}} = \Sigma\ C_i\ (V_{\mathrm{h,x}}/[\eta])_i/\Sigma\ C_i, \tag{5.18}$$

where $V_{\mathrm{h,x}} = [\eta]_{\mathrm{x}} M_{\mathrm{x}}$ is the data retrievable from the universal calibration curve. By rearranging Eq. (5.17) using Eq. (5.16) the following expression is obtained:

$$\begin{aligned} M_{\mathrm{n}} &= \Sigma\ C_i/\Sigma\ (\ln\ \eta_{\mathrm{rel}}/V_{\mathrm{h,x}})_i \quad \text{or} \\ M_{\mathrm{n}} &= \text{sample amount}\ /\Sigma\ (\ln\ \eta_{\mathrm{rel}}/V_{\mathrm{h,x}})_i \end{aligned} \tag{5.19}$$

The sample amount can be easily determined from the injection volume and the sample concentration, and no information from a concentration detector is required. With this approach, the M_n value of any polymer sample can be determined by SEC using only the viscosity detector. Other molar mass averages, however, cannot be determined. The advantage of the Goldwasser M_n method is that it can access much wider molar mass ranges than other existing methods like osmometry or endgroup methods.

Due to some problems encountered with SEC-LS and SEC-viscometry, a triple-detector SEC technology is frequently used, where three on-line detectors are used together in a single SEC system. In addition to the concentration detector, an on-line viscometer and a LS instrument are coupled to the SEC (TriSEC). With TriSEC, absolute molar mass determination is possible for polymers that are very different in chemical composition and molecular conformation. The usefulness of the TriSEC approach has been demonstrated in a number of applications. Pang and Rudin showed that only by using both viscometer and LS detection accurate molar mass distributions are obtained [29]. Wintermantel et al. have developed a custom-made multidetector instrument and demonstrated that it has great potential not only for absolute molar mass determinations but also for structure characterization of linear flexible, semiflexible, and branched polymers [30]. Degoulet et al. characterized polydisperse solutions of branched PMMA [31], while Jackson et al. investigated linear chains of varying flexibility in order to prove universal calibration [32]. Yau and Arora discussed the advantages of TriSEC for the determination of Mark-Houwink coefficients, long-chain branching, and polymer architecture [33].

5.2.2.1 Analysis of Polystyrene Stars by SEC-Viscometry [98]

Aim

Star-shaped polymers exhibit unusual flow and viscosity properties compared to linear analogues. Such polymers may be constructed with several arms radiating from a central core, either by preparing the individual arms and attaching them a central molecule or by growing the polymer arms from a central core. A number of commercial polymers can be constructed with a star-branched morphology relatively easily, their characterization, however, is still a challenge. In the present application, online coupled SEC-viscometry shall be used for the analysis of star-shaped PSs.

Materials

- **Calibration Standards**. The samples under investigation are star-shaped PSs with different numbers of arms. Three samples were investigated with 5, 14 and 21 arms, respectively.

Equipment

- **Chromatographic System**. PL-GPC 220.
- **Columns**. Two PLgel Mixed-C columns with average pore sizes of 5 μm.
- **Mobile Phase**. THF, HPLC grade.
- **Detectors**. RI and viscometer of the integrated PL-GPC 220.
- **Column Temperature**. 40 °C.
- **Sample Concentration**. 1.0 g/L.
- **Injection Volume**. 100 μL.

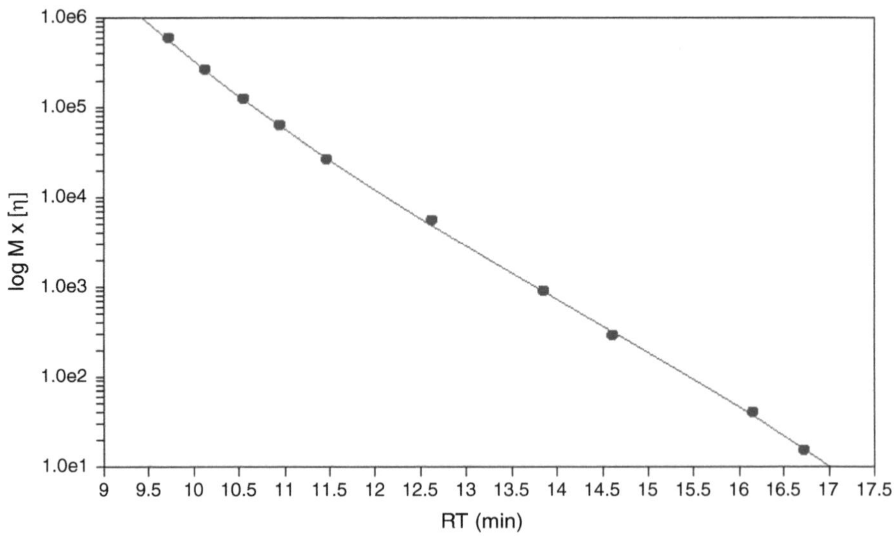

Fig. 5.20 Universal calibration curve based on linear PS (adapted from [98])

Preparatory Investigations

A universal calibration curve was generated as described in Sect. 5.2.2. A set of linear PS standards with narrow molar mass dispersity has been used. The universal calibration curve is presented in Fig. 5.20.

Separations

The samples were measured using the present SEC system with dual RI and viscometer detection. The chromatograms based on the two detector traces are shown in Fig. 5.21.

Evaluation

Based on the universal calibration the molar mass averages and the weight-average intrinsic viscosity were calculated, see Table 5.9.

Using these data Mark-Houwink plots can be generated. These are compared to the Mark-Houwink plot of a broad linear PS, see Fig. 5.22. As expected, an increasing number of arms results in a decreasing intrinsic viscosity, as can be seen for the 5-arm and the 14-arm PS. The 21-arm PS showed a rather unusual viscosity behaviour: it varied strongly with molar mass.

Based on the linear regions in the Mark-Houwink plots for the stars, the intrinsic viscosity contraction factor g' can calculated as a function of molar mass using the following equation:

$$g' = [\eta]_{\text{star}}/[\eta]_{\text{linear}}. \tag{5.20}$$

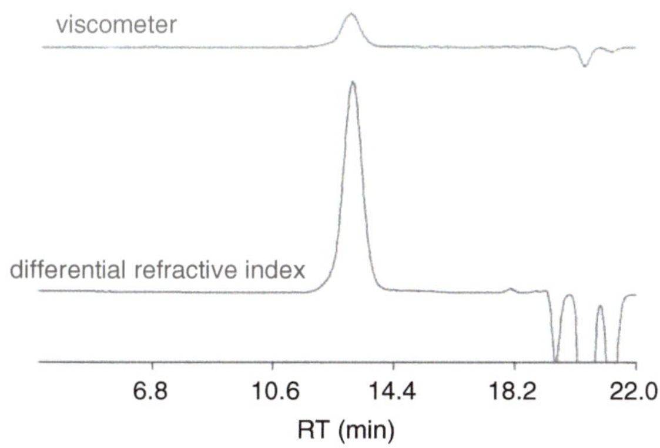

Fig. 5.21 SEC separation of the 14-arm star-shaped PS with RI and viscometer detection (adapted from [98])

Table 5.9 Molar masses and viscosities of the star-shaped PS

Sample	M_n (kg/mol)	M_w (kg/mol)	M_z (kg/mol)	M_v (kg/mol)	M_w/M_n	$[\eta]$
5-Arm	10.5	64.8	98.6	46.3	6.2	0.28
14-Arm	26.8	29.3	32.4	28.7	1.1	0.10
21-Arm	111.4	157.9	201.2	141.3	1.4	0.21

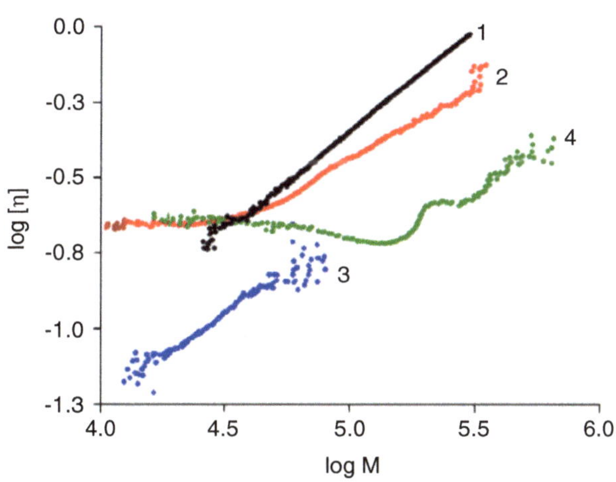

Fig. 5.22 Mark-Houwink plots for a linear PS (black, 1) and the star-shaped PS with 5 arms (red, 2), 14 arms (blue, 3) and 21 arms (green, 4) (adapted from Ref. [98])

From the g' data, the radius of the gyration contraction factor g was calculated using the following empirical relationship, where $a = 1.104$, $p = 7$ and $b = 0.906$ [99]:

$$g' = [a + (1 - a)g^p]g^b. \tag{5.21}$$

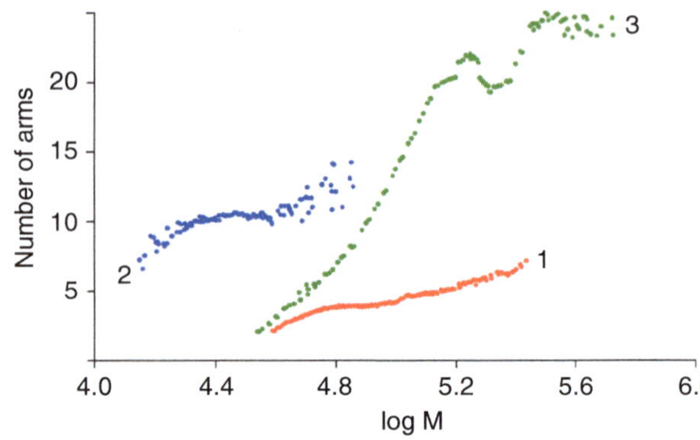

Fig. 5.23 Functionality plots for the star-shaped PS with 5 arms (red, 1), 14 arms (blue, 2) and 21 arms (green, 3) (adapted from Ref. [98])

Using the calculated *g* values, the functionality '*f*' for the stars, which is the theoretical number of arms, was calculated using the following model based on the assumption that the arms were 'random', that is, polydisperse in molar mass. For random stars with f number of polydisperse arms [100, 101]:

$$g = 3f\ /(f+1)^2. \tag{5.22}$$

Figure 5.23 shows an overlay of *f* for the stars as a function of the log molar mass. The random model gave a prediction of the functionality *f* which was in good agreement with the value expected from the synthesis. However, for all of the star-shaped PS, especially the 21-arm PS, the calculated value of *f* increased sharply with molar mass indicating that a considerable portion of the sample at low molar mass contained components with fewer than expected arms. Due to the low separation efficiency of SEC the macromolecules with different numbers of arms cannot be fully separated from each other. This problem will be addressed more in detail in Sect. 6.5.3.

References

1. Pasch H, Trathnigg B (1998) HPLC of polymers. Springer, Berlin
2. Glöckner G (1991) Gradient HPLC and chromatographic cross-fractionation. Springer, Berlin
3. Yau WW, Kirkland JJ, Bly DD (1979) Modern size exclusion chromatography. Practice of gel permeation and gel filtration chromatography. Wiley Interscience, New York, NY
4. Janca J (1984) Steric exclusion liquid chromatography. Dekker, New York, NY
5. Glöckner G (1987) Polymer characterization by liquid chromatography. Elsevier, Amsterdam
6. Mori S, Barth HG (1999) Size exclusion chromatography. Springer, Berlin
7. Provder T (ed) (1980) Size exclusion chromatography. ACS Symp Ser 138. ACS, Washington, DC
8. Provder T (ed) (1984) Size exclusion chromatography. ACS Symp Ser 245. ACS, Washington, DC
9. Provder T (ed) (1987) Detector and data analysis in size exclusion chromatography. ACS Symp Ser 352. ACS, Washington, DC

10. Provder T (ed) (1993) Size exclusion chromatography. ACS Symp Ser 521. ACS, Washington, DC
11. Provder T, Barth HG, Urban MW (eds) (1995) Chromatographic characterization of polymers. Adv Chem Ser 247. ACS, Washington, DC
12. Barth HG, Boyes BE, Anal Chem 62 (1990) 381R, 64 (1992) 428R, 66 (1994) 595R, 68 (1996) 445R, 70 (1998) 251R
13. Striegel AM (2000) In: Striegel AM (ed) Multiple detection in size exclusion chromatography. ACS Symp Ser 893. ACS, Washington, DC
14. Striegel AM, Yau WW, Kirkland JJ, Bly DD (2009) Modern size-exclusion liquid chromatography. Practice of gel permeation and gel filtration chromatography. Wiley, Hoboken, NJ
15. Pasch H (2000) Adv Polym Sci 150:1
16. Mori S, Uno Y (1987) Anal Chem 59:90
17. Maliakal AS, O'Shaughnessy B, Turro NJ (2005) In: Striegel AM (ed) Multiple detection in size exclusion chromatography. ACS Symp Ser 893. ACS, Washington, DC
18. Kendrick BS (2005) In: Striegel AM (ed) Multiple detection in size exclusion chromatography. ACS Symp Ser 893. ACS, Washington, DC
19. Yokoyama WH, Knuckles BE (2005) In: Striegel AM (ed) Multiple detection in size exclusion chromatography. ACS Symp Ser 893. ACS, Washington, DC
20. Trathnigg B (1990) J Liq Chromatogr 13:1731
21. Trathnigg B (1991) J Chromatogr 552:505
22. Trathnigg B, Yan X (1992) Chromatographia 33:467
23. Podzimek S (1994) J Appl Polym Sci 54:91
24. Jackson C, Barth HG (1995) In: Wu CS (ed) Molecular weight sensitive detectors for size exclusion chromatography, Chapter 4. Marcel Dekker, New York, NY
25. Reed WF (2005) In: Striegel AM (ed) Multiple detection in size exclusion chromatography. ACS Symp Ser 893. ACS, Washington, DC
26. Cotts PM (2005) In: Striegel AM (ed) Multiple detection in size exclusion chromatography. ACS Symp Ser 893. ACS, Washington, DC
27. Striegel AM (2005) In: Striegel AM (ed) Multiple detection in size exclusion chromatography. ACS Symp Ser 893. ACS, Washington, DC
28. Podzimek S (2005) In: Striegel AM (ed) Multiple detection in size exclusion chromatography. ACS Symp Ser 893. ACS, Washington, DC
29. Pang S, Rudin A (1992) Polymer 33:1949
30. Wintermantel M, Antonietti M, Schmidt M (1933) J Appl Polym Sci Appl Polym Symp 52:91
31. Degoulet C, Nicolai T, Durand D, Busnel JP (1995) Macromolecules 28:6819
32. Jackson C, Chen YJ, Mays JW (1996) J Appl Polym Sci 61:865
33. Yau WW, Arora KS (1994) Polym Mater Sci Eng 69:210
34. Jackson C, Barth HG (1994) Trends Polym Sci 2:203
35. Yau WW (1990) Chemtracts-Macromol Chem 1:1
36. Runyon JR, Barnes DE, Rudel JF, Tung LH (1969) J Appl Polym Sci 13:2359
37. Ogawa T (1979) J Appl Polym Sci 23:3515
38. Revillon A (1980) J Liq Chromatogr 3:1137
39. Tung LH (1979) J Appl Polym Sci 24:953
40. Trathnigg B, Jorde C (1987) J Chromatogr 385:17
41. Trathnigg B, Feichtenhofer S, Kollroser M (1997) J Chromatogr A 786:75
42. Johann C, Kilz P (1989) In: Proceedings of the 1st international conference on molar mass characterization, Bradford, UK
43. Dawkins JV (1995) Compositional heterogeneity of copolymers by coupled techniques with chromatographic columns and multidetectors. In: Provder T, Barth HG, Urban MW (eds) Chromatographic characterization of polymers. Hyphenated and multidimensional techniques, Chapter 15. Adv Chem Ser 247. American Chemical Society, Washington, DC
44. Meehan E, O'Donohue S, McConville JA (1993) Polym Mater Sci Eng 69:269

45. Gores F, Kilz P (1993) Copolymer characterization using conventional SEC and molar mass-sensitive detectors. In: Provder T (ed) Chromatography of polymers, Chapter 10. ACS Symp Ser 521. American Chemical Society, Washington, DC
46. Lee HC, Ree M, Chang T (1995) Polymer 36:2215
47. Pasch H, Trathnigg B, Augenstein M (1994) Macromol Chem Phys 195:743
48. Dayal U, Mehta SK (1994) J Liq Chromatogr 17:303
49. Dayal D (1994) J Appl Polym Sci 53:1557
50. Jackson C, Yau WW (1993) J Chromatogr 645:209
51. Jackson C (1994) J Chromatogr A 662:1
52. Pang S, Rudin A (1993) In: Provder T (ed) Chromatography of polymers, ACS Symp Ser 521. American Chemical Society, New York, NY
53. Podzimek S (2011) Light scattering, size exclusion chromatography and asymmetric field flow fractionation. Wiley, Hoboken, NJ
54. Gores F, Kilz P (1993) In: Provder T (ed) Chromatography of polymers, Chapter 10, ACS Symp Ser 521. American Chemical Society, Washington, DC
55. Kratochvil P (1987) In: Jenkins AD (ed) Classical light scattering from polymer solutions, Polym Sci Libr 5. Elsevier, Amsterdam
56. Jeng L, Balke ST, Mourey TH, Wheeler L, Romeo P (1993) J Appl Polym Sci 49:1359
57. Jeng L, Balke ST (1993) J Appl Polym Sci 49:1375
58. Pille L, Solomon DH (1994) Macromol Chem Phys 195:2477
59. Mourey TH, Coll H (1994) Polym Mater Sci Eng 69:217
60. Mourey TH, Coll H (1995) J Appl Polym Sci 56:65
61. Grubisic-Gallot Z, Gallot Y, Sedlacek J (1994) Macromol Chem Phys 195:781
62. Grubisic-Gallot Z, Gallot Y, Sedlacek J (1995) J Liq Chromatogr 18:2291
63. Laun S, Pasch H, Longieras N, Degoulet C (2008) Polymer 49:4502
64. Chen J, Radke W, Pasch H (2003) Macromol Symp 193:107
65. Mourey TH, Bryan TG (2002) J Chromatogr A 964:169
66. Wang PJ, Rivard RJ (1987) J Liq Chromatogr 10:3059
67. Drott EE (1977) J Chromatogr Sci 8:41
68. Drott EE (1977) In: Cazes J (ed) Liquid chromatography of polymers and related materials. Marcel Dekker, New York, NY, p 41
69. Schorn H, Kosfeld R, Hess M (1983) J Chromatogr 282:579
70. Buijtenhuijs FA, van de Riet AMC (1997) Polym Mater Sci Eng 77:40
71. Adler M, Pasch H, Meier C, Senger R, Koban HG, Augenstein M, Reinhold G (2004) e-polymers no. 055
72. Wade A, Weller PJ (eds) (1994) Handbook of pharmaceutical excipients, 2nd edn. Pharmaceutical Press, London
73. Barth HG (1987) In: Provder T (ed) ACS Symp Ser 352, Chapter 2, American Chemical Society, New York, NY
74. Barth HG, Regnier FE (1980) J Chromatogr 192:275
75. Kato T, Tokuya T, Nozaki T, Takahashi A (1984) Polymer 25:218
76. Callec G, Anderson AW, Tsao GT, Rollings JE (1984) J Polym Sci Polym Chem 22:287
77. Muller G, Yonnet C (1984) Makromol Chem Rapid Commun 5:197
78. Mori S (1989) Anal Chem 61:530
79. Dubin PL, Principi JM (1989) J Chromatogr 479:159
80. Visser S, Slangen CJ, Robben AJPM (1992) J Chromatogr 599:205
81. Ahmed F, Modrek B (1992) J Chromatogr 599:25
82. Wittgren B, Welinder A, Porsch B (2003) J Chromatogr A 1002:101
83. Styring M, Armonas JE, Hamielec AE (1987) In: Provder T (ed) Detection and data analysis in size exclusion chromatography, ACS Symp Ser 352. American Chemical Society, Washington, DC

84. Haney MA, Armonas JE, Rosen L (1987) In: Provder T (ed) Detection and data analysis in size exclusion chromatography, ACS Symp Ser 352. American Chemical Society, Washington, DC
85. Brower L, Trowbridge D, Kim D, Mukherjee P, Seeger R, McIntyre D (1987) In: Provder T (ed) Detection and data analysis in size exclusion chromatography, ACS Symp Ser 352. American Chemical Society, Washington, DC
86. Mori S (1993) J Chromatogr 637:129
87. Cheung P, Lew R, Balke ST, Mourey TH (1993) J Appl Polym Sci 47:1701
88. Lesec J, Millequant M, Havard T (1993) In: Provder T (ed) Chromatography of polymers, ACS Symp Ser 521. American Chemical Society, Washington, DC
89. Benoit H, Rempp P, Grubisic Z (1967) J Polym Sci B5:753
90. Kurata M, Tsunashima Y, Iwama, M, Kamada K (1975) In: Brandrup J, Immergut I (eds) Polymer handbook. Wiley, New York, NY, p IV-1
91. Sanayei RA, Suddaby KG, Rudin A (1993) Makromol Chem 194:1953
92. Puskas JE, Hutchinson R (1993) Rubber Chem Technol 66:742
93. Xie J (1994) Polymer 35:2385
94. Lew R, Suwanda D, Balke ST, Mourey TH (1993) J Appl Polym Sci Appl Polym Symp 52:125
95. Harrison CA, Mourey TH (1995) J Appl Polym Sci 56:211
96. Chiantore O, Lazzari M, Caldari S, Zecchi E (1998) Proceedings of the 11th International Symposium on Polymer Analysis and Characterization, Santa Margerita Ligure, Italy
97. Goldwasser JM (1989) In: Proceedings of the International GPC Symposium, Newton, MA
98. http://www.chromatographyonline.com/lcgc/data/articlestandard/lcgceurope/502004/136920/article.pdf
99. Weissmuller M, Burchard W (1997) Polym Int 44:380
100. Burchard W (1983) Adv Polym Sci 48:1
101. Burchard W (1977) Macromolecules 10:919

6 Two-Dimensional Liquid Chromatography

6.1 Introduction

State-of-the-art polymeric materials possess property distributions in more than one parameter of molecular heterogeneity. Copolymers, for example, are distributed in molar mass and chemical composition, while telechelics and macromonomers are distributed frequently in molar mass and functionality. It is obvious that n independent properties require n-dimensional analytical methods for accurate (independent) characterization of the different structural parameters. In chromatography, the separation efficiency of any single separation method is limited by the efficiency and selectivity of the separation mode, i.e. the plate count of the column and the phase system selected. Adding more columns will not overcome the need to identify more components in a complex sample, due to the limitation of peak capacities. The peak capacity in an isocratic separation can be described, following Grushka [1], as in Eq. (6.1):

$$n = 1 + \frac{\sqrt{N}}{4} \cdot \ln \frac{V_p}{V_0}, \tag{6.1}$$

where n is the peak capacity, N is the plate number, V_p is the pore volume, and V_0 is the interparticle volume.

The corresponding peak capacity in a n-dimensional separation is enormously higher due to the fact that each dimension contributes to the total peak capacity as a factor and not as an additive term for single dimension methods as described in Eq. (6.2):

$$n_{total} = \Pi\, n_i \cdot \sin^{(i-1)} \vartheta_i, \tag{6.2}$$

where n_{total} represents the total peak capacity, n_i the peak capacity in dimension i and ϑ_i is the "angle" between two dimensions. The angle between dimensions is determined by the independence of the methods; a 90° angle is obtained by two

H. Pasch and B. Trathnigg, *Multidimensional HPLC of Polymers*, Springer Laboratory,
DOI 10.1007/978-3-642-36080-0_6,

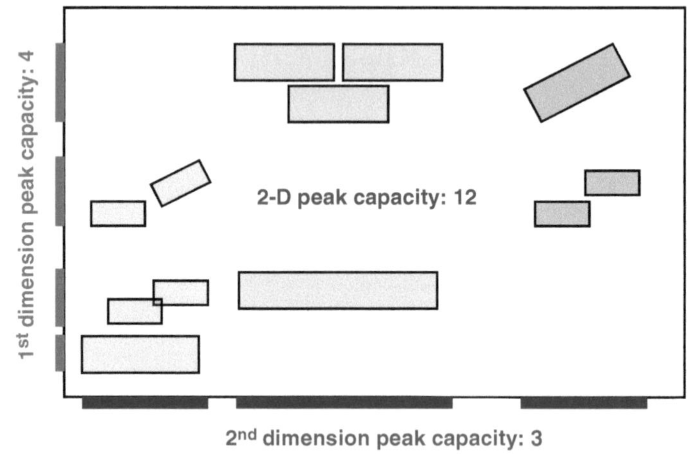

Fig. 6.1 Schematic contour map representation of resolution enhancement and peak capacity in 2D separations (peaks in each dimension are indicated by *colour bars* at the *axes*)

methods, which are completely independent of each other and will, for example, separate two properties solely on a single parameter without influencing themselves.

In the case of a two-dimensional (2D) system the peak capacity is given by Eq. (6.3):

$$n_{2D} = n_1 \cdot n_2 \cdot \sin\vartheta = \left(1 + \frac{\sqrt{N_1}}{4} \cdot \ln\frac{V_{p,1}}{V_{0,1}}\right) \cdot \left(1 + \frac{\sqrt{N_2}}{4} \cdot \ln\frac{V_{p,2}}{V_{0,2}}\right) \cdot \sin\vartheta. \quad (6.3)$$

This effect is schematically illustrated in Fig. 6.1.

Multidimensional chromatography separations can be done in planar systems or coupled-column systems. Examples of planar systems include two-dimensional thin-layer chromatography (TLC) [2, 3], where successive one-dimensional TLC experiments are performed at 90° angles with different solvents, and 2D electrophoresis, where gel electrophoresis is run in the first dimension followed by isoelectric focusing in the second dimension [4–6]. Hybrids of these systems where chromatography and electrophoresis are used in each spatial dimension were reported more than 40 years ago [7]. Belenkii et al. reported on the analysis of block copolymers by TLC [8, 9]. Diblock copolymers of styrene and t-butyl methacrylate were separated first with regard to chemical composition by TLC at critical conditions followed by a SEC-type separation to determine the molar masses of the components.

The main problem using planar methods is the difficulty in detection and collection of fractions among other less critical problems, such as homogeneous preparation of chromatographic media. However, the detection problem exists also for the coupled-column methods, mainly because of fraction dilution by each stage in a multidimensional seaparation system. Another aspect is the adjustment of chromatographic time bases between the different dimensions so that first dimension peaks may be sampled an adequate number of times by the next dimension separation system. This aspect has been studied in detail by Murphy et al. [10].

In 2D column chromatography systems an aliquot from a column or channel is transfered into the next separation method in a sequential and repetitive manner. Storage of the accumulating eluent is typically provided by sampling loops connected to an automated valve. Many variations on this theme exist which use various chromatographic and electrophoretic methods for one of the dimensions. In addition, the simpler "heart cutting" mode of operation takes the eluent from a first dimension peak or a few peaks and manually injects this into another column during the first dimension elution process. A partial compilation of these techniques is given in [8, 11–17].

The use of different modes of liquid chromatography facilitates the separation of complex samples selectively with respect to different properties like hydrodynamic volume, molar mass, chemical composition, architecture or functionality. Using these techniques in combination, multi-dimensional information on different aspects of molecular heterogeneity can be obtained. If, for example, two different chromatographic techniques are combined in a "cross-fractionation" mode, information on chemical composition distribution (CCD) and molar mass distribution (MMD) can be obtained. Literally, the term "chromatographic cross-fractionation" refers to any combination of chromatographic methods capable of evaluating the distribution in size and composition of copolymers. An excellent overview on different techniques and applications involving the combination of size exclusion chromatography (SEC) and gradient HPLC was published by Glöckner in [18].

In the SEC mode, the separation occurs according to the molecular size of a macromolecule in solution, which is dependent on its chain length, chemical composition, solvent and temperature. Thus, molecules of the same chain length but different composition may have different hydrodynamic volumes. Since SEC separates according to hydrodynamic volume, SEC in different eluents can separate a copolymer in two diverging directions. This principle of "orthogonal chromatography" was suggested by Balke and Patel [19–21]. The authors coupled two SEC instruments together so that the eluent from the first one flowed through the injection valve of the second one. At any desired retention time the flow through SEC 1 could be stopped and an injection made into SEC 2. The first instrument was operated with THF as the eluent and polystyrene gel as the packing, whereas for SEC 2 polyether bonded-phase columns and THF-heptane were used. Both instruments utilized SEC columns. However, whereas the first SEC was operating so as to achieve conventional molecular size separation, the second SEC was used to fractionate by composition, utilizing a mixed solvent to encourage adsorption and partition effects in addition to size exclusion. Consequently, independent information on both MMD and CCD could not be obtained from such an experiment.

Since "orthogonality" requires that each separation technique is totally selective towards an investigated property, it seems to be more advantageous to use a sequence of methods, in which the first dimension separates according to chemical composition. In this way quantitative information on CCD can be obtained and the resulting fractions eluting from the first dimension are chemically homogeneous.

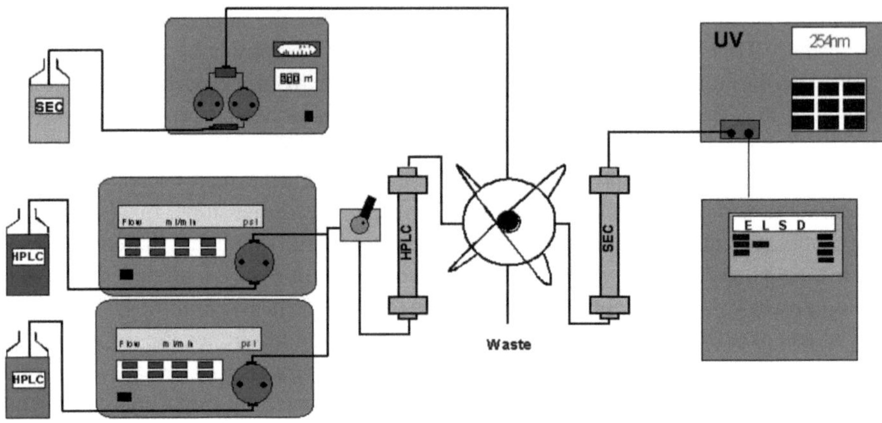

Fig. 6.2 General experimental set-up for a 2D chromatographic system (reprinted with permission from [24]. Copyright (2004) American Chemical Society)

These homogeneous fractions can then be analyzed independently in SEC mode in the second dimension to get the required MMD information. In such cases, SEC separation is strictly separating according to molar mass, and quantitative MMD information can be obtained.

6.2 Experimental Aspects of 2D Separations

Aspects of 2D Separations system is actually not as difficult as one might think at first. As long as well-known separation methods exist for each dimension the experimental aspects can be handled quite easily in most cases. Off-line systems just require a fraction collection device and something or someone who re-injects the fractions into the next chromatographic dimension. In online 2D systems the transfer of fractions is preferentially done by automatic injection valves, as proposed by Kilz et al. [16, 22, 23]. Figure 6.2 shows a general set-up for an automated 2D chromatography system.

The system is composed of two chromatographs, including pumps, injector and column for the first dimension, and a pump, column and detectors for the second dimension. The focal point in 2D chromatography separations is the transfer of fractions eluting from the first dimension into the second dimension. This can be done in various ways. The most simplistic approach is collecting fractions from one separation and manually transferring them into the second separation system. Obviously, this approach is prone to many errors, labour intensive and quite time-consuming.

A more efficient way of fraction transfer can be achieved by using electrically (or pneumatically) actuated valves equipped with two injection loops. Such a set-up allows one fraction to be injected and analyzed from one loop while the next

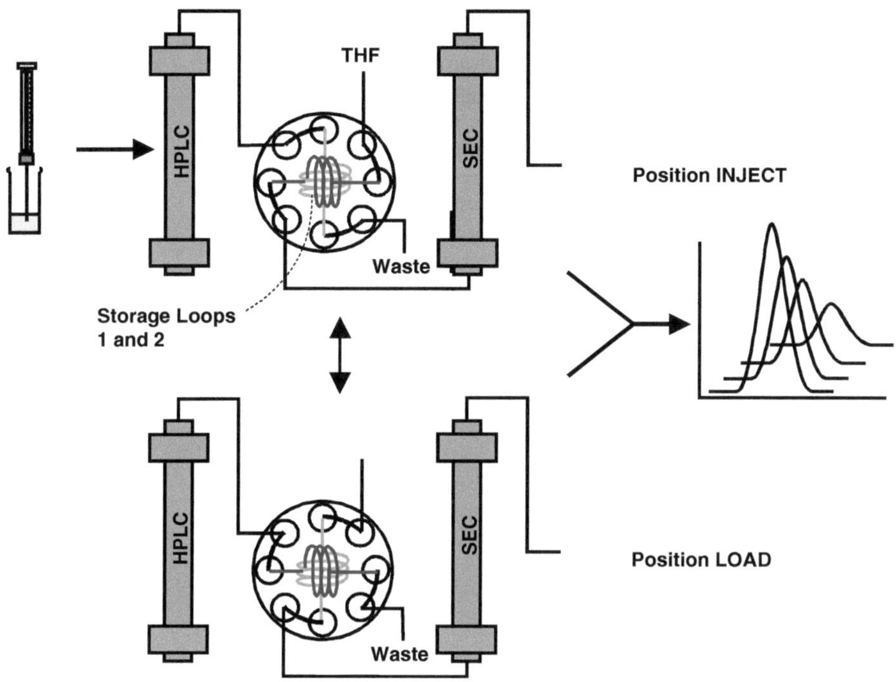

Fig. 6.3 Fraction transfer between chromatographic dimensions using a dual loop 8-port valve (reprinted from [25] with permission of Springer Science + Business Media)

fraction is collected at the same time in the second loop. The operation of such an automatic dual-loop system is schematically presented in Fig. 6.3.

When starting the separation in the first dimension, loop 1 is in the "LOAD" position, whereas loop 2 is in the "INJECT" position. Each loop has a volume of 50–200 μL, and exactly this volume is cut for each fraction from the first separation. When the first fraction leaves chromatograph 1, it enters loop 1 and fills it. When loop 1 is completely filled, the dual-loop valve switches to the opposite position, i.e. loop 1 is then connected with chromatograph 2 in the "INJECT" position, whereas loop 2 is connected to chromatograph 1 in the "LOAD" position. Now, loop 2 is filled with the next fraction from chromatograph 1 while the content of loop 1 is injected in chromatograph 2 for analysis. Total mass transfer from the first to the second dimension can be guaranteed by proper selection of flow rates in both dimensions [26]. This is a very beneficial situation compared to heart-cut transfers because by-products and trace impurities can be separated even if they are not visible in the first dimension separation. Table 6.1 summarizes potential fraction transfer options.

There are some other important aspects that have to be considered for optimum 2D experiment design, including the selection of separation techniques, the sequence of separation methods, and the detectability in the second dimension.

Table 6.1 Summary of 2D transfer injection options

Transfer	Mode	Advantages	Disadvantages	Example
Manual	Off-line	Very simple Fast set-up	Time-consuming Not for routine work Not precise No correlation of fraction elution to transfer time	Test tube
Automatic	Off-line	Simple Easy Fast set-up	Not quantitative Less precise No correlation of fraction elution to transfer time	Fraction collector Storage valve
Single-loop	On-line	Correct concentrations Correct transfer times Automation	Not quantitative Transfer not quantitative Set-up time	Injection valve (with actuation)
Dual-loop	On-line	Correct concentrations Correct transfer times Quantitative transfer Automation	Set-up time Special valve	8-Port actuated valve Combination of two 6-port valves

6.2.1 Sequence of Separation Methods

The sequence of separation methods is an important aspect in achieving the best resolution and most accurate determination of property distributions. It is advisable to use the method with the highest selectivity for the separation of one property as the first dimension. This ensures highest purity of eluting fractions being transferred into the subsequent separation. In the case of gradient HPLC and SEC as separation methods, early publications [19–21, 27, 28] used SEC as the first separation method, because it took much longer than a subsequent HPLC analysis. This is not the best set-up, however, because the SEC fractions are only monodisperse in hydrodynamic volume, but not in molar mass, chemical composition, etc. On the other hand, HPLC separations can be fine-tuned using gradients to fractionate only according to a single property, which can then be characterized for molar mass without any bias.

In many cases, interaction chromatography as the first dimension separation method is the best and most adjustable choice. From an experimental point of view, high flexibility is required for the first chromatographic dimension. In general, this is also easier achieved when running the interaction chromatography mode in the first dimension, because

(a) More parameters (mobile phase, mobile phase composition, mobile phase modifiers, stationary phase, temperature etc.) can be used to adjust the separation according to the chemical nature of the sample.
(b) Better fine-tuning in interaction chromatography allows for more homogeneous fractions.
(c) Sample load on such columns can be much higher as compared to SEC columns.

6.2.2 Detectability and Sensitivity in the Second Dimension

Because of the consecutive dilution of fractions, detectability and sensitivity become important criteria in 2D experiment design. If by-products and trace impurities have to be detected, only the most sensitive and/or selective detection methods can be employed. ELSD detection, despite several draw-backs, has been used mostly due to its high sensitivity for compounds that will not evaporate or sublime under detection conditions. Fluorescence and diode array UV/VIS are also sensitive detection methods, which can pick up samples at nanogram level. Mass spectrometers have a high potential in this respect too, however, they are currently not developed to such a state where they would be generally usable. The main problem here is not sensitivity but the analysis of high molar mass analytes without fragmentation.

Only in rare cases has RI detection, otherwise very popular in SEC, been used in multidimensional separations, because of its low sensitivity and strong dependence on mobile phase composition.

As a general rule, the higher the injection band dilution of a given separation method the more sensitive a subsequent detection method has to be. Such type of model calculations can be done easily; refer to a paper by Schure [29] for further details.

6.2.3 Other Experimental Factors Affecting 2D Separations

Depending on the specific type of the multidimensional experimental set-up, there are a number of other parameters to take into consideration. Two are listed here, but because they are specific to the method combination, this list reflects only those techniques that are most commonly used.

Influence of eluent transfer from first to second dimension: A very important aspect in multidimensional chromatography design is the compatibility of mobile phases that are transferred between the different dimensions. It is a necessity that the mobile phases in two consecutive stages in multidimensional separations are completely miscible. Otherwise the separation in the second method is dramatically influenced and the fraction transfer is restricted or completely hindered. In gradient systems, this requirement has to be verified for the total composition range. In SEC separations the transfer of mixed mobile phases can affect molar mass calibration. In order to get accurate molar mass results, the calibration curves have to be measured using the extremes of mobile phase composition and tested for changes in elution behaviour and pore size influence in the SEC column packing. The better the thermodynamic property of the SEC eluent, the less influence is expected on the SEC calibration, when the transfer of mobile phase from the previous dimension occurs. It has been shown to be advantageous to use the SEC eluent as one component of the mobile phase in the previous dimension to avoid potential interference and mobile phase incompatibility.

Time consumption: Time is an important issue when designing multidimensional experiments. Set-up time itself plays only a lesser role, but the time needed for the multidimensional separations themselves can be considerable. This is especially true for 2D separations using quantitative mass transfer via tandem-loop transfer valves. Heart-cut experiments require much less time and are often sufficient to check the applicability of the approach. Reducing the time consumption for multi-dimensional experiments is currently a heavily investigated topic. Several approaches are being investigated, and investigators are optimistic that experiment times will be reduced by a factor of about 10 for complete mass transfer experiments using optimized column sets and flow conditions.

Another time requirement in multidimensional separations is the time needed to reduce the amount of data and present them in an instructive way. With several dozen transfers between dimensions, data reduction and presentation can be very time consuming and has been a real burden for those who performed the first cross-fractionation experiments [19–21, 28]. There is a clear need for specialized multi-dimensional software, which does all the data acquisition, fraction transfer, valve switching, data reduction, data consolidation and presentation of results. Although there is a number of 2D chromatography systems available the most widely used is the system of Polymer Standards Service (Mainz, Germany) [26, 30]. A few laboratories use in-house solutions, which are specific to their own chromatography and data accumulation hardware and specific also to result calculation and report generation.

6.3 Separation Techniques for the First and Second Dimensions

In 2D chromatography, different modes of LC are combined. Depending on the individual technique, separation can be carried out with regard to molecular size, chemical composition, or architecture. An in-depth description of different LC techniques is given in the introductory chapter.

The most often used set-up for 2D chromatography is the combination of interaction chromatography and size exclusion chromatography. SEC is the standard technique for determining molar mass distributions by separating according to hydrodynamic volume. For complex polymers it must be considered, however, that hydrodynamic volume is not only a function of chain length (molar mass) but also of chemical composition and architecture. It is, therefore, not selective towards only one parameter of molecular heterogeneity.

Interaction chromatography, on the other hand, can be performed in a huge variety of different experimental set-ups. Normal phase or reversed phase systems using isocratic or gradient elution can be used. There is an abundance of stationary phases with different types of surface modifications of different polarities. This flexibility in experimental parameters is a very important criterion for using interaction chromatography as a first dimension method, since it can be fine-tuned to separate according to a given property more easily than most other chromatographic techniques.

Gradient HPLC has been useful for the characterization of copolymers [31–35]. In such experiments careful choice of separation conditions is imperative. Otherwise, low resolution for the polymeric sample will obstruct the separation. On the other hand, the separation in HPLC, dominated by enthalpic interactions, perfectly complements the entropic nature of the SEC retention mechanism in the characterization of complex polymer formulations.

Interaction chromatography is based on enthalpic interactions between the solute and the surface of the stationary phase. In pure interactive LC separations there are no entropic contributions to retention. The enthalpy change of the analyte corresponds to dispersion, polarization and charge-transfer interactions, as well as to H-bonding and ion exchange. The absence of entropic contributions to the separation is only possible if the stationary phase consists of non-porous beads or if the analyte molecules cannot penetrate into any of the pores of the stationary phase because of their size or interaction energy (e.g. ionic repulsion). In general, it will not be possible to avoid entropy changes in HPLC experiments with samples of different molar masses or sizes. In such cases it is best to select either a column that has very small or very large pores, which will force the molecules to be totally excluded from the stationary phase or to be totally permeating the pores of the packing. In both situations entropic contributions to the separation can be minimized.

In interactive and SEC modes either the enthalpy or the entropy dominates the separation. This is not the case in liquid chromatography at critical conditions (LCCC) where entropic and enthalpic interactions compensate each other [36–41]. The Gibbs free energy of the macromolecule remains constant when it penetrates the pores of the stationary phase ($\Delta G = 0$). The distribution coefficient K_d is unity, regardless of the size of the macromolecules, and all macromolecules of equal chemical structure elute from the chromatographic column in one peak. The term "chromatographic invisibility" is used to refer to this phenomenon. This means that the chromatographic behaviour is not directed by the size, but by the heterogeneities (chemical structure, architecture, endgroups, etc.) in the macromolecular chains [36–39]. This is particularly interesting for 2D-LC applications because in LCCC a nearly molar mass independent elution is achieved.

In general, as the Gibbs free energy is influenced by the length of the polymer chain and its chemical structure, the contributions of G_i for the polymer chain and G_j for the heterogeneity may be introduced as stated in Eq. (6.4):

$$\Delta G = \Sigma n_i \, \Delta G_i + \Sigma n_j \, \Delta G_j \,. \tag{6.4}$$

For a perfectly uniform homopolymer chain the free energy change is determined by the contribution of the repeating units of the polymer chain (Eq. 6.5):

$$\Delta G = \Sigma n_i \, \Delta G_i \,. \tag{6.5}$$

At the critical point of adsorption of the polymer chain of a complex polymer, however, the contribution G_i becomes zero and the chromatographic behaviour is exclusively directed by imperfections in the macromolecular chain (Eq. 6.6):

$$\Delta G = \Sigma n_j \Delta G_j . \tag{6.6}$$

This chromatographic effect can be employed to determine imperfections in the polymer chain selectively and without any contribution by the repeating units themselves. LCCC has been successfully used for the determination of the functionality type distribution of telechelics and macromonomers [42–46], for the analysis of block copolymers [47–49], macrocyclic polymers [50], and polymer blends [51–54].

Thus, LCCC represents a chromatographic separation technique yielding fractions that are homogeneous in one property (e.g. chemical composition), but polydisperse in a different property (e.g. molar mass). These fractions can readily be analyzed by SEC, which for chemically homogeneous fractions provides true molar mass distributions without interference of CCD or FTD. Therefore, the combination of LCCC and SEC in a 2D chromatography experiment can be regarded as "orthogonal" chromatography in the strict sense provided that LCCC is used as the first dimension separation mode. Consequently, for functional homopolymers being distributed in functionality and molar mass, the coupling of LCCC with SEC can yield combined information on functionality type distribution (FTD) and MMD.

Although much of the early work on multidimensional chromatography was focused on using SEC in the first dimension, it is now widely accepted that an interactive type of separation should be used as the first step. On the other hand, SEC is preferentially used as the second method to retrieve molar mass information.

There are a number of other chromatographic separation methods that can only be used in the last stage of a multidimensional experiment. These include capillary electrophoresis (CE) and supercritical fluid chromatography (SFC). CE is a very efficient microseparation method (typically $N > 100{,}000$), which uses a strong electrical field to create an electro-osmotic flow in which the species will migrate. The reason for that is that the surface of the silicate glass capillary contains negatively charged functional groups that attract positively charged counterions. The positively charged ions migrate towards the negative electrode and carry solvent molecules in the same direction. This overall solvent movement is called electro-osmotic flow. During a separation, uncharged molecules move at the same velocity as the electro-osmotic flow (with very little separation). Positively charged ions move faster and negatively charged ions move slower.

SFC is a relatively 'exotic' chromatographic technique, which has been commercialized in the early 1980s. The more frequently used technique is supercritical fluid extraction. In SFC, the sample is carried through a capillary or packed column by a supercritical fluid (typically carbon dioxide). The properties of the mobile phase can be modified easily by polar additives and/or pressure programming, just as in gradient HPLC, to optimize selectivity. All three basic modes of chromatography (interaction, size exclusion and critical conditions) have been verified in SFC separations [55]. SFC is a very efficient separation technique, which has most of its applications in low molar mass separations.

SFC has several advantages over conventional chromatographic techniques. Separations can be done considerably faster than in HPLC, because the diffusion of solutes in supercritical fluids is about ten times greater than that in liquids (and about three times less than in gases). This results in a decrease in resistance to mass transfer in the column and allows for fast high resolution separations. Compared with GC, capillary SFC can provide high resolution chromatography at much lower temperatures. This allows fast analysis of thermolabile and non-volatile compounds. These advantages make SFC a good choice for multidimensional chromatography setups. Since SFC involves the use of fluids at high pressures, it can only be used in the last separation step in multi-dimensional separations.

6.4 Data Aquisition and Processing

In 2D chromatography, the most reliable operation of the instrument can be achieved, when hardware operations as well as data aquisition are organized by a unified software package. Therefore, all detectors and injectors, including the storage loops, are connected to a suitable data station. For the present investigations the 2D-CHROM hardware and software (and later versions such as PSS WinGPC UniChrom) of Polymer Standards Service, Mainz, Germany, is used. It allows for operating the switching valves of the storage loops by suitable time events given, to conduct data aquisition and all calculations.

Calibration and quantification. In addition to qualitative information on the molecular heterogeneity of a complex copolymer, quantitative data on the different distribution functions must be obtained. Since the first step is preferentially HPLC, a calibration with respect to chemical composition or functionality has to be conducted. The quantitative determination of FTD is rather straightforward, because a separation into homogeneous functionality fractions is obtained by LCCC. These fractions must be identified and can then be quantified via the corresponding peak areas taking into account the respective detector response factors.

The quantitative determination of CCD depends on the separation procedure. If a separation into single oligomers is obtained, these may be quantified via the corresponding peak areas. The determination of the chemical composition of diblock and triblock copolymers can be carried out according to procedures that are described for the quantification in LCCC [25]. The separation of statistical copolymers with respect to CCD is possible by gradient HPLC. In this case, the quantitative evaluation of the copolymer chromatograms requires knowledge of the influence of polymer composition on elution time, the influence of molar mass on elution time, and the influence of composition on detector signal intensity.

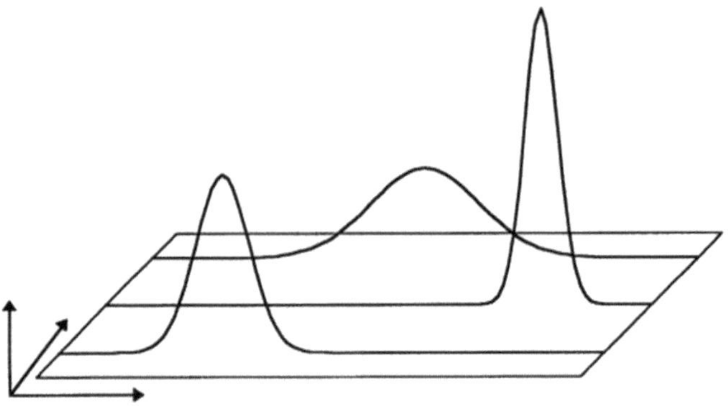

Fig. 6.4 Discontinuous representation of a two-dimensional separation

This information can be obtained by calibration with a series of samples graded in composition.

The calibration of the second separation step, which is in most cases SEC, can be carried out similar to ordinary SEC experiments. However, a number of important features of two-dimensional chromatography must be considered: every calibration procedure relates a certain elution volume to a certain molar mass. This calibration is valid as long as the chromatographic procedure and apparatus are not changed. If now, instead of an ordinary injection of a polymer sample into the SEC eluent, an injection of a polymer sample in a mixed solvent via an automated switching valve is made, the hydrodynamic volume and the elution volume of the sample will change. Therefore, in order to obtain reproducible and reliable results, the SEC calibration standard must pass the first separation step (HPLC) and enter the SEC system via automated injection. The calibration of the SEC system has to be carried out in the following way:

1. The calibration standard is injected into the first chromatographic system and the retention volume V_1 is determined only for this system.
2. The calibration standard is injected into the 2D system and after V_1 the automated injection valve is switched to inject the calibration standard into the SEC system. The injection time is taken as the starting point ($V_e = 0$) and the elution volume V_2 for the SEC system is determined.
3. V_2 is plotted against the molar mass of the calibration standard to yield the calibration curve of the SEC system.

Presentation of results: The visualization of the results of a 2D separation is possible in different ways. A discontinuous presentation is obtained, when the SEC chromatograms of all fractions are plotted along the elution volume axis of the first separation step, see Fig. 6.4.

A much clearer presentation of the separation is obtained, when a continuous plot is used, see Fig. 6.5. In this case the separation in the first and second modes are plotted along the Y- and X-axis, respectively. The concentration profile is presented by a color code in the Z-axis.

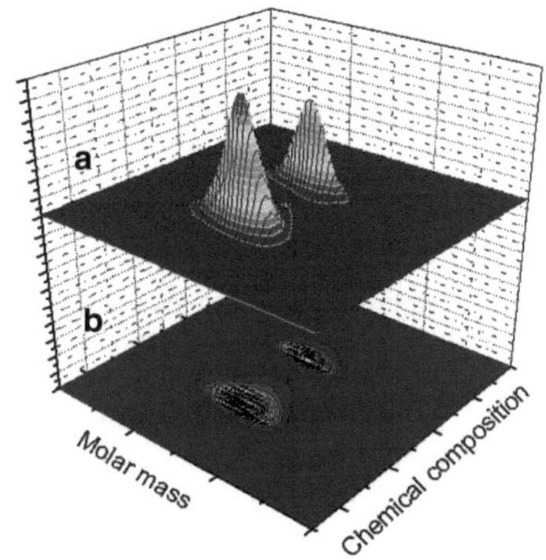

Fig. 6.5 Continuous representation of a two-dimensional separation, (**a**) surface plot with intensity lines, (**b**) contour plot

6.5 Coupling of Gradient HPLC and SEC

Much work on chromatographic cross-fractionation was carried out with respect to combination of SEC and gradient HPLC. In most of the early applications SEC was used as the first separation step, followed by HPLC. In a number of early papers the cross-fractionation of model mixtures was discussed. Investigations of this kind demonstrated the efficiency of gradient HPLC for separation by chemical composition. Mixtures of random copolymers of styrene and acrylonitrile were separated by Glöckner et al. [28]. In the first dimension a SEC separation was carried out using THF as the eluent and polystyrene gel as the stationary phase. In total, about 10 fractions were collected and subjected to the second dimension, which was gradient HPLC on a cyano-modified phase using i-octane-THF as the mobile phase. Model mixtures of random copolymers of styrene and 2-methoxyethyl methacrylate were separated in a similar way, the mobile phase of the HPLC mode being iso-octane/methanol in this case [56]. This procedure was also applied to real-world copolymers [28]. Graft copolymers of methyl methacrylate onto ethylene-propylene-diene rubber (EPDM) were analyzed by Augenstein and Stickler [57], whereas Mori reported on the fractionation of block copolymers of styrene and vinyl acetate [58]. For all these experiments the same limitation with respect to the SEC part holds true: when SEC is used as the first dimension, true molar mass distributions are not obtained.

From the theoretical point of view, a more feasible way of analyzing copolymers is the pre-fractionation through HPLC in the first dimension and subsequent analysis of the fractions by SEC [59, 60]. HPLC was found to be rather insensitive towards molar mass effects from a certain molar mass upwards and yielded very uniform fractions with respect to chemical composition.

One of the very first applications of two-dimensional gradient HPLC-SEC was published by Kilz et al. describing the analysis of styrene-butadiene star polymers [61]. 4-arm star polymers based on poly(styrene-block-butadiene) were prepared by

Table 6.2 Molar masses of the star copolymers

Sample	Butadiene content (wt%)	M_w (g/mol)	M_n (g/mol)
1	17	87,000	35,000
2	39	88,000	31,000
3	63	79,000	32,000
4	78	77,000	29,000

anionic polymerization to give samples with well known structure and molar mass control. In a first reaction step, a poly(styrene-block-butadiene) with a reactive chain end at the butadiene was prepared. This precursor reacted with a tetrafunctional terminating agent to give a mixture of linear (of molar mass *M*), 2-arm (2*M*), 3-arm (3*M*) and 4-arm (4*M*) species. Four samples with varying butadiene content (about 20, 40, 60, 80%) were prepared in this way, see Table 6.2. A mixture of these samples was used for the 2D experiment. Accordingly, a complex mixture of 16 components, resulting from the combination of four different butadiene contents and four different molar masses (*M*, 2*M*, 3*M*, 4*M*) had to be separated with respect to chemical composition and molar mass.

Initially, the 16-component star block copolymer was investigated by SEC. As can be seen in Fig. 6.6a, four peaks were obtained. They correspond to the four molar masses of the sample consisting of oligomers with one to four arms. The molar masses were in the ratio *M*–2*M*–3*M*–4*M*. Despite the high resolution, the chromatogram did not give any indication of the very complex chemical structure of the sample. Even when pure fractions with different chemical compositions were investigated, the retention behaviour did not show significant changes as compared to the sample mixture. In each case a tetramodal molar mass distribution was obtained indicating the different topological species. The SEC separation alone did not show any difference in chemical composition of the samples, which vary from 20 to 80% butadiene content.

Running the sample mixture in gradient HPLC gave poorly resolved peaks, which might suggest different composition, but gave no clear indication of different molar mass and topology, see Fig. 6.6b.

The combination of the two methods in the 2D setup dramatically increased the resolution of the separation system and gave a clear picture of the complex nature of the 16-component sample. A 2D representation of the gradient HPLC-SEC separation is given in Fig. 6.7. Each tracing represents a fraction transferred from HPLC to SEC and gives the result of the SEC analysis. Based on the composition of the sample, a contour map with the coordinates chemical composition and molar mass is expected to show 16 spots, equivalent to the 16 components. Each spot would represent a component which is defined by a single composition and molar mass. The experimental evidence of the improved resolution in the 2D analysis is given in Fig. 6.8. This contour plot was calculated from experimental data based on 28 transfer injections.

The contour plot clearly revealed the chemical heterogeneity (*Y*-axis, chemical composition) and the molar mass distribution (*X*-axis) of the mixture. The relative

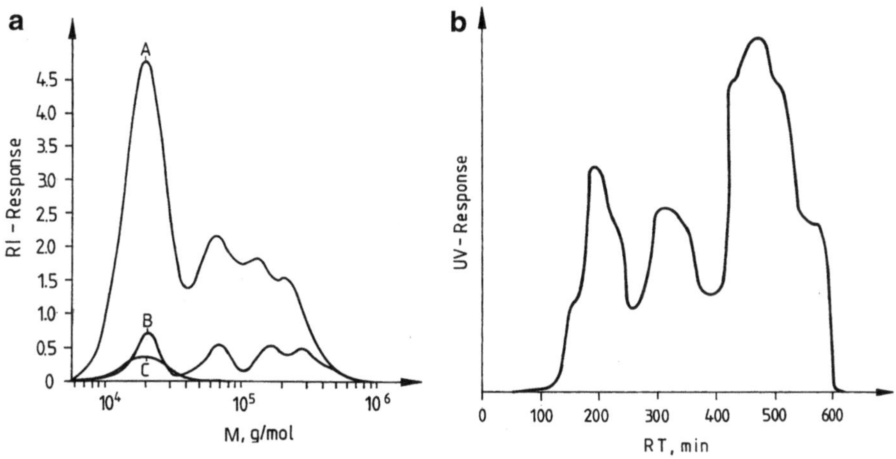

Fig. 6.6 SEC chromatograms of the 16-component sample (A) and the star polymers 1 (C) and 4 (B) (**a**) and gradient HPLC of the 16-component sample (**b**); (**a**) stationary phase: SDV PSS 10^3 and 10^5 Å, mobile phase: THF, RI detection; (**b**) stationary phase: silica gel, mobile phase: i-octane-THF, linear, 20–100 % THF (reprinted with permission from [61]. Copyright (1995) American Chemical Society)

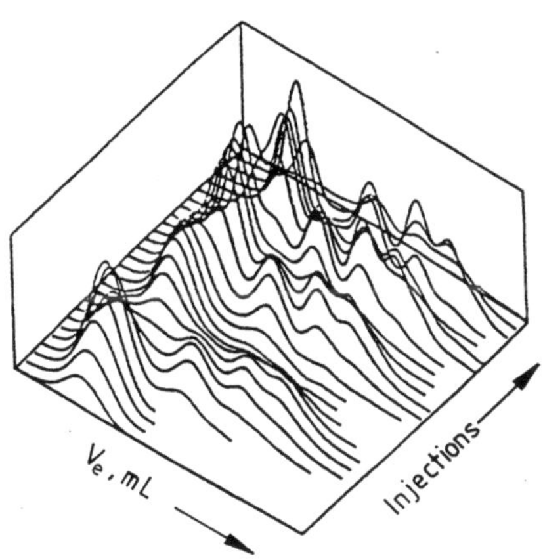

Fig. 6.7 Three-dimensional plot of the HPLC-SEC analysis of the 16-component star copolymer (reprinted with permission from [61]. Copyright (1995) American Chemical Society)

concentrations of the components were indicated by colours. sixteen major peaks were resolved with high selectivity. These correspond directly to the components. For example, peak 1 corresponds to the component with the lowest butadiene content and the lowest molar mass (molar mass *M*) whereas peak 13 relates to the component with the lowest butadiene content but a molar mass of 4*M*. Accordingly, peak 16 is due to the component with the highest butadiene content and a molar mass of 4*M*. A certain molar mass dependence of the HPLC separation is indicated

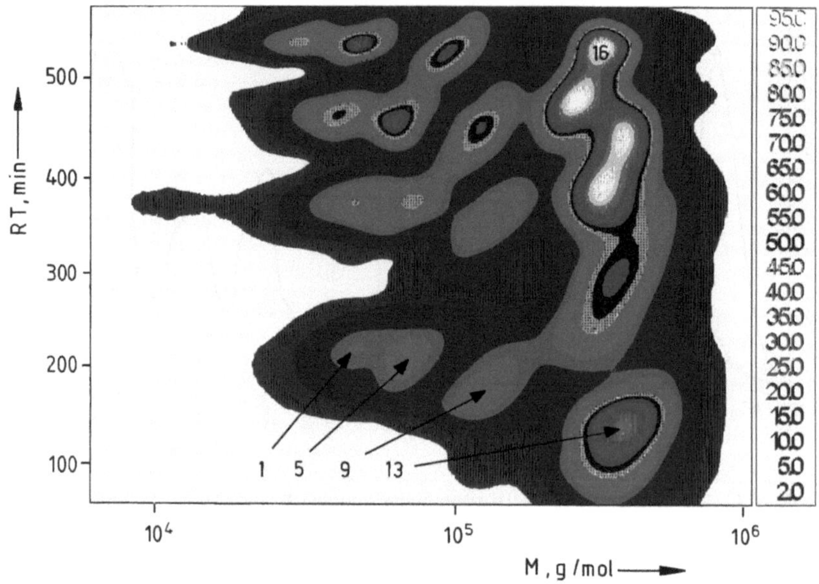

Fig. 6.8 Contour plot of the two-dimensional separation of a 16-component styrene-butadiene star copolymer (reprinted with permission from [61]. Copyright (1995) American Chemical Society)

by a drift of the peaks for components of similar chemical composition, see peaks 1–5–9–13, for example. This kind of behaviour is typical for polymers, because the pores of the HPLC stationary phase lead to size exclusion (entropic) effects which overlap with the adsorptive interactions at the surface of the stationary phase. Consequently, 2D separations of this type will in general be not orthogonal but skewed, depending on the pore size distribution of the stationary phase. The quantitative amount of butadiene in each peak could be determined via an appropriate calibration with samples of known composition. The molar masses could be calculated based on a conventional SEC calibration of chromatograph 2.

6.5.1 Analysis of the Grafting Reaction of Methyl Methacrylate onto EPDM [62]

Aim

Graft copolymers with an elastomeric backbone and thermoplastic grafts serve as impact modifiers in rubber-modified thermoplastics. This type of graft copolymer A-g-B is prepared by polymerizing a monomer B radically in the presence of a polymer A. Grafts B grow from macroradicals A. As a result of the grafting reaction, a complex product is obtained comprising the graft copolymer A-g-B, residual non-grafted polymer backbone A and homopolymer B. Accordingly, the reaction product is distributed both in molar mass (MMD) and chemical composition (CCD). In the past, the analysis of such products has always been difficult and tedious. To evaluate the two-dimensional parameter field of the composition and the MMD of such copolymers, classical and chromatographic cross-fractionation can be used. The classical approach relies on the dependence of copolymer solubility on composition and chain length. Far from an ideal separation, precipitation fractionation yields fractions that vary both in chemical composition and molar mass.

The present application describes the analysis of the grafting reaction of methyl methacrylate onto EPDM by on-line coupled 2D chromatography. In order to be selective towards chemical composition, interaction chromatography is used in the first dimension. In the second dimension, SEC provides information on MMD.

Materials

- **Calibration Standards.** Narrow-disperse PMMA in the molar mass range of 1,000–1,000,000 g/mol (PSS GmbH, Mainz, Germany).
- **Polymers.** EPDM (M_n 91,000 g/mol) was grafted with MMA (1:2 by weight) in concentrated toluene solution, at 80 °C with dibenzoyl peroxide as a radical initiator (0.2 mol%) and *n*-dodecyl mercaptane (0.2 mol%) as a transfer agent. After different reaction times samples were taken and analyzed: sample (reaction time): 2 (150 min); 5 (190 min); 6 (200 min); 9 (420 min). The graft products were precipitated with methanol.

Equipment

- **Chromatographic System.** A modular chromatographic system comprising two chromatographs connected via one eight-port injection valve and two storage loops was used. The chromatograph for the first separation step (chromatograph 1) comprised a Rheodyne six-port injection valve with a 100 μL injection loop and a gradient SD-200 Rainin pump. One electrically driven eight-port injection valve (Valco EHC8W) was used to connect the two chromatographs. In addition, they were connected to two storage loops of a volume of 200 μL each. The chromatograph for the second separation step (chromatograph 2) comprised a Waters model 510 pump. The operation of the coupled injection valves was controlled by the software, which was used for data collection and processing. In the present case the software package "PSS-2D-GPC-Software" of Polymer Standards Service, Mainz, Germany, was used. Molar mass calibration is based on polymethyl methacrylate.
- **Columns.** Chromatograph 1: Knauer Nucleosil CN 300+500 Å, 7 μm average particle size (Knauer, Berlin, Germany). Column size was 250 × 4 mm I.D. For the two-dimensional separation in the first dimension: Phenomenex CN 100 Å, 5 μm average particle size (Phenomenex, Torrance, USA). Column size was

125 × 4 mm i.d. The cartridge temperature for the column was 40 °C. Chromatograph 2: PSS SDV lin XL, 10 μm average particle size and column size of 50 × 20 mm i.d. (PSS GmbH, Mainz, Germany).
- **Mobile Phase.** Chromatograph 1: THF-cyclohexane for LCCC and THF-i-octane for gradient HPLC. The gradient started from 99:1 (v/v), being held constant for 1 min, being changed linearly to 34:66 i-octane/THF within 4 min, being held constant for 6 min and being changed linearly to 0:100 i-octane/THF within 2 min. The flow rate was 1 mL/min. Chromatograph 2: THF, all solvents were HPLC grade.
- **Detectors.** Waters 486 tunable UV detector at 254 nm and evaporative light scattering detector (ELSD) model ELSD 500 of Altech both after chromatograph 2.
- **Column Temperature.** 25 °C
- **Sample Concentration.** 20–40 mg/mL. All samples are dissolved in the mobile phase of chromatograph 1.
- **Injection Volume.** 50 μL

Preparatory Investigations

Grafting-from processes never yield pure graft copolymers. They always result in reaction mixtures of complex chemical composition. In the present case, in addition to the graft copolymer EPDM-g-PMMA residual non-grafted EPDM and PMMA homopolymer were expected in the reaction mixture. For a detailed analysis, these components must be separated from each other by appropriate chromatographic techniques. For a first information, SEC was used to separate with regard to molecular size. The SEC chromatogram of a typical reaction product is shown in Fig. 6.9. It indicates that the molar mass differences of the components are not large enough to produce separate elution peaks. However, an indication of the distribution of methyl methacrylate across the elution volume axis can be obtained by comparing the traces of the ELSD and UV (254 nm) detectors. Since PMMA has a higher UV response at 254 nm than EPDM, it becomes apparent that the lower elution volume part (higher molar mass) of the chromatogram is rich in PMMA. This part assumingly belongs to the graft copolymer, however, a definite information cannot be obtained due to incomplete separation.

For the separation of complex polymers according to chemical composition different techniques of interaction chromatography can be used. LCCC is one useful technique for separating segmented copolymers. Operating at the critical conditions of a particular homopolymer, this homopolymer can be separated from other chemically different polymers or copolymers containing this homopolymer as a block or graft. In the present case, for the separation of the components of the grafting reaction, chromatographic conditions adjusted to the critical point of PMMA are used. The critical point of adsorption of PMMA can be established on a polar stationary phase such as cyano-modified silica gel (Nucleosil CN) and

Fig. 6.9 SEC chromatogram of the graft product sample 5; stationary phase: PSS SDV linear, eluent: THF, detection: (*continous line*) UV 254 nm, (*dashed line*) ELSD (reprinted from [62] with permission of Wiley-VCH)

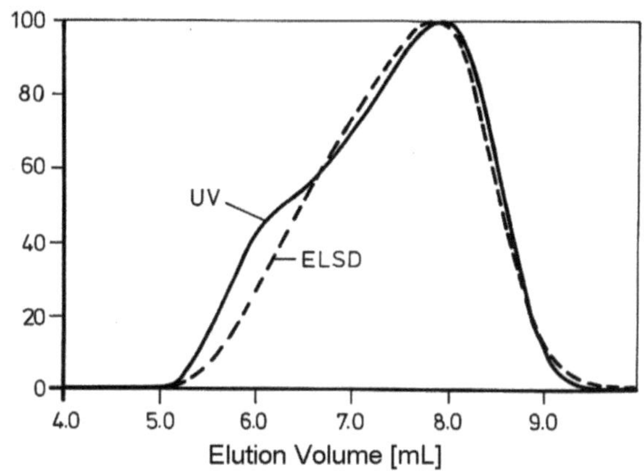

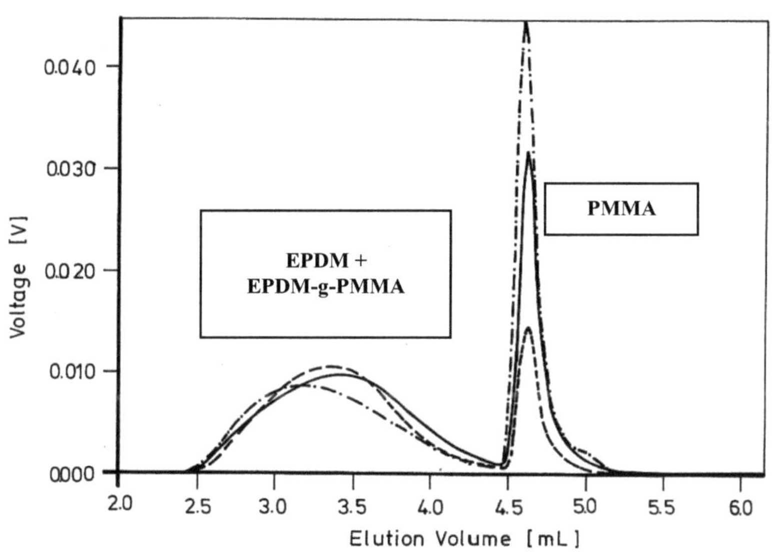

Fig. 6.10 LCCC chromatograms of graft products samples 2 (– – –), 5 (‾‾), and 9 (–·–); stationary phase: Nucleosil CN 300 + 500 Å, eluent: THF-cyclohexane 63:37% by volume, detection: ELSD (reprinted from [62] with permission of Wiley-VCH)

THF-cyclohexane as the mobile phase. The critical point corresponds to an eluent composition of THF-cyclohexane 63:37% by volume.

The LCCC chromatograms of different reaction mixtures are given in Fig. 6.10. Two well separated elution peaks are obtained. The peak at an elution volume of approximately 4.6 mL can be assigned to the PMMA homopolymer and, therefore, the other elution peak corresponds to the graft copolymer fraction and residual non-grafted EPDM. Taking the ELSD signal for the concentration profile it is obvious that a large amount of PMMA homopolymer is formed as a by-product of the

Table 6.3 Amounts of PMMA homopolymer determined by LCCC after different grafting times

Fraction	PMMA in the crude product (wt%)
1	14
3	21
4	25
5	35
7	30
8	40

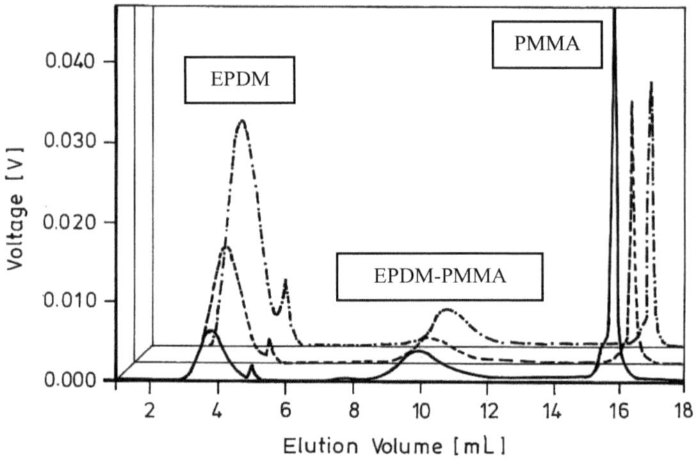

Fig. 6.11 Gradient HPLC chromatograms of samples 2 (–·–·–), 6 (– – –), and 9 (——); stationary phase: Nucleosil CN 300 + 500 Å, eluent: THF/i-octane, detection: ELSD (reprinted from [62] with permission of Wiley-VCH)

grafting reaction. The relative amount of PMMA as a function of the reaction time is given in Table 6.3. The quantification is carried out by calibrating the ELSD signal of the PMMA peak.

Separations

Another most useful technique for the separation of copolymers according to chemical composition is gradient HPLC. The combined effects of adsorption and precipitation in high performance precipitation LC (HPPLC) can be used to separate complex reaction products into the components.

In the present case the same column set as for LC-CC is used for gradient HPLC. A stepwise gradient of THF-i-octane is used starting with 99% by volume of i-octane and going to 100% by volume THF. The resulting chromatograms of three samples are presented in Fig. 6.11. A perfect separation into three fractions is obtained for all samples. The assignment of the peaks is carried out by comparison with the chromatographic behaviour of EPDM and PMMA, assuming that the elution order is a function of component polarity. EPDM as the least polar component is eluted first followed by the graft copolymer EPDM-g-PMMA and the most polar PMMA homopolymer. The small peak at a retention time of 5 min is due to the fact that EPDM is not fully retained on the column after injection.

Now that all components of the reaction mixture are separated, they can be quantified from their signal intensities. Although the ELSD detector is regarded a universal mass detector, it is well known that the response of chemically different species can be different. This is also the case for the present application. It is, therefore, necessary to calibrate the ELSD by determining the concentration versus response behaviour for PMMA and EPDM. The amount of graft copolymer EPDM-g-PMMA can then be calculated from the total amount and the amount thereof and the amounts of EPDM and PMMA.

While the gradient HPLC experiments yield detailed information on the chemical composition of the graft products, information on molar masses must be obtained by SEC. For an optimization of the conditions of the grafting reaction it is of particular interest to determine the MMD of each of the product components. Using on-line coupled 2D chromatography dual information on chemical composition and molar mass can be obtained. In the first chromatographic dimension separation is conducted with regard to chemical composition using gradient HPLC, while in the second dimension SEC separation is carried out.

The results of the 2D separations of two graft products are presented as contour diagrams in Fig. 6.12. The ordinate represents the separation in the first dimension, while the abscissa indicates the SEC separation of the fractions. The contour plot indicates three fractions which are different in chemical composition and molar mass. The assignment of the fractions is straightforward and based on the HPLC separation. Fractions 1–3 correspond to EPDM, EPDM-g-PMMA, and PMMA, respectively.

Evaluation

The contour plot clearly indicates that the molar masses of the three components are in similar ranges. Therefore, obviously, a SEC type separation alone (as in Fig. 6.9) could not resolve the different components of the graft product. The relative concentrations of the components are obtained from the intensities of the contour plot peaks. For sample 2, EPDM has the highest concentration, while the concentrations of the reaction products EPDM-g-PMMA and PMMA are comparable, see Fig. 6.12a. A sample taken at a later stage of the grafting reaction is presented in Fig. 6.12b. In this sample, the relative concentration of EPDM is much lower and PMMA constitutes the major product. A summary of the composition and the molar masses of these samples is given in Table 6.4.

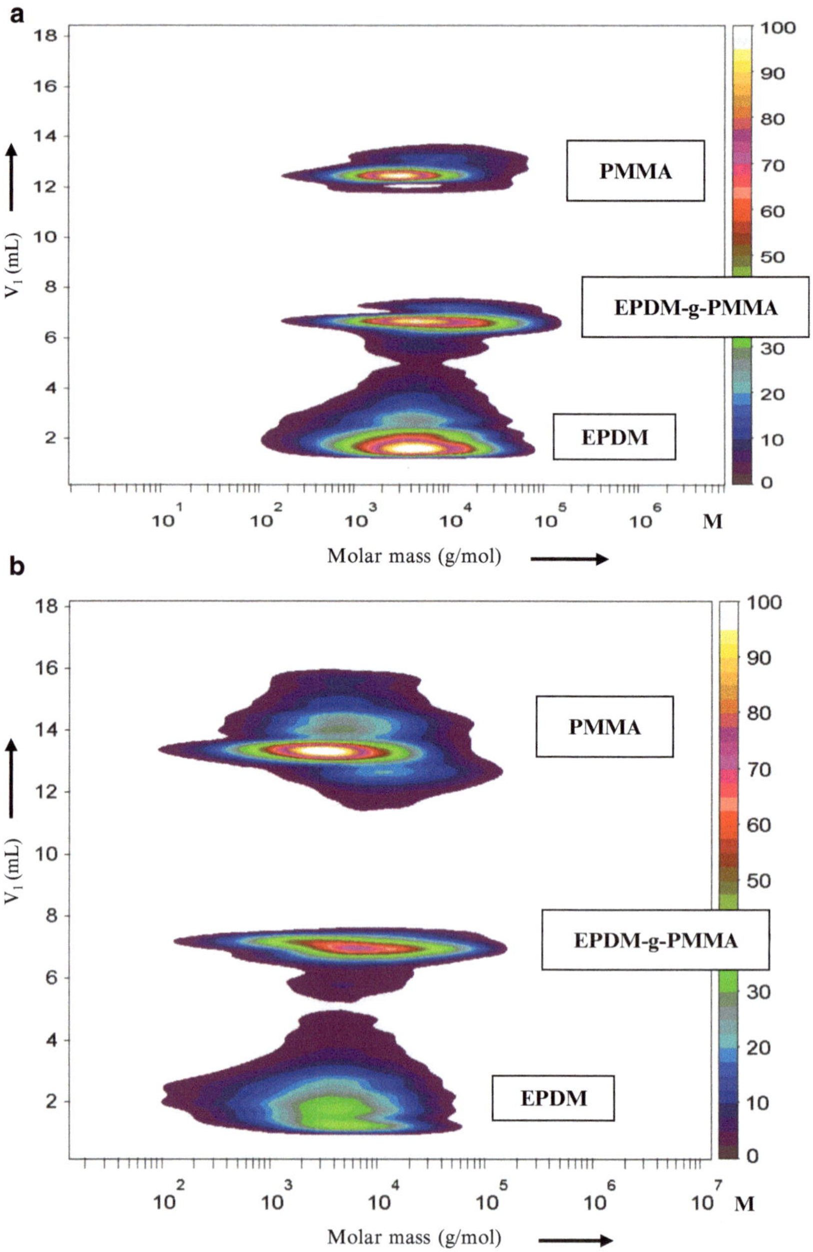

Fig. 6.12 Contour plot of the two-dimensional separation of the graft product samples 2 (**a**) and 6 (**b**); first dimension: gradient HPLC, second dimension: SEC, detection: ELSD (reprinted from [62] with permission of Wiley-VCH)

Table 6.4 Composition of the graft products determined by 2D chromatography

Sample	Component	M_w (g/mol)	Amount (Peak volume %)
2	EPDM	184,200	47
	EPDM-g-PMMA	389,200	25
	PMMA	156,900	21
6	EPDM	165,900	27
	EPDM-g-PMMA	359,700	24
	PMMA	167,500	42

6.5.2 Analysis of the Grafting Reaction of Styrene onto Polybutadiene [63]

Aim

The combination of styrene and butadiene in different copolymer structures is important for a huge variety of technical polymeric materials. To name a few, styrene-butadiene rubbers are the basis for the manufacturing of automotive tyres while the modification of polystyrene (PS) with styrene-butadiene copolymers results in high-impact polystyrene. One way to obtain styrene-butadiene copolymers is the radical grafting of polybutadiene (PB) with styrene.

The present application describes the analysis of the grafting reaction of styrene onto polybutadiene by on-line coupled 2D chromatography. In order to be selective towards chemical composition, gradient HPLC is used in the first dimension. In the second dimension, SEC provides information on molar mass distribution.

Materials

- **Calibration Standards.** Narrow-disperse PS in the molar mass range of 1,000–1,000,000 g/mol (PSS GmbH, Mainz, Germany).
- **Polymers.** PB (Sigma-Aldrich, St. Louis, USA) was purified by reprecipitation and grafted with styrene (1:6 by weight) in concentrated toluene solution, at 80 °C with dibenzoyl peroxide as a radical initiator (0.2 mol%). After different reaction times samples were taken and analyzed: sample (reaction time): 1 (30 min); 2 (60 min); 3 (120 min); 4 (180 min), 5 (240 min); 6 (360 min); 7 (480 min). The graft product was precipitated with methanol.

Equipment

- **Chromatographic System.** A modular chromatographic system comprising two chromatographs connected via one eight-port injection valve and two storage loops was used. The chromatograph for the first separation step (chromatograph 1) comprised a Rheodyne six-port injection valve with a 100 μL injection loop and a gradient SD-200 Rainin pump. One electrically driven eight-port injection valve (Valco EHC8W) was used to connect the two chromatographs. In addition, they were connected to two storage loops of a volume of 200 μL each. The chromatograph for the second separation step (chromatograph 2) comprised a Waters model 510 pump. The operation of the coupled injection valves was controlled by the software, which was used for data collection and processing. In the present case the software package "PSS-2D-GPC-Software" of Polymer Standards Service, Mainz, Germany, was used. Molar mass calibration is based on PS.

- **Columns.** Chromatograph 1: PSS ANIT 100 Å, 5 μm average particle size (PSS GmbH, Mainz, Germany). Column size was 50 × 8 mm i.d. Chromatograph 2: PSS SDV lin XL, 10 μm average particle size and column size of 50 × 20 mm i.d. (PSS GmbH, Mainz, Germany).
- **Mobile Phase.** Chromatograph 1: Cyclohexane-chloroform. The gradient was starting from 100% cyclohexane being held constant for 2 min, being changed linearly to 84:16 cyclohexane-chloroform within 2 min, being held constant for 7 min and being changed linearly to 0:100 cyclohexane-chloroform within 2 min. The flow rate was 1 mL/min. Chromatograph 2: THF, all solvents were HPLC grade.
- **Detectors.** Waters 486 tunable UV detector at 260 nm and evaporative light scattering detector (ELSD) model ELSD 500 of Altech both after chromatograph 2.
- **Column Temperature.** 25 °C
- **Sample Concentration.** 20–40 mg/mL. All samples are dissolved in cyclohexane.
- **Injection Volume.** 50 μL

Preparatory Investigations

Grafting-from processes never yield pure graft copolymers. They always result in reaction mixtures of complex chemical composition. In the present case, in addition to the graft copolymer PB-g-PS residual non-grafted PB and PS homopolymer were expected in the reaction mixture. For a first information, size exclusion chromatography (SEC) was used to separate with regard to molecular size. The SEC chromatogram of a typical reaction product is shown in Fig. 6.13. It indicates that the molar mass differences of the components are not large enough to obtain baseline-separated elution peaks. However, an indication of the distribution of styrene across the elution volume axis can be obtained by comparing the traces of the ELSD and UV (260 nm) detectors. Since PS has a higher UV response at 260 nm than PB, it becomes apparent that the higher elution volume part (lower molar mass) of the chromatogram is rich in PS. This part assumingly belongs to PS homopolymer, however, a definite information cannot be obtained due to incomplete separation.

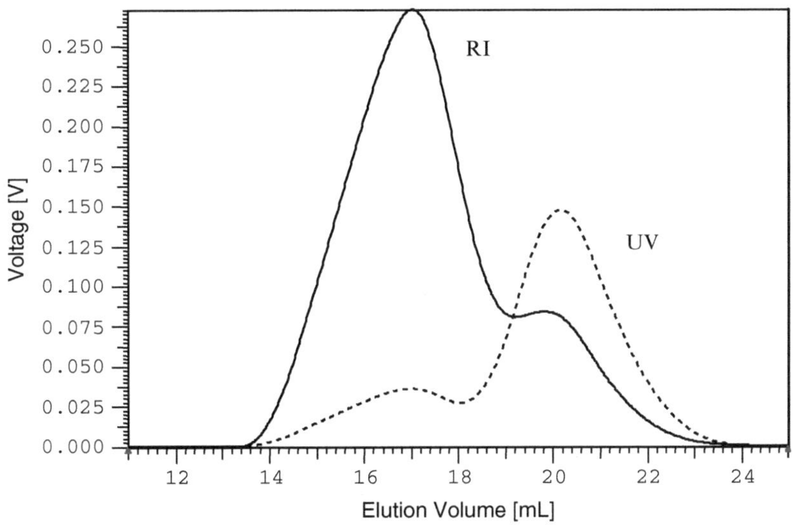

Fig. 6.13 SEC chromatogram of the graft product sample 1; stationary phase: PSS SDV linear, eluent: THF, detection: (– – – –) UV 260 nm, (——) ELSD

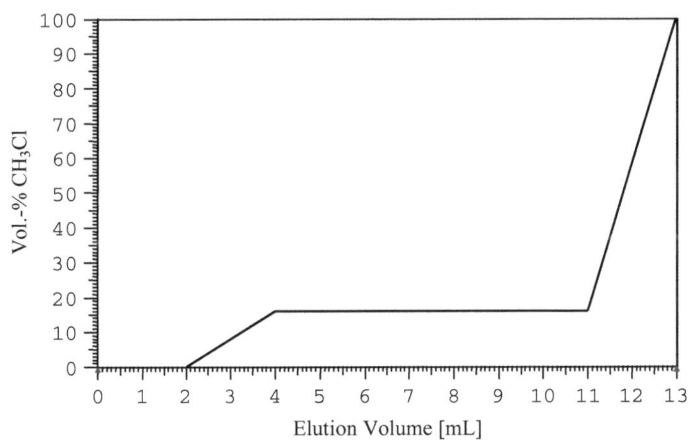

Fig. 6.14 Gradient for the separation of the graft products samples stationary phase: PSS ANIT 100 Å, eluent: cyclohexane-chloroform

Separations

As has been shown in the previous application, gradient HPLC is the most efficient way to separate the products of a grafting reaction. In the present case a stepwise gradient of cyclohexane-chloroform is used as is shown in Fig. 6.14. The resulting chromatograms of four samples are presented in Fig. 6.15. A perfect separation into three fractions is obtained for all samples. The assignment of the peaks is carried out by comparison with the chromatographic behaviour

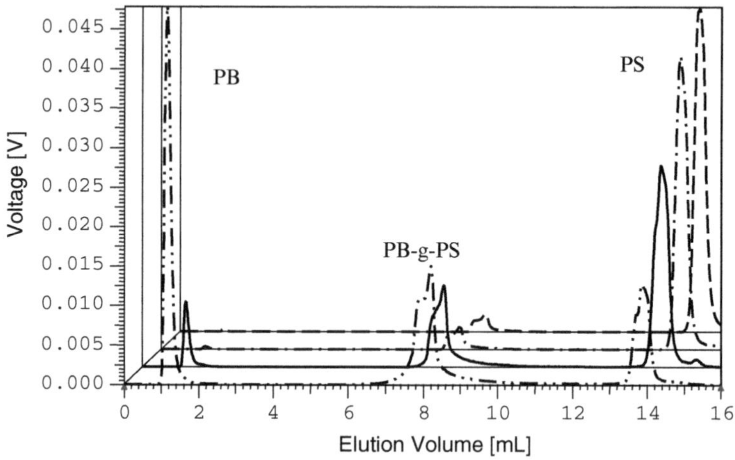

Fig. 6.15 Gradient HPLC chromatograms of samples 1 (-·······-), 3 (——), 5 (-·-·-·-) and 7 (– – – –); stationary phase: PSS ANIT 100 Å, eluent: cyclohexane-chloroform, detection: ELSD

of PB and PS, assuming that the elution order is a function of component polarity. PB as the least polar component is eluted first followed by the graft copolymer PB-g-PS and the most polar PS homopolymer.

After separating all components of the reaction mixture, these can be quantified from their signal intensities. Although the ELSD detector is regarded a universal mass detector, it is well known that the response of chemically different species can be different. This is also the case for the present application. It is, therefore, necessary to calibrate the ELSD by determining the concentration versus response behaviour for PS and PB. The amount of graft copolymer PB-g-PS can then be calculated from the total amount of sample and the amounts of PB and PS. The calibration curves for the homopolymers are shown in Fig. 6.16. As can be seen, the responses of both polymers are quite similar. The compositions of the samples as determined from the ELSD traces are summarized in Table 6.5.

Using on-line coupled 2D chromatography dual information on chemical composition and molar mass can be obtained. In the first chromatographic dimension separation is conducted with regard to chemical composition using gradient HPLC, while in the second dimension SEC separation is carried out.

The results of the 2D separations of the initial PB and two graft products are presented as contour diagrams in Fig. 6.17. The ordinate represents the separation in the first dimension, while the abscissa indicates the SEC separation of the fractions. The contour plot indicates three fractions that are different in chemical composition and molar mass. The assignment of the fractions is straightforward and based on the HPLC separation. Fractions 1–3 correspond to PB, PB-g-PS, and PS, respectively.

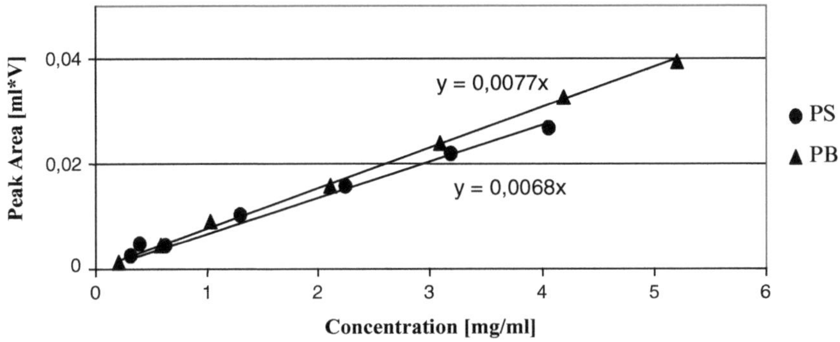

Fig. 6.16 Concentration calibration of PB and PS, detection: ELSD

Table 6.5 Concentrations of the components in the reaction mixtures as determined by gradient HPLC

Sample	Reaction time (min)	PB (wt%)	PB-g-PS (wt%)	PS (wt%)
1	30	40	31	29
2	60	22	36	42
3	120	7	30	63
4	180	2	24	74
5	240	1	24	75
6	360	–	18	82
7	480	–	15	85

Evaluation

The contour plots clearly indicate that the molar masses of the three components are in similar ranges. Although the molar mass of the polystyrene is gradually lower, a SEC type separation alone (as in Fig. 6.13) could not resolve the different components of the graft product. The relative concentrations of the components are obtained from the intensities of the contour plot peaks. For sample 1 polybutadiene has the highest concentration, while the concentrations of the reaction products PB-g-PS and PS are lower, see Fig. 6.17b. A sample taken at a later stage of the grafting reaction is presented in Fig. 6.17c. In this sample, the relative concentration of PB is very low indicating a nearly complete consumption. The homopolymer PS constitutes the major product. A summary of the compositions and the molar masses of the graft copolymer samples is given in Table 6.6.

6.5.3 Branching Analysis of Star-Shaped Polystyrenes [64, 65]

Aim

Branched polymers and polymers with complex architectures are of significant importance, e.g. as viscosity modifiers in mineral oils or as cross-linking agents

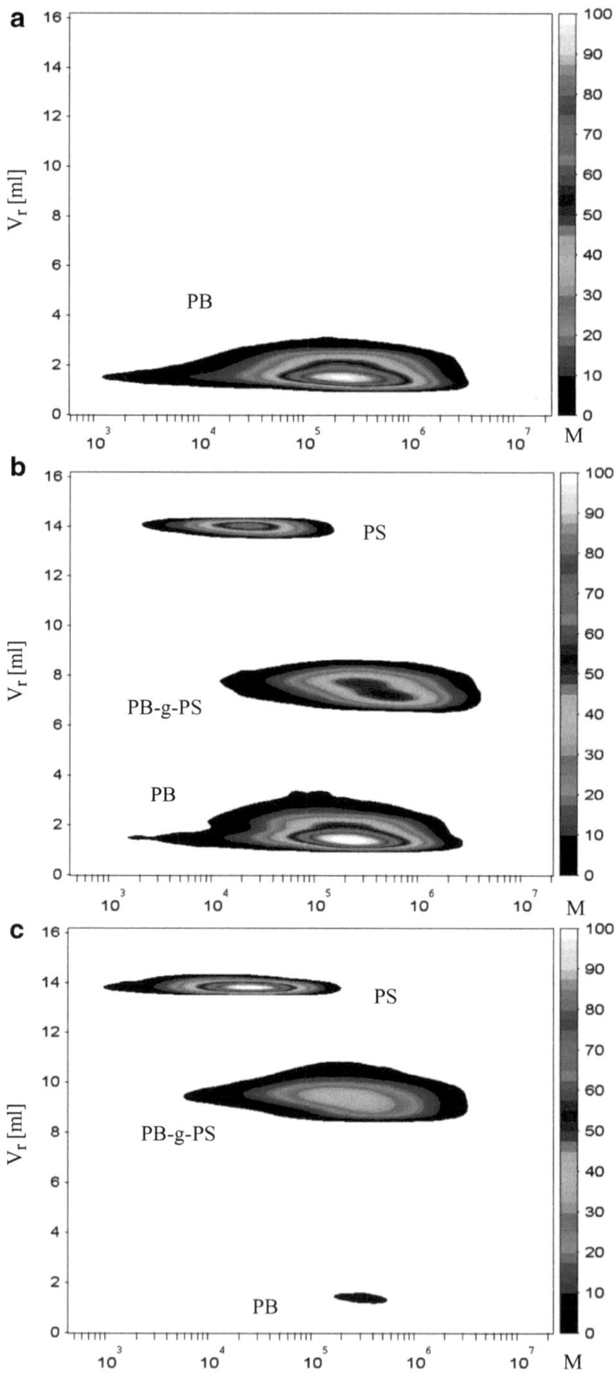

Fig. 6.17 Contour plots of the two-dimensional separation of the initial PB (**a**) and the graft product samples 1 (**b**) and 3 (**c**); first dimension: gradient HPLC, second dimension: SEC, detection: ELSD

Table 6.6 Composition analysis of the graft products determined by 2D chromatography

Sample	Reaction time (min)	Structure	M_n (g/mol)	M_w (g/mol)	M_w/M_n	Amount (Peak volume %)
PB			138,000	367,100	2.7	100
		PB	140,300	338,100	2.4	51
1	30	PB-g-PS	200,700	566,100	2.8	36
		PS	15,700	28,300	1.8	13
		PB	141,800	248,000	1.8	2
3	120	PB-g-PS	117,000	431,800	3.7	61
		PS	13,900	28,200	2.1	37
		PB	–	–	–	–
7	480	PB-g-PS	174,400	444,900	2.6	42
		PS	17,100	35,000	2.0	58

in polyurethanes. The rheological properties of polymers change significantly even at small amounts of branching thus influencing the final mechanical properties of the material. The introduction of branches results in a reduction of the melt viscosity, thus, in an improved processability without loss of mechanical properties. The characterization of branched polymers is still a challenge. Branching results in a contraction of the size of the molecules relative to that of the linear analogues of the same molar mass as has been theoretically shown by Zimm and Stockmayer in 1949. Even more than 50 years later, branching analysis is still mainly done by viscosity or light scattering experiments on the basis of the Zimm–Stockmayer theory.

From the analytical point of view, it is difficult to determine the number and length of branches in a polymeric material directly and quantitatively. In the present application it shall be investigated if liquid chromatographic techniques can be used to separate polymers with regard to branching. As an ultimate goal conditions shall be found where linear and branched molecules of the same chemical composition are separated. Further, branched molecules shall be separated according to the number of branches.

Materials

- **Calibration Standards.** Linear PS in the molar mass range of 1,000–1,000,000 g/mol (PSS GmbH, Mainz, Germany).
- **Polymers.** Linear and star-shaped PS in the molar mass range of 10,000–1,000,000 g/mol. The samples were prepared by anionic polymerization as described in [63, 64]. For star polymer formation PS macromonomers were copolymerized with divinylbenzene. The molar mass of the macromonomers (molar mass of one arm in the resulting star polymer) was varied between 4,800 and 42,000 g/mol

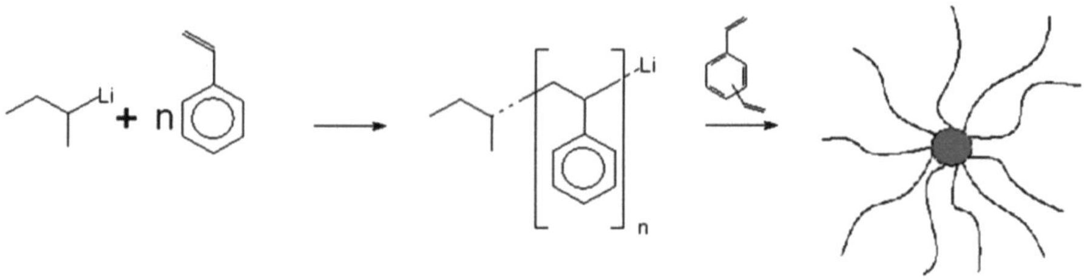

Equipment

- **Chromatographic System.** Gradient chromatography and fractionations were performed on a modular HPLC system consisting of a Shimadzu degasser DGU 14A, a low pressure gradient former FCV10ALVp, a HPLC-pump LC10ADVp, an autosampler TSP AS100 equipped with a 100 μL loop, and a Techlab K4 column oven. Data acquisition and fraction collector control was performed using PSS WINGPC software (PSS GmbH, Mainz, Germany). For the separations, a method of temperature gradient interaction chromatography was used. For the 2D experiments the temperature program involved a 10.5 min isothermal step at 20 °C followed by a 110 min linear temperature rise to 32 °C and a hold step for 10 min. To ensure complete elution the temperature was raised linearly to 57.5 °C within 10 min.
- **Columns.** Chromatograph 1: Nucleosil C18, 5 μm average particle size, 300 Å average pore size, 250 × 4.6 mm i.d. (Macherey&Nagel, Dueren, Germany), Chromatograph 2: two columns of PSS SDV $10^4 + 10^5$ Å, 5 μm average particle size and column size of 50 × 20 mm i.d (PSS GmbH, Mainz, Germany).
- **Mobile Phase.** Chromatograph 1: Acetonitrile(ACN)-tetrahydrofuran(THF) 48:52% by weight. A temperature gradient was used. The flow rate was 0.2 mL/min. Chromatograph 2: THF, all solvents were HPLC grade.
- **Detectors.** Knauer UV/VIS filter photometer at 254 nm and evaporative light scattering detector (ELSD) model ELSD 500 of Altech both after chromatograph 2.
- **Column Temperature.** Temperature gradient
- **Sample Concentration.** 0.5–5 mg/mL. All samples were dissolved in the mobile phase.
- **Injection Volume.** 100 μL

Preparatory Investigations

Each elution volume in SEC may correspond to two different molecular species, a linear fraction of lower and a branched fraction of higher molar mass. The molar masses of the different species can be found from the calibration curves of the respective structures. If in a different mode of chromatography these co-eluting species elute at different elution volumes (on the right side of Fig. 6.18 a stronger

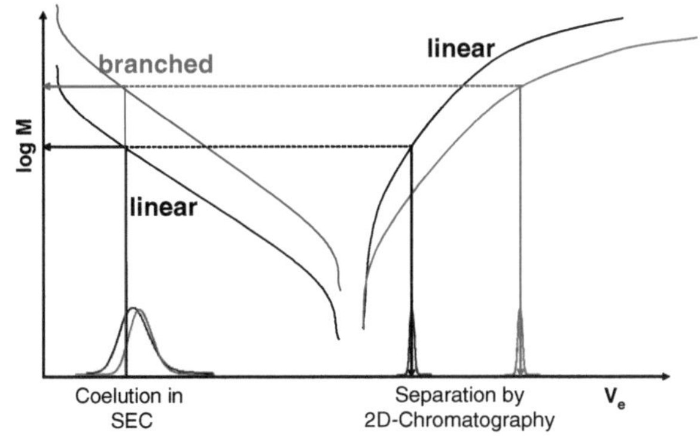

Fig. 6.18 Schematic representation of a two-dimensional separation of linear and branched polymers, for adsorbing conditions a stronger retention for the branched species is assumed (reprinted from [64] with permission of European Polymer Federation)

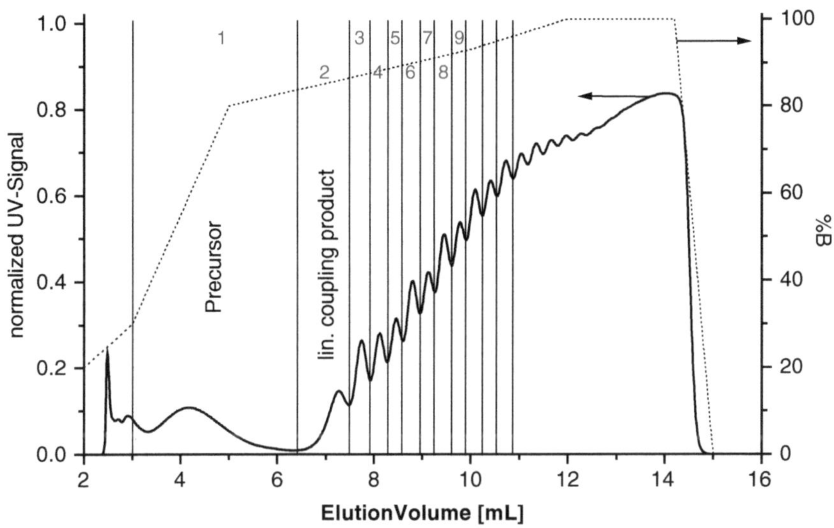

Fig. 6.19 Gradient HPLC separation of a star polymer according to number of arms (M_{arm} = 4,800 g/mol), stationary phase: Nucleosil 300 C18, mobile phase: eluent A: THF/ACN 35:65 wt%, Eluent B: THF/ACN 51:49 wt%., T = 35 °C, UV detection (reprinted from [64] with permission of European Polymer Federation)

retention of the branched species is assumed), then the separation of the heterogeneous fraction taken at a certain SEC elution volume will result in two peaks, one for the linear and one for the branched species, respectively.

For a proper molar mass analysis of the branched (star-shaped) PS, they were separated and fractionated by gradient HPLC, see Fig. 6.19 for a sample with an arm length of 4,800 g/mol. The fractions were subjected to SEC-MALLS measurements to determine the true molar masses of the fractions. Each fraction corresponded to a star polymer with a fixed number of arms. Fraction 1 was found

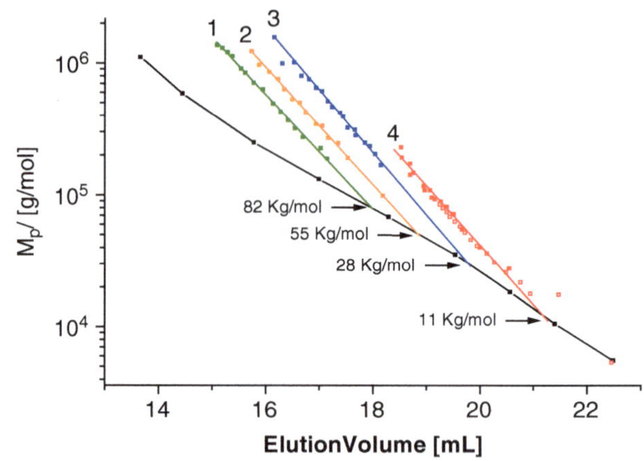

Fig. 6.20 Comparison of calibration curves for linear (*circle*) and star polymers having arm lengths of M_{arm} = 34,000(*green square*), 22,500(*orange square*),10,500(*blue square*) and 4,800 (*red square*) g/mol (reprinted from Ref. [64] with permission of European Polymer Federation)

to correspond to the precursor macromonomer, fraction 2 to the linear coupling product, fractions 3 and 4 to the tri-arm and tetra-arm polymers, respectively, etc.

The SEC calibration curves obtained for star polymers with different arm molar masses are presented in Fig. 6.20. As can be seen, they significantly deviate from the calibration curve for linear PS. It becomes clear that the use of the standard calibration curve significantly underestimates the true molar masses of the star polymers. From Fig. 6.20 it is possible to relate the molar mass of a linear PS having the same elution volume to each molar mass of a star polymer having a given arm length. This correlation will be used to characterize the elution behaviour of star polymers under gradient conditions.

Separations

As already stated, LAC and LCCC are two additional modes of polymer chromatography beside the well known SEC. In order to separate star polymers and linear polymers according to Fig. 6.18 a chromatographic mode has to be found which can be combined with SEC in a 2D experiment. For linear homopolymers it is known that gradient chromatography results in a molar mass independent elution under certain conditions. Therefore, the effect of polymer architecture on the elution behaviour in gradient chromatography was investigated. In order to do so, a temperature gradient was developed that resulted in a sufficiently strong retention and thus a sufficient separation of all star polymers. This temperature gradient HPLC was used as first dimension and all fractions of this separation were injected into the second dimension of the on-line two-dimensional chromatographic system. The second dimension was run under SEC conditions for PS and was first calibrated using linear PS standards. Each star polymer fraction transferred from the first dimension resulted in an SEC chromatogram, which could be evaluated using the

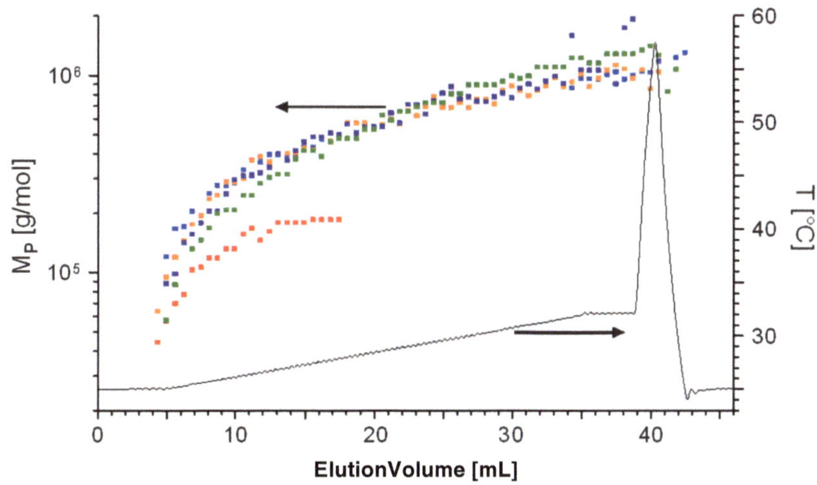

Fig. 6.21 Comparison of calibration curves in temperature gradient chromatography for star polymers having different arm molar masses $M_{arm} = 4{,}800$(*red square*), 10,500(*blue square*), 22,500(*orange square*), 34,000(*Violet square*) and 42,000(*green square*) g/mol, mobile phase: THF-AN 52:48 % by weight (reprinted from [64] with permission of European Polymer Federation) (Color figure online)

previously established PS calibration curve. Since for a given arm molar mass the relation between the true and the apparent molar mass is given in Fig. 6.20, it was possible to assign the true molar mass of the star polymer to each gradient elution volume. The elution volume versus molar mass calibration is shown in Fig. 6.21.

It becomes clear that all curves follow a common line within the experimental error. The only exception is the curve for the lowest arm molar mass of $M_{arm} = 4{,}800$ g/mol. Since a given molar mass of the star polymer corresponds to a different number of arms, depending on the arm length, no significant effect of polymer topology becomes visible under these conditions, i.e. star polymers above a certain arm molar mass elute in the order of increasing molar mass in gradient chromatography.

Assuming a molar mass dependent and, therefore, topology independent elution in gradient chromatography a 2D experiment on a mixture of linear and star-shaped PS should result in a separation into a linear and a star-shaped fraction according to Fig. 6.18. This is indeed the case, as can be seen in Fig. 6.22 for a mixture of a linear PS and a star polymer having an arm molar mass of 4,800 g/mol. The abscissa shows the retention volume in the second dimension (SEC), while the ordinate corresponds to the elution volume in the first dimension (temperature gradient chromatography).

Two different regions of elution can clearly be distinguished. According to the argumentation above the higher SEC elution volumes have to be assigned to the star polymers, since these should exhibit a smaller dimension in solution at the same molar mass (gradient retention), i.e. a higher SEC retention volume. Extrapolation over the peak maxima of the different fractions merge at an elution volume of

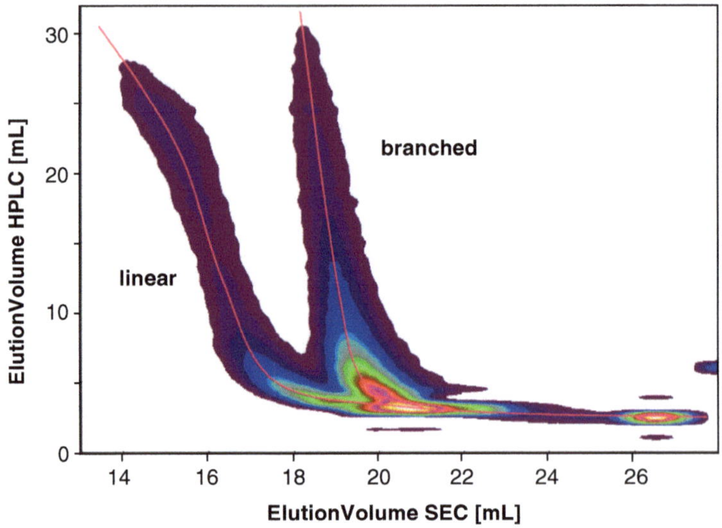

Fig. 6.22 On-line 2D separation of a mixture of a star polymer having an arm molar mass of 4,800 g/mol and polydisperse linear polystyrene (reprinted from [64] with permission of European Polymer Federation)

approximately 21 mL. This elution volume corresponds to a linear PS of a molar mass of $M_p = 13{,}500$ g/mol. Since the arm molar mass of the star is 4,800 g/mol, the deviation of the star retention curve from that of the linear one is in rather good agreement with the expected molar mass of the first branched species.

Evaluation

As has been shown, the combination of temperature gradient HPLC and SEC in a 2D set-up is a very powerful tool to separate linear from branched polymer species. The separation is excellent for branched polymers with rather short branches, e.g. for arm lengths of roughly 5,000 g/mol. In this case, a total separation between branched and linear species is obtained irrespective of the number of branches (arms).

In order to evaluate the resolution of the method as a function of the arm length, similar experiments have been conducted for star polymers with increasing arm lengths. In Fig. 6.23 the HPLC elution volume is plotted as a function of the SEC elution volume for star polymers with different arm lengths.

As is obvious, there is a distinct influence of the arm length on the separation efficiency for linear and branched molecules. For short arms, e.g. 5,000 or 11,000 g/mol, the separation is very good while for long arms, e.g. 52,000 g/mol, the star polymer behaviour approaches the behaviour of the linear polymer. Accordingly, for branched polymers with very long branches a separation between linear and branched species cannot be achieved.

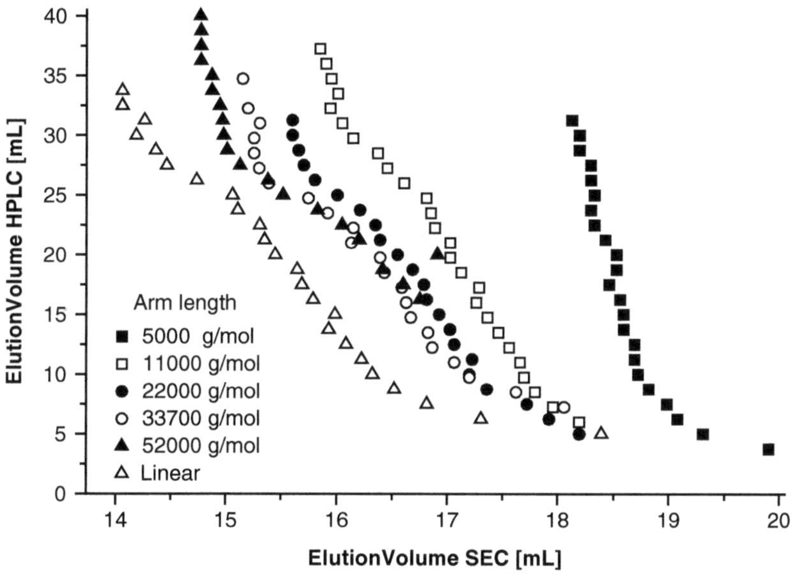

Fig. 6.23 Dependence of HPLC elution volume vs. SEC elution volume for linear polystyrene and star polystyrenes with different arm lengths

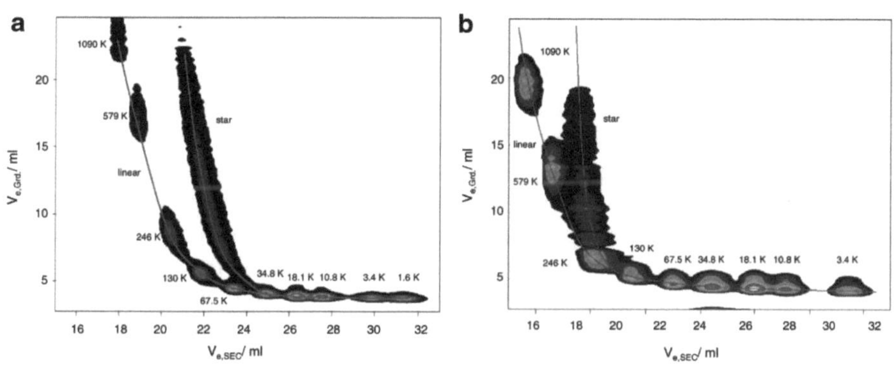

Fig. 6.24 Off-line 2D separation of a mixture of a star polymers having an arm molar mass of 10,500 g/mol (A) or 42,000 g/mol (B) and linear polystyrene calibration standards (reprinted from [65] with permission of Elsevier)

A very clear presentation of the limitations of the method is given in Fig. 6.24. While for star polymers with an arm length of 10,500 g/mol a decent separation between the linears and the stars is obtained, this is not the case for an arm length of 42,000 g/mol.

Based on the theory of interactive chromatography of polymers, Gorbunov and Vakhrushev analyzed the separation of polydisperse linear and star-shaped polymers by topology using two-dimensional chromatography [66]. They showed that the molar mass distribution is an important factor that influences the separation. Simulations showed that linear and star components can be reasonably well separated by 2D chromatography only if the polydispersity of the arms is less then 1.1. The separation of linears and stars having more than three arms is possible for much higher values of polydispersity. The pore size of the SEC stationary phase has a significant effect on the 2D separation pattern. Best results can be expected when the average pore radius is of the same order as the average radius of gyration of the macromolecules in the sample. According to their theory and simulations, the effect of the adsorption interaction in the 2D set-up is of little significance; therefore in order to reduce the time of the separation one can use gradient HPLC instead of adsorption chromatography at fixed interaction conditions. It was shown theoretically that the separation of symmetric and very asymmetric stars is possible. It should be noted, that the theory is based on the ideal chain model which does not account for excluded volume effects, and therefore it is applicable rather to Θ-solvents than to good solvent systems. Although the calculations cannot be directly compared to the present separations, the experimental findings are in a good qualitative agreement with Gorbunov's theory and calculations.

6.6 Coupling of LCCC and SEC

Liquid chromatography at critical conditions (LCCC) relates to a chromatographic situation where the entropic and enthalpic interactions of the macromolecules and the column packing compensate each other. The Gibbs free energy of the macromolecule does not change when entering the pores of the stationary phase. The distribution coefficient K_d is unity, regardless of the size of the macromolecules, and all macromolecules of equal chemical structure elute from the chromatographic column in one peak. To describe this phenomenon the term "chromatographic invisibility" is used, meaning that the chromatographic behaviour is not directed by the size but by the inhomogeneities (chemical structure) of the macromolecules. Under such chromatographic conditions it is possible to determine the heterogeneities of the polymer chain selectively and without any influence of the polymer chain length.

Thus, LCCC represents a chromatographic separation technique yielding fractions that are homogeneous with respect to chemical composition but distributed in molar mass. These fractions can readily be analysed by SEC which for chemically homogeneous fractions provides true MMDs without interference of CCD or FTD. Therefore, the combination of LCCC and SEC in a 2D setup can truly be regarded as "orthogonal" chromatography provided that LCCC comprises the first dimension. Consequently, for functional homopolymers being distributed in functionality and molar mass, coupled LCCC versus SEC can yield combined information on FTD and MMD.

Table 6.7 Molecular parameters of AH-polyesters as supplied by the manufacturer

Sample	M_n (g/mol)	OH groups/molecule
PE1	930	1.83
PE2	1,300	1.78
PE3	1,650	1.69
PE4	2,400	1.61
PE5	1,020	1.98
PE6	1,400	1.86
PE7	1,810	1.86
PE8	2,220	1.73
PE9	5,940	1.64

6.6.1 Characterization of Aliphatic Polyesters [61, 67]

Aim

Polyesters of adipic acid and 1,6-hexanediol are manufactured for a wide variety of applications with an output of thousands of tons per year. They are intermediates for the manufacture of polyurethanes, and their functionality type distribution is a major parameter in affecting the quality of the final products. In particular, non-reactive cyclic species are responsible for the "fogging effect" in polyurethane foams.

The determination of the FTD of polyesters is reported by a number of authors. Vakhtina [68] separated aliphatic polyesters by thin layer chromatography, and Filatova et al. investigated polyesters by gradient elution LC and LCCC [69, 70]. The aim of the present experiment is the separation of technical polyesters of adipic acid and 1,6-hexanediol according to functionality, the identification of the functionality fractions, and the quantitative determination of their MMDs.

Materials

- **Calibration Standards.** Since narrow dispersed adipic acid-hexanediol (AH) polyesters samples are not available, technical samples in the molar mass range of 1,000–6,000 g/mol are used.
- **Polymers.** Technical AH-polyesters of the following average structure

$$HO-\left[-OC-(CH_2)_4-COO-(CH_2)_6-O-\right]_n-H$$

The average molar masses and functionalities of the samples are summarized in Table 6.7.

Additional model polyesters with specific endgroups were prepared and will be discussed in the following part.

Equipment

- **Chromatographic System.** Modular HPLC system comprising conventional HPLC pumps, a Rheodyne six-port injection valve and any conventional HPLC column oven.
- **Columns.** Critical chromatography: Silica gel of Tessek (Prague, Czech Republic), 7 μm average particle size and 120 Å average pore diameter. Column size was 250 × 4 mm i.d. SEC: Two-column set of PL-gel (Polymer Laboratories, UK), 5 μm average particle size and 50 Å and 100 Å average pore diameter. Column size was 300 × 8 mm i.d.
- **Mobile Phase.** LCCC: mixtures of acetone and hexane, SEC: acetone, all solvents are HPLC grade.
- **Detectors.** ERC 7511 differential refractometer
- **Column Temperature.** 30 °C
- **Sample Concentration.** 0.5–5 mg/mL. All samples are dissolved in the mobile phase.
- **Injection Volume.** 20–100 μL

Preparatory Investigations

As a first step of the experiments, the critical point of adsorption of the polymer (polyester) must be determined. As AH-polyesters may have different terminal groups such as HO–OH, HO–COOH and HOOC–COOH, they can differ in terms of polarity over a wide range. Therefore silica gel and and reversed phases may be tested as stationary phases. However, reversed phases were found to be unsuitable due to insufficient selectivity. In addition, the mobile phase composition corresponding to the critical point was found to be close to the precipitation point, causing partial precipitation of the higher molar mass samples on the column. With silica gel and a mobile phase of acetone-hexane, good solubility of the samples is obtained. Testing silica gels of different pore sizes, optimum resolution and peak shape is obtained with an average pore diameter of about 100 Å.

The critical diagram molar mass versus retention time for the silica gel 120 Å is shown in Fig. 6.25. The critical point of adsorption is obtained at a mobile phase composition of acetone-hexane 51:49% by volume, where regardless of the molar mass, all calibration samples elute at the same elution volume. At higher concentrations of acetone in the mobile phase the SEC mode is operating, whereas at lower concentrations of acetone in the mobile phase separation corresponds to the adsorption mode.

Separations

Separations of the AH-polyesters given in Table 6.7 according to their terminal groups are carried out at the critical point of adsorption, as indicated in Fig. 6.25. The critical chromatogram of an AH-polyester with a molar mass of 2,400 g/mol is given in Fig. 6.26 as a representative example.

In all chromatograms one major peak at a retention time of about 14 min is obtained. In addition, a number of elution peaks of lower intensity indicate other

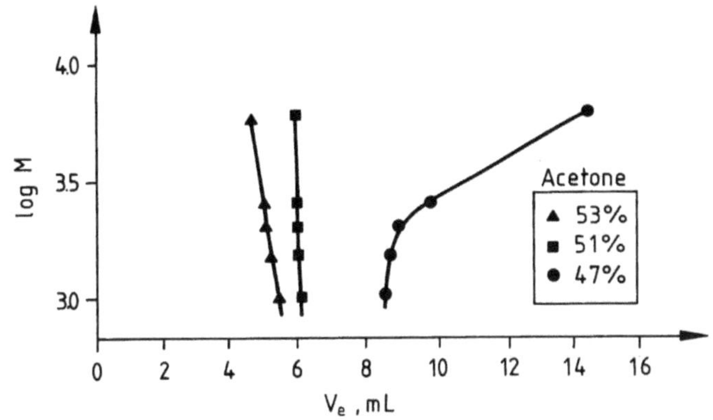

Fig. 6.25 Critical diagram molar mass versus elution volume of AH-polyesters, stationary phase: Tessek silica gel, mobile phase: acetone-hexane (reprinted from [67] with permission of Taylor & Francis)

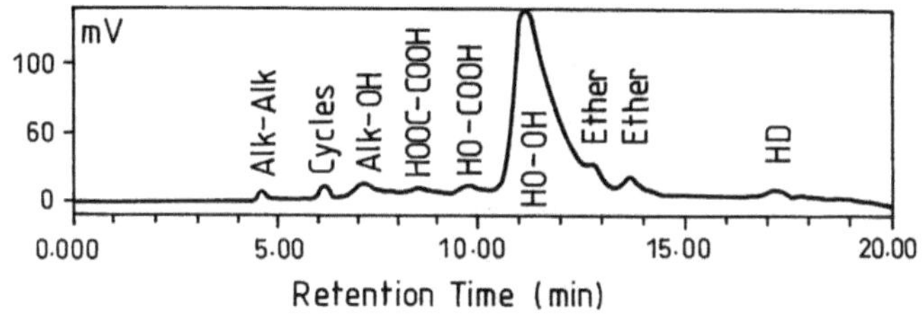

Fig. 6.26 Chromatogram of an AH-polyester (sample PE 4), stationary phase: Tessek silica gel, mobile phase: acetone-hexane 51:49 % by volume (reprinted from [67] with permission of Taylor & Francis)

functionality fractions. In total, up to nine elution peaks may be identified in the chromatograms of the polyesters. It is known from the hydroxy and acid numbers that the samples are mainly hydroxy-terminated AH-polyesters. Accordingly, the major peak (peak 6) can be assigned to the α, ω-dihydroxy species

$$HO{-}(CH_2)_6{-}O{-}\left[{-}OC{-}(CH_2)_4{-}COO{-}(CH_2)_6{-}O{-}\right]_n{-}H$$

For the assignment of the other peaks a number of model polyesters may be used, which have the same AH-polyester chain and specific endgroups. By comparison of the chromatograms of these functionally uniform AH-polyesters with the peaks in Fig. 6.26, the following assignment can be made [67], see Table 6.8.

Peaks 7 and 8 are obtained due to the formation of ether structures in the polyester samples.

$$HO{-}R^1{-}OH + HO{-}R^2{-}OH \rightarrow HO{-}R^1{-}O{-}R^2{-}OH$$

where R^1 and R^2 represent AH-polyester chains.

Table 6.8 Peak assignment of Fig. 6.26 $R^1-[-OC-(CH_2)_4-COO-(CH_2)_6-O-]_n-R^2$

Fraction	Symbol	R^1	R^2
1	Alk–Alk	$CH_2{=}CH(CH_2)_4$	X
2	cycles	- - - - - - - - -	- - - - - - - - -
3	Alk–OH	$CH_2{=}CH(CH_2)_4$	–H
4	HOOC–COOH	HO–	$–OC–(CH_2)_4–COOH$
5	HOOC–OH	HO–	–H
6	HO–OH	$HO–(CH_2)_6–O–$	–H
9	1,6-hexanediol		

$X: -OC-(CH_2)_4-COO-(CH_2)_4-CH=CH_2$

Evaluation

The amount of the different functionality fractions may be determined from the relative peak areas, taking into account the different detector responses. These data can be correlated with the hydroxy and the acid numbers of the total samples.

The MMDs of the functionality fractions may be determined by preparatively separating the fractions and subjecting them to SEC. The SEC chromatograms of fractions 1–9 are summarized in Figure 6.27. For a number of fractions oligomer separations are obtained, which can be used to calibrate the SEC system. The SEC calibration curves for the functionality fractions 1, 2, 4–6 are given in Fig. 6.28. For functionality fractions 1 and 2 virtually the same calibration curve is obtained. The calibration curves for fractions 4–6 are very similar, but differ remarkably from the calibration curve of fractions 1 and 2. This clearly indicates that differences in endgroup functionality have a strong effect on the SEC behaviour and must be considered when investigating this type of samples by SEC. The effect of endgroups on the SEC behaviour is particularly strong for oligomers as in the present case.

Separation with regard to molar mass and functionality is obtained by using 2D chromatography as the analytical technique. The contour plot in Fig. 6.29 clearly reveals all structural peculiarities of the polyester sample. The different functionality fractions can be readily identified in the ordinate direction and their MMDs can be obtained through the SEC separation in the abscissa direction.

6.6.2 Analysis of Epoxy Resins [71, 72]

Aim

Epoxy resins present one of the most important types of cross-linking polymers. High chemical corrosion resistance, good mechanical and thermal properties, and outstanding adhesion to various substrates are characteristics of the materials.

Epoxy resins are mostly prepared by the reaction of epichlorohydrin and bisphenol A (BPA), see for example the reaction scheme of the TAFFY process

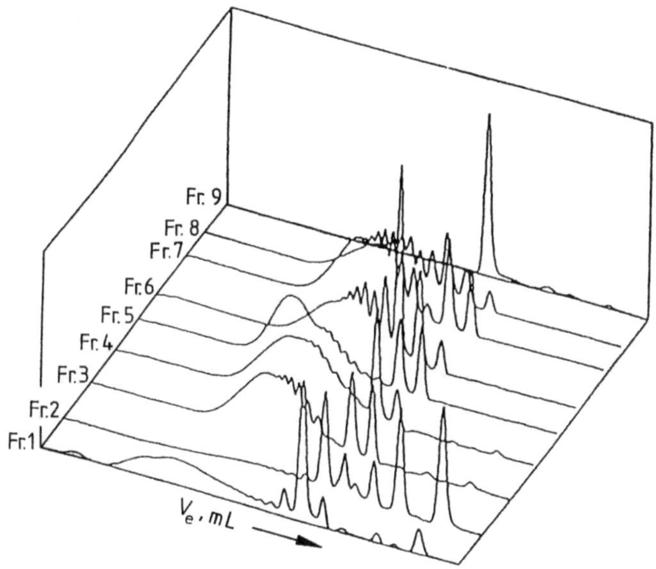

Fig. 6.27 SEC chromatograms of fractions 1–9 taken from separation of AH-polyester, stationary phase: PL-gel, 300 × 8 mm I.D., mobile phase: acetone (reprinted from [67] with permission of Taylor & Francis)

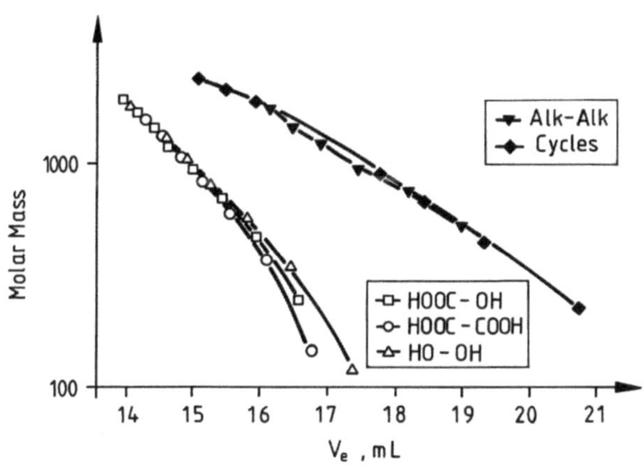

Fig. 6.28 SEC calibration curves for different functionality fractions of AH-polyesters (reprinted from [67] with permission of Taylor & Francis)

in Fig. 6.30. As a result of this reaction oligomers are formed which mainly contain glycidyl endgroups. Due to side reactions such as the hydrolysis of the epoxy groups or incomplete dehydrohalogenation, other endgroups may be also formed. In further reactions, branching can take place. Accordingly, epoxy resins may exhibit a functionality type distribution and a topological distribution in addition to the usual MMD.

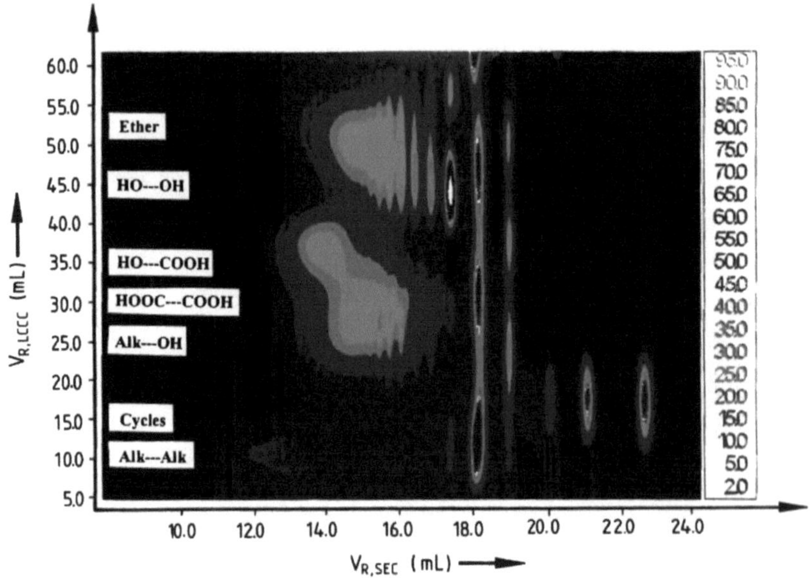

Fig. 6.29 Contour plot of the two-dimensional separation of AH-polyester PE4 (reprinted from [61] with permission of Taylor & Francis)

Materials

- **Calibration Standards.** Since narrow-disperse epoxy resin samples are not available, technical samples in the molar mass range of 800–5,000 g/mol are used.
- **Polymers.** Technical Bisphenol-A based epoxy resins of the average structure given in the reaction scheme. The average molar masses of the samples are summarized in Table 6.9.

Equipment

- **Chromatographic System.** Modular HPLC system comprising two chromatographs connected via one eight-port injection valve and two storage loops. The chromatograph for the first separation step (chromatograph 1) comprised a Rheodyne six-port injection valve with a 50 μL injection loop and an ISCO piston pump 100 DX. One electrically driven eight-port injection valve (Valco EHC8 W) was used to connect the two chromatographs. The two storage loops had a volume of 100 μL each. The chromatograph for the second separation step (chromatograph 2) comprised a Waters model 510 pump. The operation of the coupled eight-port injection valve was directed by the software, which was used for data collection and processing. In the present case, the software package "PSS-2D-GPC Software" of Polymer Standards Service GmbH, Mainz, Germany, was used.

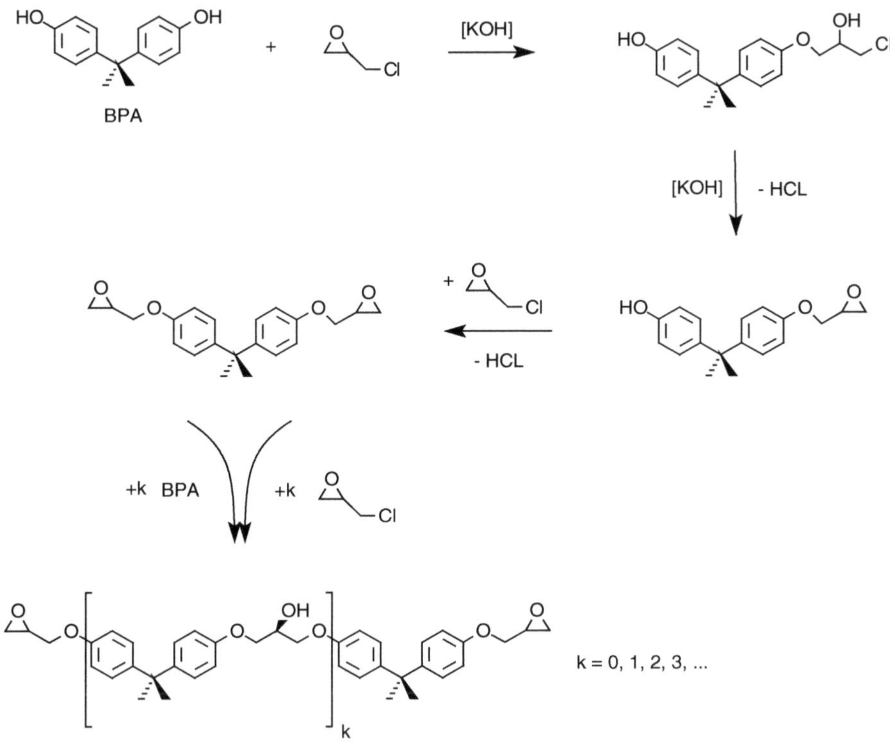

Fig. 6.30 Reaction scheme for the synthesis of Bisphenol-A based epoxy resins by the TAFFY process

Table 6.9 Molar masses of the epoxy resins as determined by SEC

Sample	M_n (g/mol)	M_w (g/mol)
1	5,020	23,800
2	3,280	11,410
3	2,890	9,560
4	4,090	16,190
5	2,650	6,890
6	3,920	16,150
7	1,060	2,330
8	860	1,510

- **Columns.** Critical chromatography (chromatograph 1): Nucleosil 50–5 of Macherey-Nagel (Düren, Germany), 5 μm average particle size and 50 Å average pore diameter. Column size was 200 × 4 mm i.d. SEC (chromatograph 2): Two-column set of PL Mixed-D and Mixed-E (Polymer Laboratories, UK), 5 μm average particle size. Column size was 300 × 8 mm i.d.

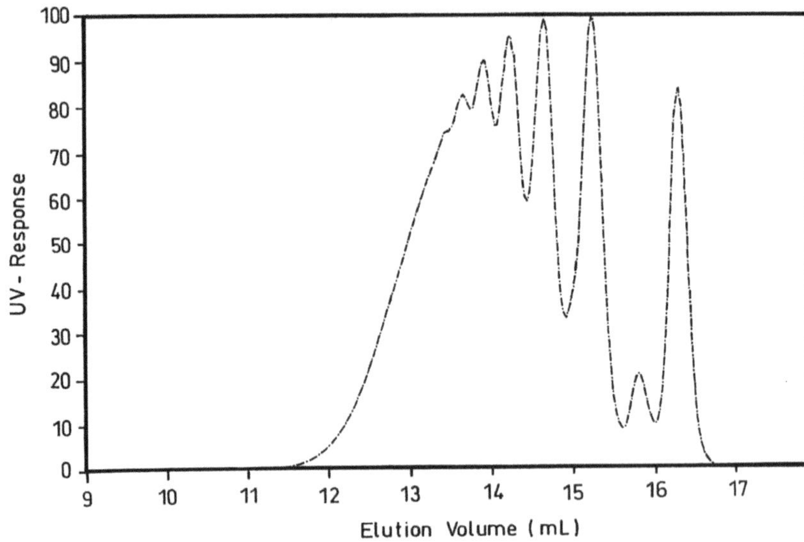

Fig. 6.31 SEC chromatogram of an epoxy resin (sample 8), stationary phase: PL Mixed-D + Mixed-E, mobile phase: THF (reprinted from [72] with permission of Wiley-VCH)

- **Mobile Phase.** LCCC: mixtures of THF and hexane, SEC: THF, all solvents were HPLC grade.
- **Detectors.** Waters model 486 UV/VIS detector at 280 nm
- **Column Temperature.** 30 °C
- **Sample Concentration.** 1–5 mg/mL. All samples are dissolved in the mobile phase.
- **Injection Volume.** 50 μL

Preparatory Investigations

For first information on molar mass the samples were analyzed by SEC, which was optimized with respect to the forthcoming two-dimensional experiments. Fast analysis with acceptable resolution was obtained with a set of two linear columns and a flow rate of 2 mL/min. The SEC chromatogram of a typical low molar mass epoxy resin is shown in Fig. 6.31. Well resolved oligomer peaks are obtained, the last peak in the chromatogram being the monomeric bisglycidyl bisphenol A.

The peaks at 15.3 mL, 14.8 mL, and 14.2 mL can be assigned to the respective dimer, trimer, and tetramer. Since no calibration standards for epoxy resins are available, a calibration curve was constructed from polystyrene calibration standards and the epoxy oligomers [71].

For the analysis of the functionality type distribution of the epoxy resins LCCC was used. In general, the critical point of adsorption is determined by running a number of calibration samples in eluents of different compositions. This was not possible with the epoxy resins since calibration samples were not available. However, it is known that the major functionality fraction in typical epoxy resins is

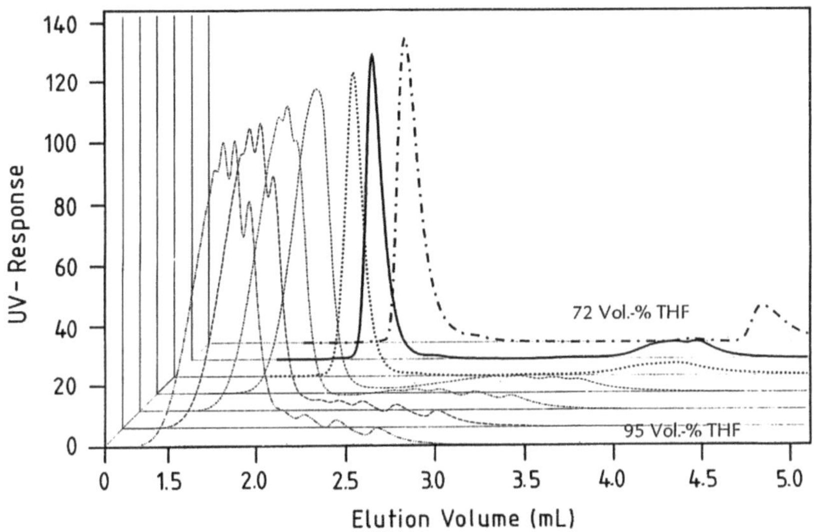

Fig. 6.32 Chromatograms of sample 8 at different eluent compositions, stationary phase: Nucleosil 50–5, mobile phase: THF-hexane 95/90/85/80/75/74/72 % (reprinted from [71] with permission of Wiley-VCH)

always the bisglycidyl fraction, constituting the major elution peak in the chromatograms. This peak was regarded as the reference peak and its chromatographic behaviour for samples of different molar masses was analyzed as a function of the eluent composition. As can be seen in Fig. 6.32, significantly different chromatograms were obtained depending on the composition of the eluent. In 95 % THF, the sample exhibited typical SEC behaviour. With increasing concentration of *n*-hexane in the mobile phase the oligomer peaks merged in one narrow elution peak. The critical point of adsorption corresponded to a mobile phase composition of THF-hexane 74:26 % by volume, see Fig. 6.33.

Separations

A typical LCCC chromatogram is given in Fig. 6.34. Three well separated elution regions P1, P2, and P3 were obtained. The broadness of P2 and P3 indicates, however, that these fractions do not behave critically or they are still heterogeneous. For a detailed analysis of the fractions, sample 8 was fractionated multiple times to obtain the narrow fractions A-H, which were subjected to MALDI-TOF mass spectrometry [71].

The MALDI-TOF analysis gave the following assignment of the chromatographic fractions, see Table 6.10. As can be seen, the first fractions contain the oligomers that are rich in epoxy groups. It is interesting that in addition to linear oligomers mono- and dibranched oligomers were identified. These oligomers obviously result from the reaction of a terminal epoxy group with the secondary hydroxy group of another oligomer. The fractions C-G correspond to oligomers that

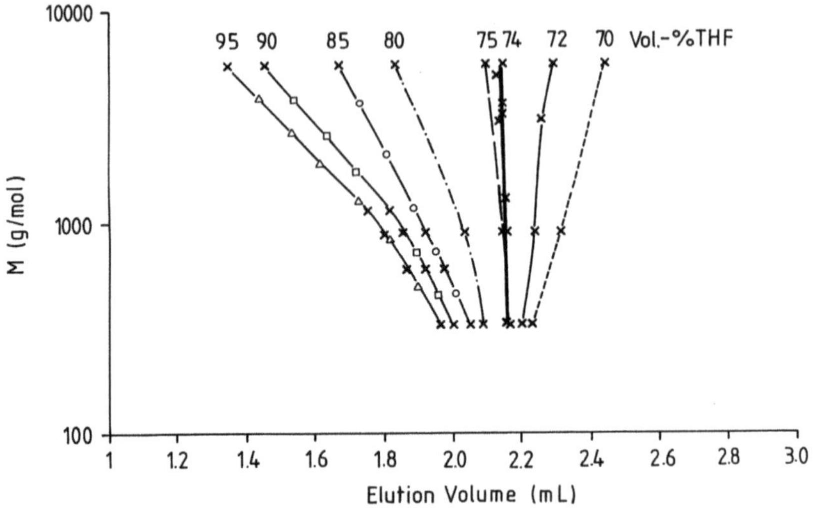

Fig. 6.33 Critical diagram of molar mass versus elution volume for the bisglycidyl BPA peak, chromatographic conditions see Fig. 6.34 (reprinted from [71] with permission of Wiley-VCH)

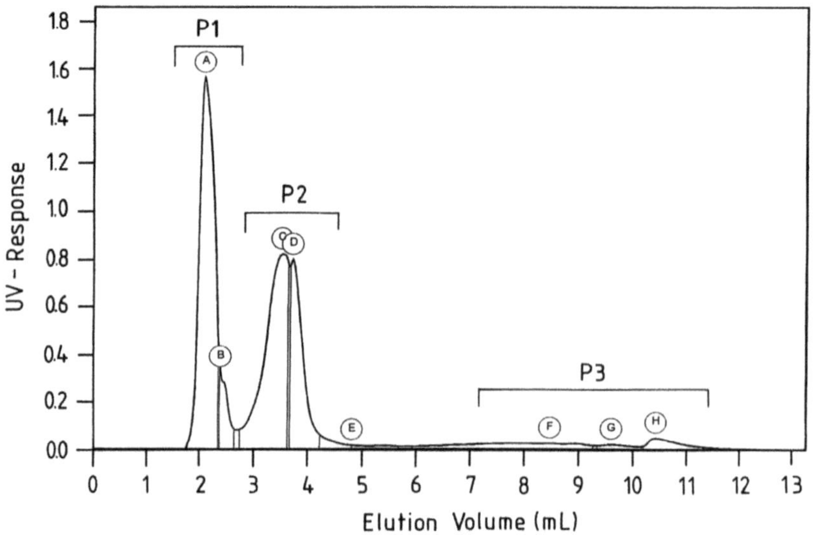

Fig. 6.34 LCCC chromatogram of sample 8, stationary phase: Nucleosil 50–5, mobile phase: THF-hexane 74:26 % by volume, detection: UV 280 nm (reprinted from [71] with permission of Wiley-VCH)

contain epoxy and diol terminal groups. In contrast, fraction H belongs to oligomers without epoxy groups. Such oligomers cannot react in a cross-linking reaction and are, therefore, unwanted by-products.

Table 6.10 Assignment of epoxy resin structures to the chromatographic fractions of sample 8[a]

Fraction	General Structure	
A,B	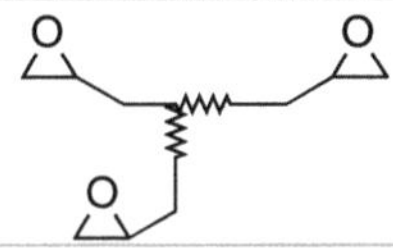	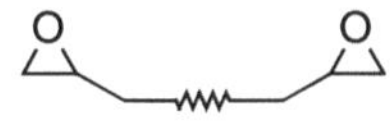
C	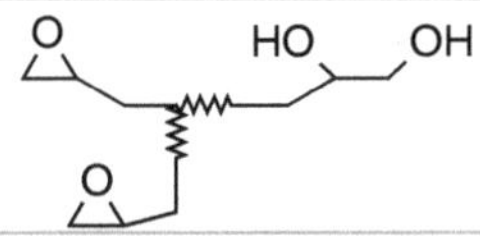	
D	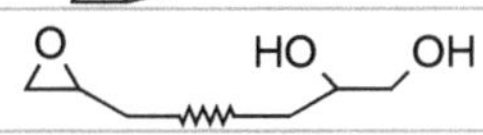	
F,G	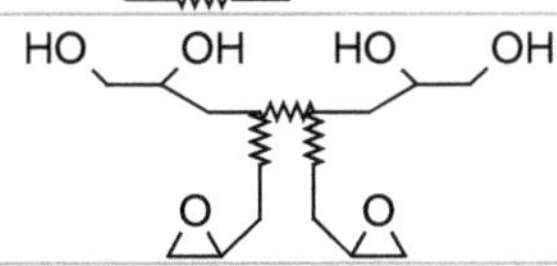	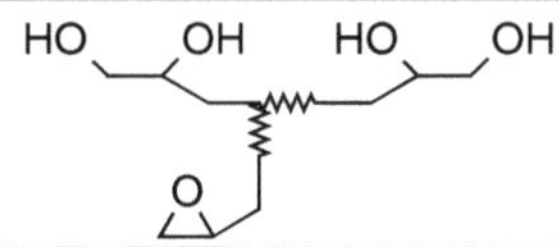
H	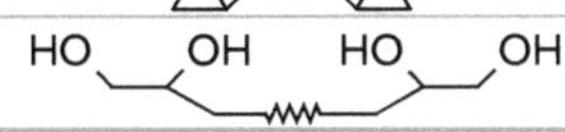	

[a]for fraction E no useful spectrum could be obtained

Using the same chromatographic conditions, LCCC and SEC were combined in an on-line coupled 2D set-up. Only the flow rates of both methods had to be adjusted to the 2D experiment. For the SEC part a flow rate of 2 mL/min was selected. With this flow rate, one SEC run required an analysis time of 5 min. Within these 5 min one loop of the dual-loop system having a volume of 100 μL had to be filled with a fraction from the LCCC. Accordingly, for the LCCC separation a flow rate of 20 μL/min had to be used. As a result of the 2D separation, a number of SEC chromatograms from chromatograph 2 are obtained, each of them characterizing a fraction of 100 μL from chromatograph 1.

The 2D experiment yields separation with respect to functionality and molar mass, and FTD and MMD can be determined quantitatively. For calculating FTD, the relative concentration of each functionality fraction must be determined. These concentrations are equivalent to the volume of each peak in the contour plot because a UV detector has been used. The UV response of the different oligomers is directly proportional to concentration. With the appropriate software, quantification can be done easily. The determination of MMD of each fraction is possible after calibrating chromatograph 2 as has been described in [71]. The calculation of MMD can then be done in the usual way, taking one single chromatogram for each functionality fraction, preferably from the region of the highest peak intensity.

The contour plot of sample 8 in Fig. 6.35 reveals a number of features that are not seen in the off-line LCCC and SEC experiments. Three regions 1–3 are obtained in the contour plot, which correspond to P1, P2 and P3 in Fig. 6.34. However, these regions are not uniform but exhibit different substructures which are coded as (a) and (b). In each of the contour plot regions single oligomer peaks are obtained at

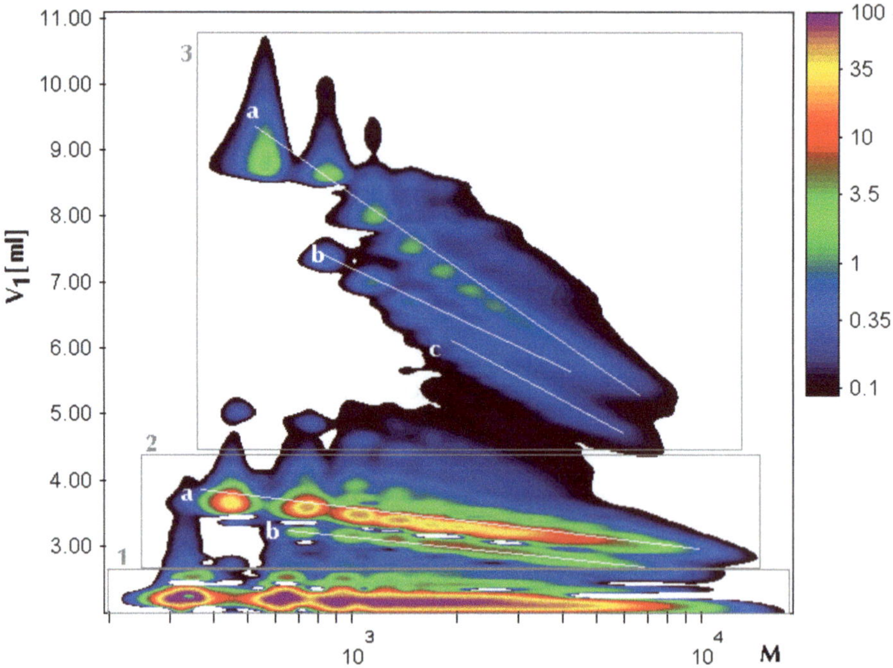

Fig. 6.35 Contour plot of the 2D separation of sample 8; first dimension: LCCC, second dimension: SEC, detection: UV 280 nm, regions 1, 2, 3 and series a, b, c see text (reprinted from [72] with permission of Wiley-VCH)

low molar masses which merge into continuous distributions at higher molar masses. The lowest molar mass peak of each distribution corresponds to the monomer, which has been verified by comparison with model compounds. Region 1 corresponds to the α,ω-bisepoxy functionality fraction, the lowest molar mass peak being the bisphenol-A diglycidyl ether. As can be seen very clearly, all oligomers of this fraction are characterized by the same elution volume (V_1) with respect to the LCCC axis. This behaviour indicates that the bisepoxy oligomers elute strictly at the critical point of adsorption. The oligomer distribution spreads strictly in parallel to the molar mass axis.

For functionality fraction 2, which is assigned to the α-epoxy-ω-diol functionality fraction, critical behaviour is not strictly obeyed. As can be seen from the contour plot, the oligomer distribution exhibits a negative slope meaning that low molar mass oligomers have higher elution volumes in the first dimension (V_1) than high molar mass oligomers. This effect is even more pronounced for functionality fraction 3, which corresponds to the α,ω-bisdiol functionality fraction. The observed behaviour can originate from different effects but has been proven to result mainly from adsorption phenomena.

Another feature of the contour plot in Fig. 6.35 is, that instead of one oligomer series two or even three oligomer series (a), (b) and (c) are obtained for one functionality fraction, series (a) being always the most abundant. This is in agreement with the results of MALDI-TOF mass spectrometry, where in addition to linear oligomers branched oligomers have been found within each functionality fraction.

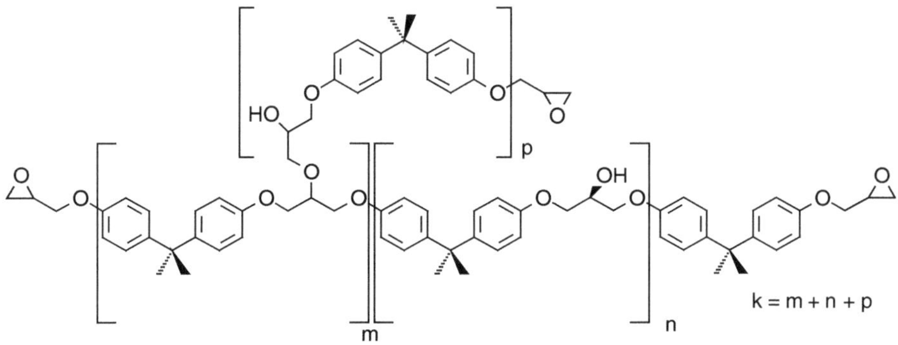

The most abundant oligomer series is the series of linear molecules ($m = 0$), corresponding to oligomer series (a) in the contour plot. Accordingly, series (b) can be assigned to mono-branched ($m = 1$) and series (c) to di-branched species ($m = 2$). This assignment is in agreement with the elution behaviour in LCCC. The branches are terminated by epoxy groups which are less polar than the diol endgroups. Accordingly, the polarity of the oligomer series decreases with an increasing number of branches.

Evaluation

As is demonstrated in Fig. 6.35, resolution of 2D chromatography by far exceeds the resolution capabilities of LCCC and SEC. Not only different functionality fractions are observed but different degrees of branching are seen as well. In addition, for each oligomer series the quantitative oligomer distribution can be obtained. As was pointed out previously, the colour code in the contour plot corresponds to concentration. Accordingly, for each structural feature, the concentration and the molar mass distribution can be given. The wealth of information, obtained by a 2D experiment is clearly demonstrated in Table 6.11.

In addition to the desired α,ω-bisepoxy functionality fraction a significant amount of oligomers of lower functionality is present in sample 8. The α,ω-bisepoxy fraction amounts only to 65% of the total sample. The MMD of the total sample obtained from the 2D experiment agrees well with the off-line SEC measurement. A comparison of the molar masses of the different fractions shows, that they are very similar for all linear molecules regardless of their functionality. As has to be expected, for the branched oligomers higher molar masses are obtained.

Table 6.11 Amounts and average molar masses of the functionality fractions of sample 8

Region	Structure	Amount (%)	M_n (g/mol)	M_w (g/mol)	M_w/M_n
1	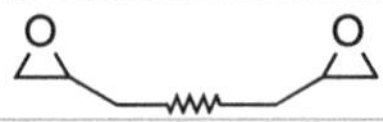	65	860	1,720	1.99
2	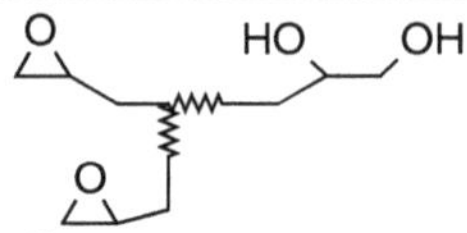	4	1,690	2,620	1.55
	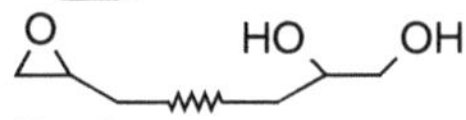	26	1,130	1,840	1.62
	Total	30	1,180	1,940	1.65
3	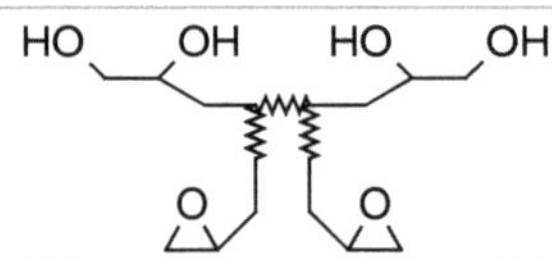	1	2,990	3,680	1.23
	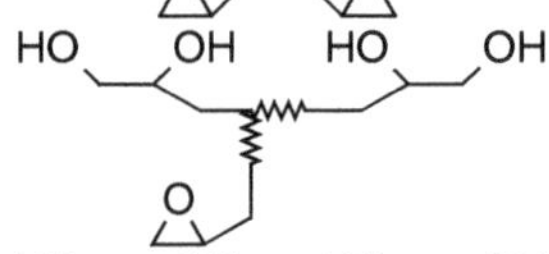	1	1,360	1,590	1.17
	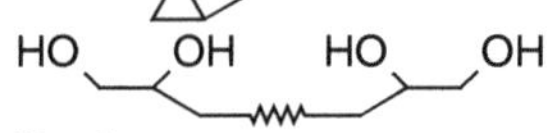	3	1,100	1,990	1.81
	Total	5	1,215	2,250	1.84
Total		100	950	1,810	1.90

It has been pointed out earlier, that epoxy resins usually are prepared by the TAFFY or the advancement process. The TAFFY process yields oligomer mixtures with even and odd degrees of polymerization while the advancement process gives mainly oligomers with odd degrees of polymerization. As can be seen for the oligomer series in Fig. 6.36, for sample 8 even and odd degrees of polymerization are obtained. Accordingly, this sample was prepared by the TAFFY process.

In contrast to sample 8, the 2D analysis of sample 7 results in a rather uniform contour plot. In this case, however, mainly odd degrees of polymerization are obtained, indicating that sample 7 was prepared by the advancement process. In addition, this sample is much more uniform with respect to the functionality. Figure 6.36 shows clearly, that the major functionality fraction is the α,ω-bisepoxy fraction, the α-epoxy-ω-diol fraction is present only in minor amounts, and the α,ω-bisdiol fraction is not present at all. The relative amounts and the molar masses of the different oligomer series are summarized in Table 6.12. They agree well with the data of the off-line SEC measurement.

A significantly higher molar mass sample with a broad MMD is sample 6. The contour plot of this sample in Fig. 6.37 shows the complexity of the sample with respect to functionality. In addition to the functionality fractions 1–3 corresponding

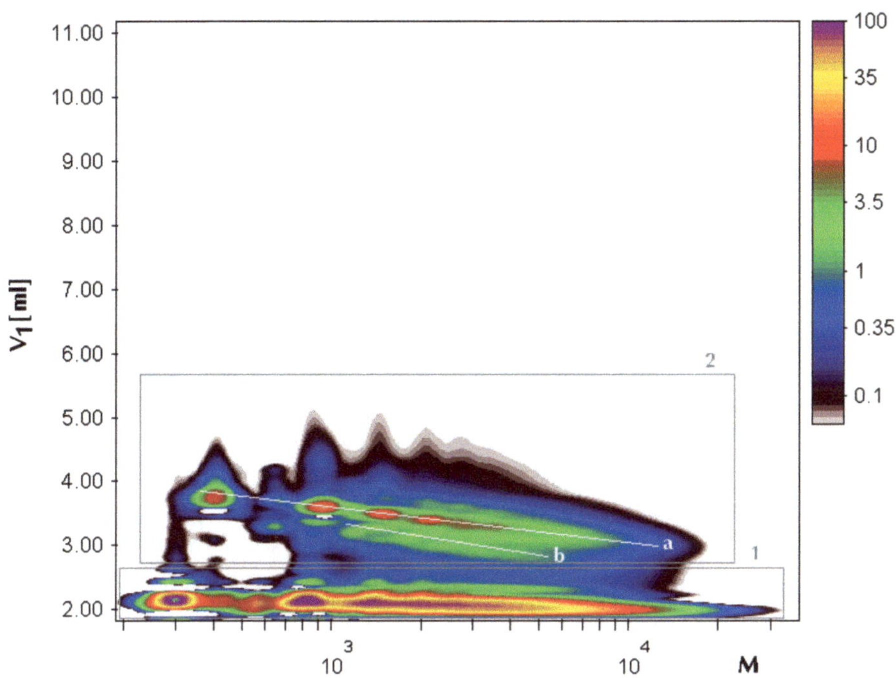

Fig. 6.36 Contour plot of the 2D separation of sample 7; first dimension: LCCC, second dimension: SEC, detection: UV 280 nm, regions 1, 2, 3 and series a, b, c see text (reprinted from [72] with permission of Wiley-VCH)

Table 6.12 Amounts and average molar masses of the functionality fractions of sample 7

Region	Structure	Amount (%)	M_n (g/mol)	M_w (g/mol)	M_w/M_n
1		87	1,013	2,412	2.38
2	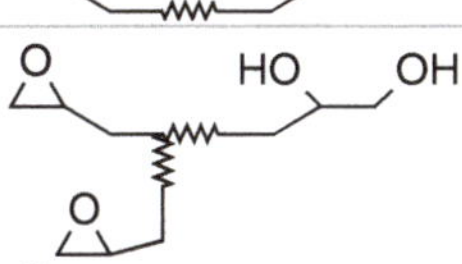	3	1,600	2,970	1.86
		10	1,290	2,715	2.11
	Total	13	1,340	2,760	2.06
Total		100	1,050	2,470	2.35

to the α,ω-bisepoxy, the α-epoxy-ω-diol, and the α,ω-bisdiol fractions a functionality fraction 4 is obtained, which is significantly higher in molar mass than the other fractions. This functionality fraction exhibits an even higher polarity than fraction 3, indicating that the oligomers contain a greater number of polar groups. A possible explanation is the formation of branched oligomers with more than two diol endgroups.

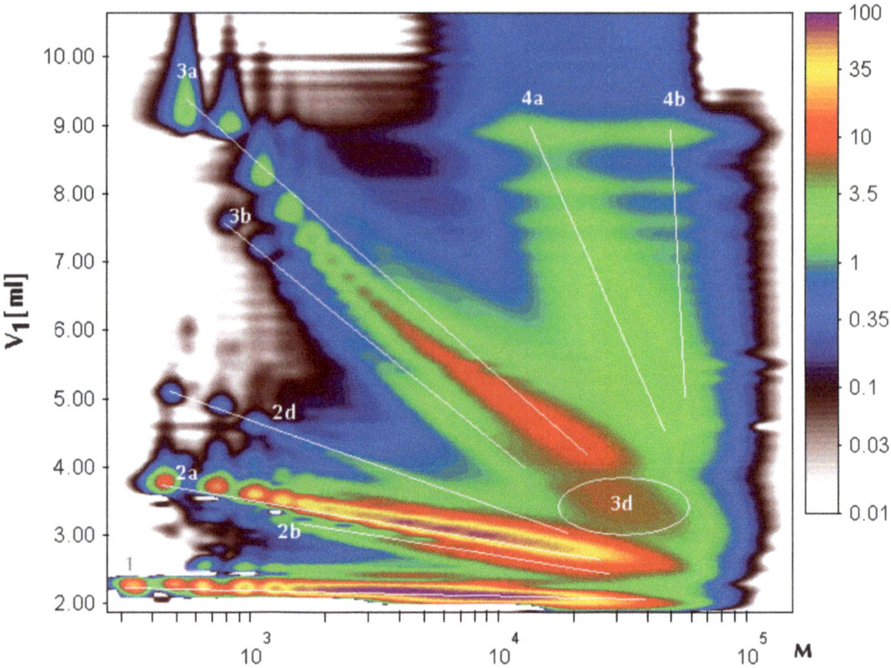

Fig. 6.37 Contour plot of the 2D separation of sample 6; first dimension: LCCC, second dimension: SEC, detection: UV 280 nm, regions 1, 2, 3 and series a, b, c see text (reprinted from [72] with permission of Wiley-VCH)

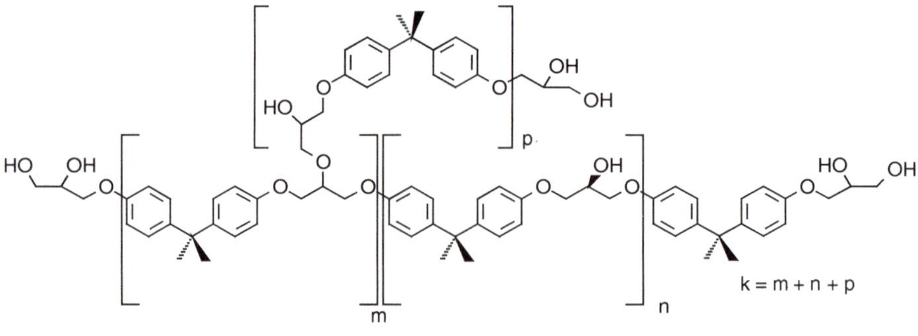

Further oligomer fractions appear in 2d and 3d (see Fig. 6.37) which could not yet be assigned to a specific structure. The quantitative composition of sample 6 is given in Table 6.13. The fact that even and odd oligomers are observed in the contour plot indicates that this sample was prepared by the TAFFY process. This is in agreement with the significant chemical heterogeneity of the sample and the rather low amount of bisepoxy oligomers.

Table 6.13 Amounts and average molar masses of the functionality fractions of sample 6

Region	Structure	Amount (%)	M_n (g/mol)	M_w (g/mol)	M_w/M_n
1	bisepoxy	32	2,760	8,160	2.96
2	epoxy-diol	37	4,120	11,180	2.71
3	bisdiol	21	6,130	17,610	2.87
4	polydiol	10	23,120	34,890	1.51
Total		100	4,100	14,000	3.42

6.6.3 Characterization of Star-Shaped Polylactides [73, 74]

Aim

Homo- and copolymers of lactides (LA) find a wide range of applications related to their ability to hydrolytical and/or biological degradation. First, biomedical applications were developed including bioresorbable surgical sutures, slow-release drug delivery systems, fractured bone fixation or tissue engineering [75–78]. Poly(L-lactide) (PLA) is also being considered as an environmentally friendly commodity thermoplastic and fiber forming material [79–81].

Various practically important properties of PLA such as degradation rate or thermo-mechanical parameters can be adjusted by the macromolecular architecture (topology), stereochemistry, structure of endgroups, and molar mass [82, 83]. Ring-opening polymerization offers methods for the controlled synthesis of PLA composed of macromolecules with tailor-made structure, e.g. star-shaped PLAs. In ring-opening polymerization the architecture of the resulting PLA macromolecules result from the structure of the alkoxide group derived either from the initiator or from the chain transfer agent. Very frequently, PLAs fitted with hydroxyl endgroups are formed. Thus, by selecting the appropriate initiator, star-shaped PLAs with different numbers of arms can be prepared. Such PLA stars shall be analysed with regard to the number of arms by means of 2D chromatography.

Materials

- **Calibration Standards.** Laboratory samples of poly-L-lactide in the molar mass range of 1,000–100,000 g/mol. Linear α-butyl-ω-hydroxy-poly(L-lactide)s (Bu-PLA-OH) of various molar masses were prepared by LA polymerization initiated by $Sn(OBu)_2$ according to [84]. The molar masses of the samples are summarized in Table 6.14.
- **Polymers.** Star-shaped poly-L-lactides with up to 13 arms, see Table 6.14. A series of star-shaped, hydroxyl group terminated poly(L-lactide)s $(R\text{-}(PLA\text{-}OH)_x)$ was prepared according to the procedure employing $Sn(Oct)_2$ as co-initiator, developed for the synthesis of linear PLAs [85, 86]. Namely, L,L-dilactide was

Table 6.14 Molar masses of the polylactides as determined by SEC

Sample	Initiator	M_n (g/mol)	M_w/M_n
1	$Sn(OBu)_2$	2,300	1.59
2	$Sn(OBu)_2$	3,150	1.16
3	$Sn(OBu)_2$	5,160	1.12
4	$Sn(OBu)_2$	8,970	1.12
5	$Sn(OBu)_2$	42,900	1.18
6	$Sn(OBu)_2$	74,000	1.52
7	$Sn(Oct)_2$/DEG[b]	9,000[a]	1.11
8	$Sn(Oct)_2$/TMP[b]	7,800[a]	1.11
9	$Sn(Oct)_2$/DTMP[b]	8,800[a]	1.15
10	$Sn(Oct)_2$/DPE[b]	11,300[a]	1.13
11	$Sn(Oct)_2$/PEHMO[b]	8,300[a]	1.19

[a]measured by vapor pressure osmometry
[b]*DEG* diethylene glycol, *TMP* trimethylol propane, *DTMP* di(trimethylol propane), *DPE* di (pentaerithritol), *PEHMO* poly(3-ethyl-3-hydroxy methyloxetane)

polymerized with $Sn(Oct)_2$ and the corresponding polyol (bearing 2, 3, 4, 6, and ≈13 primary OH groups) as components of the catalytic initiating system.

Equipment

- **Chromatographic System.** The chromatograph for the first separation step (LCCC) consisted of a Rheodyne six-port injection valve with a 100 μL injection loop and an electrically driven eight-port injection valve (Valco EHC8 W) to connect LCCC and SEC chromatographs. In addition, these chromatographs were connected to two storage loops with a volume of 200 μL each. The chromatograph for the second separation step consisted of a Waters model 510 pump. The operation of the coupled injection valves was controlled by the software, which was used for data collection and processing (PSS-2D-GPC Software of Polymer Standards Service GmbH, Germany). The flow rates for the first and second dimensions were 25 μL/min and 4 mL/min, respectively.
- **Columns.** LCCC (chromatograph 1): Nucleosil Si-100 + Si-300 of Macherey-Nagel (Düren, Germany), 5 μm average particle size and 100 or 300 Å average pore diameter. Column size was 200 × 4 mm i.d. SEC (chromatograph 2): PSS-SDV high-speed column (PSS GmbH, Mainz, Germany), 5 μm average particle size. Column size was 50 × 20 mm i.d.
- **Mobile Phase.** Chromatograph 1: Mixtures of dioxane and n-hexane, Chromatograph 2: tetrahydrofuran, all solvents were HPLC grade.

- **Detectors.** Waters model 486 UV/VIS detector at 235 nm and evaporative light scattering detector ELSD 500 of Altech
- **Column Temperature.** 50 °C
- **Sample Concentration.** 20 mg/mL. All samples are dissolved in the mobile phase.
- **Injection Volume.** 100 μL

Preparatory Investigations

For a detailed analysis of the star-shaped PLA it was necessary to prepare a series of linear polylactide (PLA) samples differing in molar masses. For this purpose the ring-opening polymerization of LA initiated with $Sn(OBu)_2$ was applied. In covalent metal alkoxide-initiated polymerization the hydroxyl head endgroups are always present. The tail endgroups are transferred directly from the initiator alkoxide group.

The prepared series of linear polylactides (entries 1–6 in Table 6.14) can be used to find suitable chromatographic conditions for the separation of more complex PLAs. In a first experimental step the critical conditions for PLA were determined. This was done by measuring the chromatographic behaviour of samples of different molar masses in mobile phases of varying composition, as has been described earlier. Following an approach developed by Schulz et al. [87] for oligomeric PLA, mixtures of 1,4-dioxane (good solvent) and *n*-hexane (poor solvent) were used as mobile phase. As the stationary phase silica gel was selected. The temperature of the LCCC analysis was set on 50 °C, governing sufficient solubility of the higher molar mass (M_n well above 10^4 g/mol) poly(L-lactide)s in the eluent. It has been found that at 50 °C critical conditions correspond to a mobile phase composition of 1,4-dioxane/*n*-hexane 56.25:43.75 vol.%.

Separations

Experimental conditions established for the linear PLA were then applied in the LCCC analysis of the star-shaped PLAs (R-(PLA-OH)$_x$) having various numbers of arms. All polyols were fitted with primary hydroxyl groups, enabling fast co-initiation and chain transfer reactions.

Star-shaped PLAs of $M_n \approx 10^4$ g/mol were prepared according to a procedure described in the experimental part. Table 6.14 reports on the corresponding molar masses of the resulting R-(PLA-OH)$_x$'s (entries 7–11). The number of arms was in agreement with the number of hydroxy groups in the initiator molecules, see e.g. for the 2-, 4- and 6-arm PLAs.

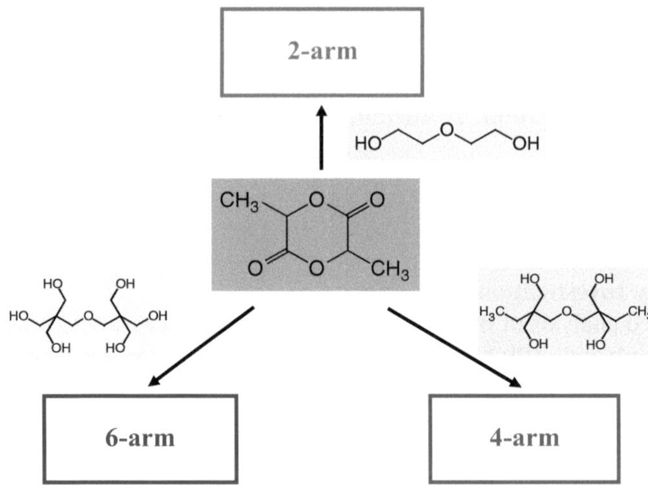

Fig. 6.38 shows a series of HPLC traces recorded for R-(PLA-OH)$_x$ bearing 1, 2, 3, 4, 6, and 13 PLA-OH arms and with M_n close to 10^4 g/mol (linear Bu-PLA-OH and HO-PLA-DEG-PLA-OH can formally be considered as star-shaped PLAs with 1 or 2 PLA-OH arms). Measurements were carried out at 50 °C using the critical composition that has been established for linear PLA (i.e. 1,4-dioxane/*n*-hexane 56.25: 43.75%). The samples have similar molar masses and, therefore, cannot be separated by SEC. In LCCC separation occurs with respect to the heterogeneity of the polymer chain. The critical conditions are optimized regarding the homopolymer chain with a fixed number and type of endgroups. If in other samples the numbers and/or types of endgroups are different, then the chromatographic behaviour of these samples will change. This can be clearly seen for the series of samples shown in Fig. 6.38.

According to Fig. 6.38 elution volumes increase with the number of PLA-OH arms in R-(PLA-OH)$_x$. The explanation for this behaviour is straightforward: the chromatographic experiments are conducted on a polar stationary phase of silica gel. The interaction between the PLA molecules and the stationary phase occurs mainly through the hydroxy endgroups which are the most polar part of the molecules. The more interacting sites (endgroups) the molecules have the stronger will be the interaction with the stationary phase and the larger will be the elution volume.

Evaluation

For polymers bearing 1, 2, 3, 4, and 6 arms this dependence is shown in Fig. 6.39. Such a behaviour results from the increasing strength of the attractive forces between the macromolecules and the column packing which increases with the increasing number of PLA arms fitted with the hydroxyl endgroups. According to the linear free relationship or Martin's rule the capacity fractor is supposed to

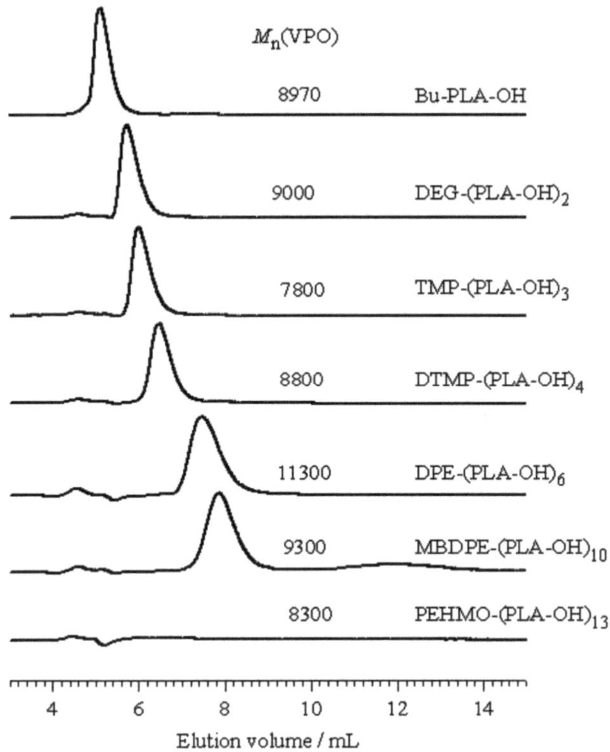

Fig. 6.38 LCCC traces of star-shaped polylactides (R-(PLA-OH)$_x$) fitted with various numbers of PLA arms; mobile phase: dioxane-hexane 56.25:43.75% by volume, detection: ELSD (reprinted from [74] with permission of Elsevier)

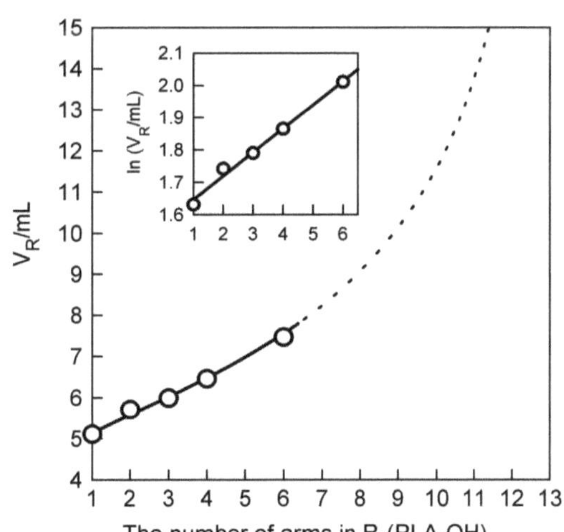

Fig. 6.39 Dependence of elution volumes measured at the peak maxima in LCCC traces shown in Fig. 6.38 on the number of PLA arms in star-shaped polylactides (R-(PLA-OH)$_x$). (reprinted from [74] with permission of Elsevier)

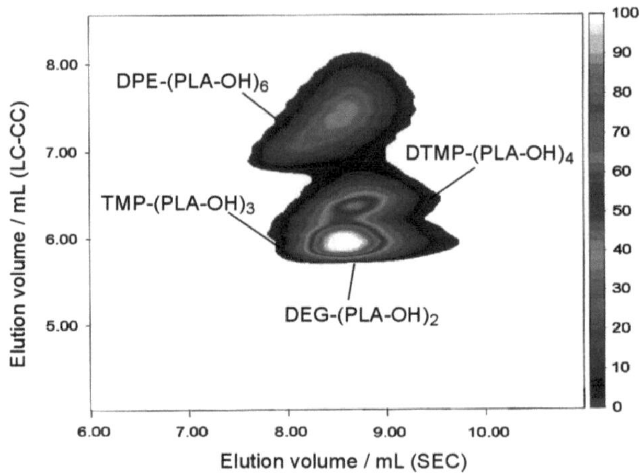

Fig. 6.40 Two-dimensional LCCC versus SEC trace of the artificial mixture composed of DEG-$(PLA\text{-}OH)_2$, TMP-$(PLA\text{-}OH)_3$, DTMP-$(PLA\text{-}OH)_4$, DPE-$(PLA\text{-}OH)_6$. (Reprinted from [74] with permission of Elsevier)

depend exponentially on the number of hydroxyl goups. This means that ln V_R is a linear function of the number of arms. This is indeed the case as can be seen from the insertion in Fig. 6.39. As can be seen in Fig. 6.38, the 13-arm PLA is very strongly retained due to the high number of terminal hydroxyl groups.

Differences in the elution volumes are sufficiently large to detect individual species for example in the LCCC trace of a mixture composed of the 2-, 3-, 4-, and 6-arms R-$(PLA\text{-}OH)_x$ all of them having similar molar masses. First the species with smaller numbers of the hydroxy endgroups are eluted followed by species having more hydroxy endgroups. It is interesting to notice that the 2-arm and the 3-arm nearly co-elute while the 4-arm is more efficiently separated from the 3-arm. A further increase of the number of arms to the 6-arm species results in a near-baseline separation.

The combination of LCCC and SEC in automated 2D chromatography enables the analysis of complex mixtures of polyaliphatic esters of various architectures with regard to functionality and molar mass. As can be seen in the contour diagram (Fig. 6.40), the complex sample is separated in both dimensions. The separation in the vertical direction corresponds to the functionality type separation. The horizontal direction summarizes the separation with regard to molar mass. The 2D trace of a mixture composed of equal masses of DEG-$(PLA\text{-}OH)_2$, TMP-$(PLA\text{-}OH)_3$, DTMP-$(PLA\text{-}OH)_4$, and DPE-$(PLA\text{-}OH)_6$ exhibits characteristic features. The separate contours related to 2-arm/3-arm, 4-arm, and 6-arm are clearly seen. The separation power increases exponentially with the increasing number of arms.

6.6.4 Analysis of PS-PDMS Block Copolymers [88, 89]

Aim

Hybrid polymers are gaining increasing interest and applications in the field of polymer science and technology. These special materials allow quite different properties to be combined in one product and overcome some shortcomings of pure materials. In addition, some specific properties can only be achieved by combining two or more blocks of distinctly different properties such as amphiphilic block copolymers which results in micelles in solvents where there is varying compatibilities to the two blocks. The applications of such hybrid materials containing polydimethylsiloxanes (PDMS) have been described in [90–96].

The different blocks of the hybrid materials in this study have distinctly different properties e.g., the glass transition temperature of polystyrene is 100 °C (brittle solid at room temperature) while that of polydimethylsiloxane is −127 °C (viscous liquid at room temperature even at higher molar masses), different surface energies etc. Due to the incompatibility of the PS and PDMS blocks in the copolymer it is possible to control the phase segregation and the morphology at the nanoscale. The phase morphology is dictated by the composition and lengths of both blocks and the amount of unwanted homopolymer (PS or PDMS) formed during synthesis.

It is the aim of this application to separate PDMS-b-PS copolymers at critical conditions for both PS and PDMS to obtain information on the block lengths and the presence of homopolymer fractions in the samples. The molar mass distributions of the respective fractions will be analysed by 2D-LC.

Materials

- **Calibration Standards.** PS and PDMS standards were purchased from Polymer Laboratories (Church Stretton, Shropshire, UK) and Polymer Standards Service (Mainz, Germany), respectively.
- **Polymers.** The PDMS-b-PS copolymers were synthesized by atom transfer radical polymerization according to reported procedures [97–100]. Briefly, silane functional PDMS was synthesized via an anionic ring opening polymerization of the cyclic D_3 PDMS monomer. The living anionic polymerization is terminated with chlorodimethylsilane to produce the end chain functionality. This chain end is modified using a hydrosilylation reaction to produce bromoisobutyrate functional PDMS. These functional PDMS molecules were the starting blocks and used as macroinitiators for the further ATRP polymerization of the styrene block. The molar masses of the samples are shown in Table 6.15.

Equipment

- **Chromatographic System.** In SEC, the instrument consisted of a Waters 1515 isocratic pump, a Waters inline degasser AF and a Waters 717 plus auto sampler with a 100 μL sample loop. The flow rate was 1 mL/min. The columns were

Table 6.15 Molar masses of the block copolymer samples as revealed by size exclusion chromatography

Sample Number	M_n	M_w	M_p	M_w/M_n
1	7,000	7,900	7,900	1.13
2	49,000	84,000	80,000	1.71
3	39,000	68,800	67,500	1.77
4	12,900	19,800	19,500	1.54
5	37,900	57,800	59,700	1.52
6	32,900	48,300	53,400	1.47
7	13,900	23,200	26,400	1.67

calibrated with polystyrene standards from Polymer Laboratories (Church Stretton, Shropshire, UK). Data processing was performed by Breeze version 3.30 SPA (Waters) software.

In LCCC, the mobile phase was delivered by a Waters 2690 separation module (Alliance) comprising of solvent mixer, vacuum degasser and auto sampler with 100 μL sample loop at a flow rate of 0.5 mL/min. Data were recorded by using PSS WinGPC software version 7. Nitrogen was used as carrier gas in the ELSD, and the pressure at the nebulizer was set to 1.0 bar. Evaporator and nebulizer temperatures were set at 80 °C and 40 °C, respectively. Mobile phases were mixed by volume.

For 2D-LC, sample fractions from the first dimension were transferred to the second-dimension column via an electronically controlled eight-port valve system (VICI Valco instruments, Houston, Texas, USA) equipped with two 100 μL sample loops. The second dimension consisted of a Waters HPLC 515 pump and a 50 × 20 mm PSS Linear M 5 μm SDV column (Polymer Standards GmbH, Mainz, Germany). Detection in the second dimension was done using an ELSD. The flow rates used in the first and second dimensions were 0.04 mL/min and 4.5 mL/min, respectively. A PS calibration curve was used for the calculation of the molar mass distributions of the second dimension.

- **Columns.** SEC: two PLgel columns (Polymer Laboratories) 5μm Mixed-C (300 × 7.5 mm) connected in series along with a PLgel guard column (50 × 7.5 mm). LCCC: (1) Jupiter Octadecyl column (Phenomenex, Torrance, CA, USA): silica-based octadecyl phase; 250 × 4.6 mm; particle diameter 5 μm; pore size 300 Å, surface area 160 m^2/g, carbon load 13.48%, (2) Nucleosil Si 300 (Macherey-Nagel, Dueren, Germany): plain silica; 250 × 4.6 mm; particle diameter 5 μm; pore size 300 Å. 2D-LC: Symmetry 300 (Waters, Ireland): silica-based octadecyl phase, 250 × 4.6 mm, particle diameter 5 μm, pore size 300 Å, surface area 113 m^2/g, carbon load 8.35 % for the first dimension, PSS Linear M 5 μm SDV for the second dimension.
- **Mobile Phase.** SEC: THF (stabilized by 0.125 % BHT), LCCC: mixed mobile phases corresponding to the respective critical points. All solvents were HPLC grade.

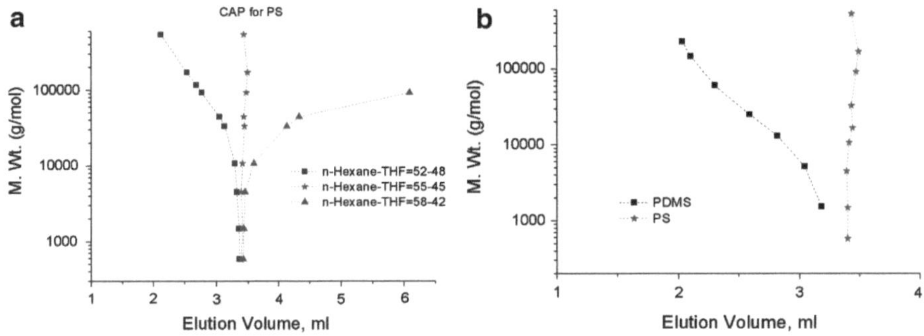

Fig. 6.41 Critical adsorption point (CAP) of polystyrene (**a**) and elution behaviour of PDMS at CAP of polystyrene (*n*-Hexane-THF: 55:45 vol.%) (**b**), stationary phase: Nucleosil Si 300 Å (reprinted from [88] with permission of Wiley-VCH)

- **Detectors.** SEC: Waters 2487 dual wavelength absorbance UV detector, Waters 2414 refractive index detector at 30.0 °C. LCCC: Agilent 1100 series variable wavelength UV detector (set at 254 nm) and PL-ELS 1000.
- **Column Temperature.** 30 °C
- **Sample Concentration.** 1.0–5.0 g/L (SEC, LCCC), 10.0 and 15.0 g/L (2D-LC), all samples were dissolved in the mobile phase.
- **Injection Volume.** 100 μL

Preparatory Investigations

As has been discussed before, LCCC is the preferred method to separate block copolymers according to the lengths of the different blocks and to detect homopolymers in the complex mixtures. Critical conditions for PS have been reported extensively in literature on different columns and with various mobile phase systems [101]. In this study, critical conditions for PS have been established on a normal phase silica column and a *n*-hexane-THF mixture was used as the mobile phase. PS elutes at roughly the same elution volume in *n*-hexane-THF 55:45 vol.% irrespective of their molar masses, see Fig. 6.41.

Under these LCCC conditions, PDMS does not exhibit enthalpic interactions with the stationary phase and elutes in the SEC mode. Figure 6.42 shows elugrams of blends of PDMS and PS homopolymers at the above mentioned conditions. PS homopolymers elute at the same elution volume irrespective of their molar masses while PDMS homopolymers show molar mass dependence as would be expected. The UV detector at 254 nm can only detect PS but not PDMS.

Critical conditions for PDMS were established on a reversed phase C_{18} using methanol-THF as the mobile phase. At a mobile phase composition of methanol-THF 41:59 vol.% PS elutes in the SEC mode while PDMS elutes irrespective of molar mass, see Fig. 6.43. Figure 6.44 shows elugrams of PDMS-PS blends at these conditions. The PDMS homopolymers elute at the same elution volume while PS

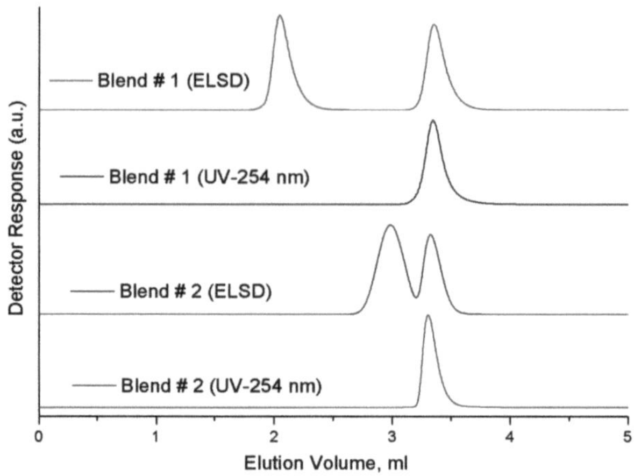

Fig. 6.42 Elugrams of blends of PS and PDMS at CAP of PS, blend #1 = PS 170800 + PDMS 147000; blend #2 = PS 4490 + PDMS 5170 (molar masses in g/mol), detectors: ELS and UV-254 (reprinted from [88] with permission of Wiley-VCH)

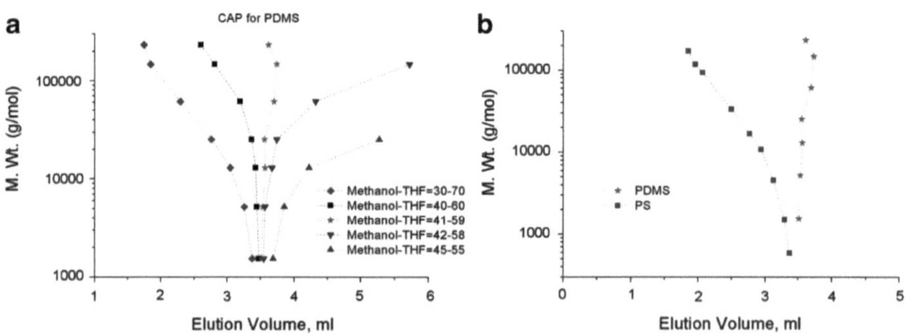

Fig. 6.43 Critical adsorption point of PDMS (**a**) and elution behaviour of PS at CAP of PDMS (methanol-THF 41:59 vol.%) (**b**), stationary phase: Jupiter Octadecyl 300 Å (reprinted from [88] with permission of Wiley-VCH)

shows a molar mass dependence. As expected the UV detector at 254 nm only detects PS but not PDMS.

Separations

The separation of the block copolymer samples at critical conditions for PS is presented in Fig. 6.45. The lengths of the PDMS block in the block copolymers are determined from these elugrams by calibration with PDMS standards. The samples show bimodal elution profiles not seen in SEC. The bimodality can be caused by (A) the presence of PDMS homopolymer in the block copolymer sample (as a result of failure in the chain extension reaction) or (B) the formation of PDMS blocks of different sizes in the block copolymer.

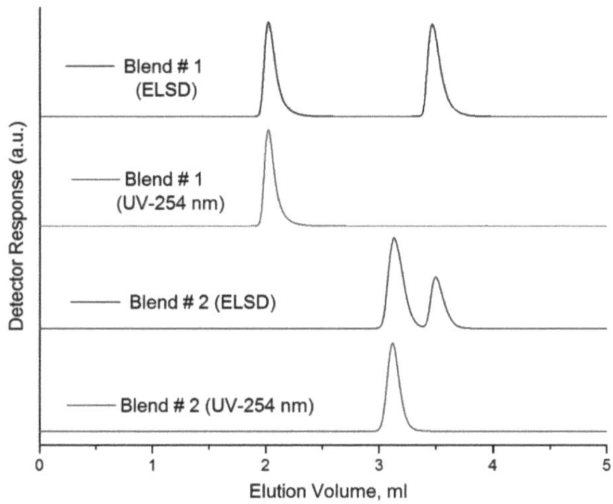

Fig. 6.44 Elugrams of blends of PS and PDMS at CAP of PDMS, blend #1 = PS 170800 + PDMS 147000; blend #2 = PS 4490 + PDMS 5170 (molar masses in g/mol), detectors: ELS and UV-254 (reprinted from [88] with permission of Wiley-VCH)

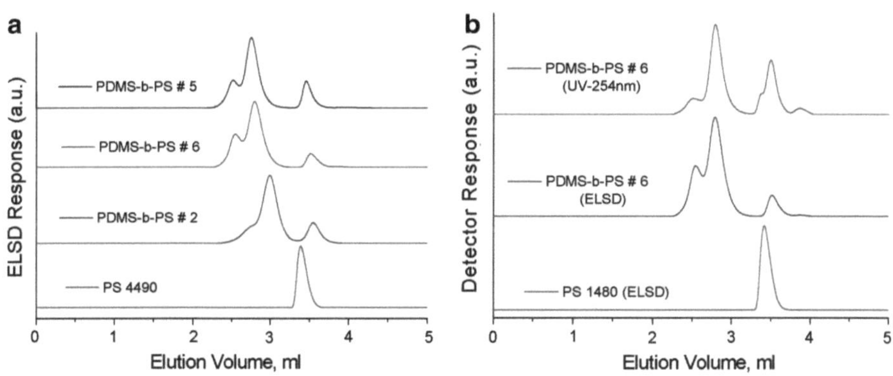

Fig. 6.45 Elugrams of PDMS-b-PS copolymers at CAP of PS, detector: ELS (**a**) and sample #6 with dual detection, detectors: UV-254 and ELS (**b**), chromatographic conditions see Fig. 6.41 (reprinted from [88] with permission of Wiley-VCH)

At an elution volume of roughly 3.5 mL the fractions of PS homopolymer are eluting at the critical point. The origin of the bimodality of the copolymer peak can be investigated by using selective detection such as UV (254 nm) which detects only the UV-active PS but not the UV-transparent PDMS in addition to the ELSD. Figure 6.45b shows the elugram of sample #6 with dual detection (ELS and UV 254 nm). It is obvious that both parts of the peak show UV activity but with

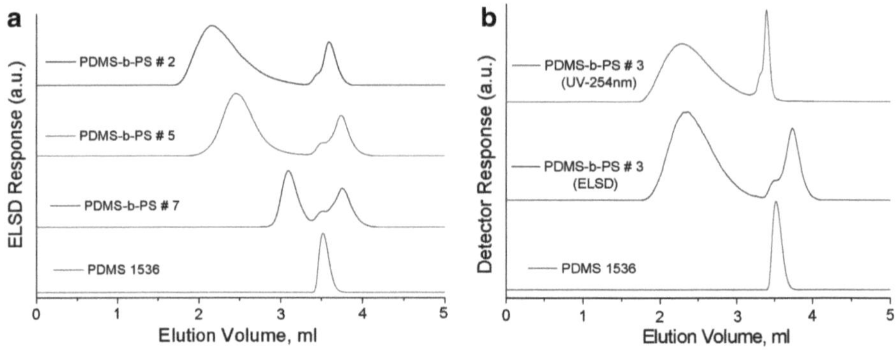

Fig. 6.46 Elugrams of PDMS-b-PS copolymers at CAP of PDMS, detector: ELS (**a**) and sample #3 with dual detection, detectors: UV-254 and ELS (**b**), chromatographic conditions see Fig. 6.43 (reprinted from [88] with permission of Wiley-VCH)

different intensities, roughly indicating the amount of PS in the block copolymer. The earlier eluting fraction corresponding to a higher PDMS molar mass has less PS in comparison to the later eluting fraction corresponding to a lower PDMS molar mass. The shoulder in the UV trace of the PS homopolymer fraction might be due to reagents that elute at the void volume of the column and are volatile and, therefore, could not be detected by ELSD.

The critical conditions for PDMS can be used to separate the block copolymers according to the PS block length. In addition, the PDMS homopolymer can also be separated from the block copolymer, see Fig. 6.46. From a PS calibration curve under these conditions, the molar masses of the PS blocks in the block copolymers were calculated.

The PDMS homopolymer fractions in the samples show at least two distinct distributions. This may be as a result of different chain end functionality. Figure 6.46b shows the elugrams of sample #3 with dual detection at critical conditions of PDMS. It is clear from this Figure that the block copolymer shows an UV activity which is not the case for the homopolymer at 3.7 mL, thus indicating the absence of PS in the homopolymer fraction. In the UV trace, there is a strong signal at the void volume of the column that is not detected by ELSD. This can again be due to reagents that are UV-active but volatile and have no interaction with the stationary phase.

The separations thus far give an idea of the chemical heterogeneity of the samples. The LCCC conditions for PS allow the separation of PS homopolymers from the block copolymers and PDMS homopolymers, while at the same time separating the block copolymers with regard to the molar mass distribution of the PDMS block. By 2D-LC the molar masses of both copolymer and homopolymer fractions can be revealed. 2D-LC contour plots of the samples under study are

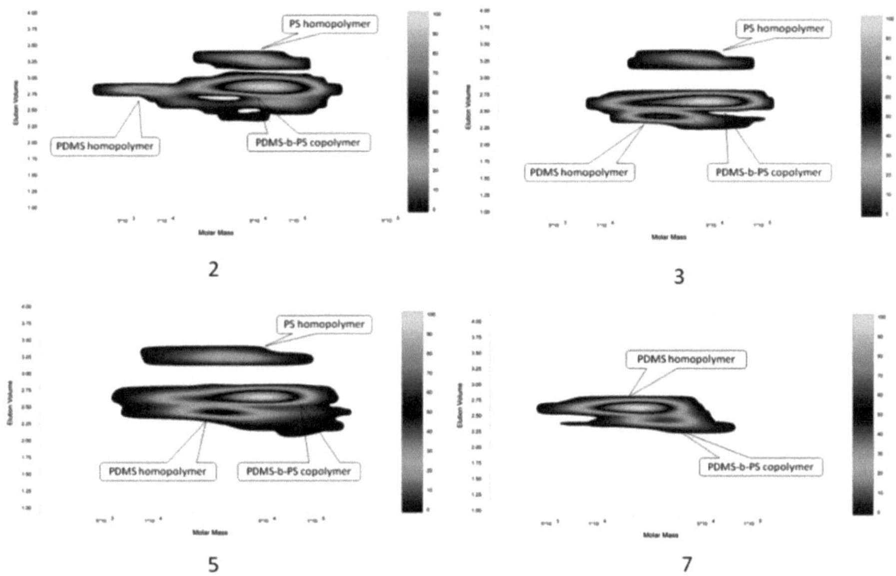

Fig. 6.47 Two-dimensional LCCC versus SEC traces of samples #2, #3, #5 and #7, first dimension: critical conditions for PS (reprinted from [89] with permission of Springer Science + Business Media)

shown in Fig. 6.47. The first dimension in these plots is LCCC of PS (*Y*-axis), with SEC forming the second dimension (*X*-axis).

The elugrams in Fig. 6.47 illustrate the first-dimension separation of PS homopolymers from PDMS-PS block copolymers. For all four samples, PS homopolymers are separated from the block copolymers and PDMS homopolymers. The separation of the PDMS homopolymers from the block copolymers in the first-dimension (*Y*-axis) can be attributed to different endgroups for the PDMS homopolymers and the PDMS block in the copolymer. These molecular species are further separated in the second-dimension (*X*-axis) due to differences in molar mass between the PDMS homopolymers and the copolymers. This results in three distinct components which are visible in the two-dimensional contour plot, namely the PS homopolymers, PDMS homopolymers, and the block copolymers. The molar masses of all separated species can be obtained from these contour plots. The SEC calibration curves of PS and PDMS are quite similar and, therefore, a PS calibration curve for molar mass calculations in the second dimension was used.

The samples were then analyzed by 2D-LC using critical conditions for PDMS in the first dimension, which should separate PDMS homopolymers from the block copolymers and PS homopolymers. Under these conditions, the block copolymers are separated according to the MMD of the PS block. Figure 6.48 shows the 2D-LC analysis of some of the block copolymers. It can be seen that PDMS homopolymers

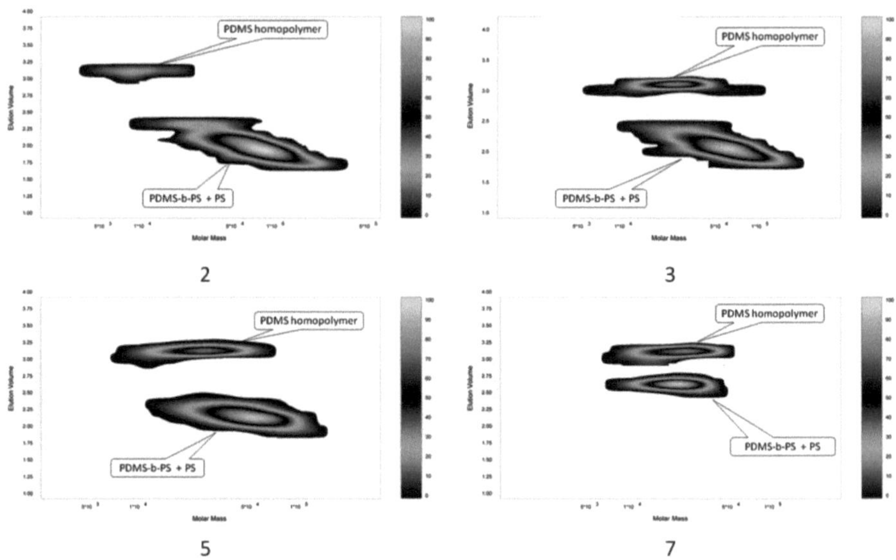

Fig. 6.48 Two-dimensional LCCC versus SEC traces of samples #2, #3, #5 and #7, first dimension: critical conditions for PDMS (reprinted from [89] with permission of Springer Science + Business Media)

are separated from the block copolymers and PS homopolymers quite well, and that the block copolymers are separated according to the MMD of the PS block. From these contour plots, MMD (relative to PS) of block copolymers and PDMS homopolymers are obtained.

Evaluation

As can be seen from the one- and two-dimensional separations, there is still partial co-elution of specific components. At critical conditions for PS, the block copolymers and PDMS are co-eluting while at the critical conditions for PDMS, the block copolymers and PS are co-eluting.

For a comprehensive analysis of all components, semi-preparative fractionations were conducted at both LCCC conditions and the fractions then analysed at alternate LCCC conditions. For example, the first fractionation was conducted at critical conditions for PS producing a fraction of PS homopolymer and a fraction of (block copolymer + PDMS homopolymer). This fraction was then separated at critical conditions of PDMS producing a fraction of pure block copolymer. This fraction was analysed by FTIR to provide the chemical composition of the pure block copolymer. Similar fractionations were conducted for all samples and the following quantitative informations were obtained: (1) percentage and molar mass of PS homopolymer, (2) percentage and molar mass of PDMS

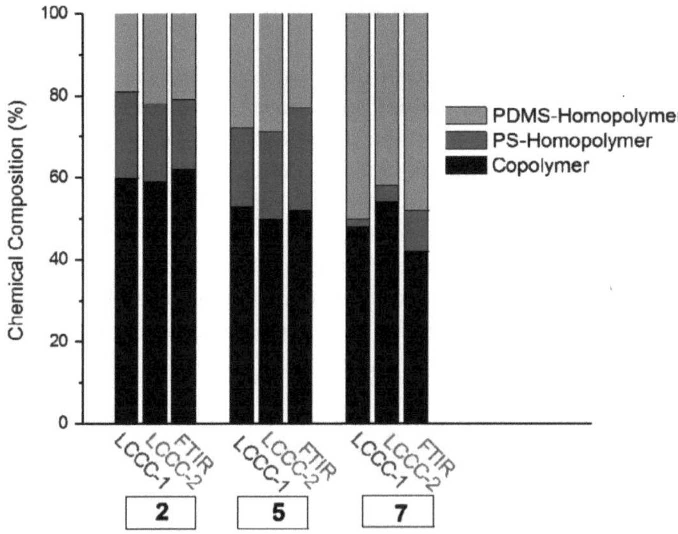

Fig. 6.49 Chemical composition of samples #2, #5 and #7 as revealed by LCCC and FTIR (LCCC-1 is based on peak areas while LCCC-2 is based on calibration of ELSD) (reprinted from [89] with permission of Springer Science + Business Media)

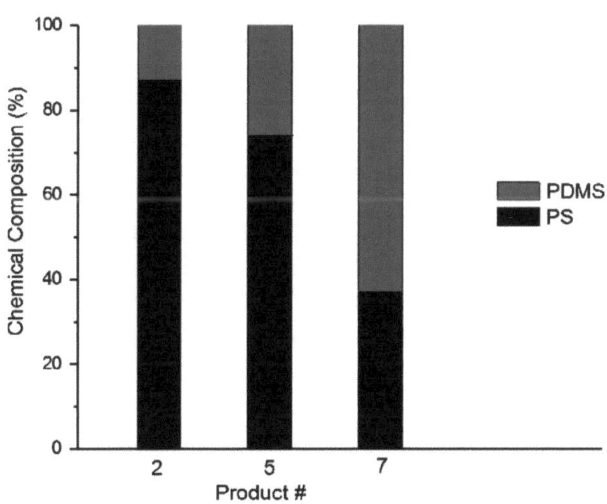

Fig. 6.50 Chemical composition of the true block copolymer fractions in samples #2, #5 and #7 as revealed by FTIR analysis (reprinted from [89] with permission of Springer Science + Business Media)

homopolymer, (3) percentage and molar mass of true block copolymer, (4) chemical composition of true block copolymer. A graphical presentation of the chemical composition information is given in Figs. 6.49 and 6.50.

6.7 Other Method Combinations in 2D-LC

In principle, there is quite a range of different fractionation methods that can be coupled in 2D set-ups. These include field flow fractionation and crystallization fractionation methods which will not be addressed in this chapter. Regarding liquid chromatographic methods, temperature gradient interaction chromatography (TGIC) is an interesting option for the first dimension while liquid adsorption chromatography (LAC) is an option to be used in the first or second dimension. Although both of these methods are feasible methods there is not much work that has been done in this field. In this section two applications will be described for (A) the coupling of TGIC with gradient HPLC and (B) the coupling of LCCC and LAC.

6.7.1 Analysis of Comb-Shaped PS by TGIC-Gradient HPLC [102]

Aim

Chain branching in macromolecules influences the rheological and mechanical properties of polymeric materials significantly, and model branched polymers (e.g., star-shaped or comb-shaped polymers) have been studied extensively to understand their single chain properties and rheological behaviour [103, 104]. Most model branched polymers are prepared by anionic polymerization, however, despite the use of the best synthetic methods available and the time-consuming post-fractionation, it is difficult to obtain structurally well-defined materials of high purity. Furthermore, the purity of such branched polymers has frequently not been confirmed unambiguously due to the lack of precise characterization methods.

The aim of the present application was the rigorous characterization of a comb-shaped PS, where deuterated PS (d-PS) branches are attached to a hydrogenous PS (h-PS) backbone. Such comb-shaped polymers resemble the structure of branched polymers with 3-way branching points in commercially important polymers and have been used extensively as model polymers having long chain branches.

Polymers

- The comb-shaped PS was prepared by grafting d-PS anions to a partially chloromethylated h-PS backbone. Both backbone and branch PS were prepared separately by anionic polymerization using high vacuum techniques. The chloromethylation of the h-PS backbone was carried out by the reaction of the h-PS with in-situ prepared chloromethyl methyl ether. Details of the synthetic procedure were reported previously [105]. The comb and precursor PS samples were characterized by SEC with light scattering detection as summarized in Table 6.16.

Table 6.16 Molar mass characteristics of the comb-shaped PS and its precursors[a], molar masses in kg/mol

Backbone		Branch		Comb		
M_n[b]	M_w/M_n	M_n[b]	M_w/M_n	M_n[b]	M_w/M_n	n_{branch}
190	1.05	83	1.02	540	1.01	4.2

[a] n_{branch} is estimated from the number-average molar mass determined by SEC measurements.
[b] number-average molar mass determined by SEC-LS measurements.

Equipment

- **Chromatographic System.** For the 2D-LC analysis, three HPLC pumps (Pump 1: Waters 515 HPLC pump, Pump 2: Bischoff compact pump 2250, and Pump 3: Spectra System P4000) and two UV detectors (TSP, UV100) were used. The two LC systems were connected via two electronically controlled 10-port 2-position valves (Alltech, SelectPro) equipped with two 100 μL storage loops and with two short diol bonded silica trapping columns (Nucleosil diol, 7 μm, 100 Å, 50 × 4.6 mm). The experimental setup has been reported in [106]. To trap the polymers from the first dimension LC effluent, ACN with 4% iso-octane was used. The mixed solvent did not cause a breakthrough problem in the second LC separation; furthermore, full desorption took place by the second dimension LC eluent. The flow rate was 0.05 mL/min for the first dimension and 1.5 mL/min for the second dimension. The temperature of the column of the first dimension was controlled by a circulating fluid from a programmable bath/circulator (ThermoHaake, C25P) through a homemade column jacket.
- **Columns.** First dimension: bare silica column (Nucleosil, 5 μm, 500 Å pore size, 50 × 4.6 mm), second dimension: C18 bonded silica column (Nucleosil C18, 7 μm, 500 Å pore size, 150 × 4.6 mm)
- **Mobile Phase.** First dimension: 52:48 vol.% of isooctane (J.T. Baker, HPLC grade) and THF (Samchun, HPLC grade), second dimension: solvent gradient run with two mixtures of CH_2Cl_2/CH_3CN (Samchun, HPLC grade) differing in their composition (A: 57:43, B: 59:41 vol.%) were used at a flow rate of 1.5 mL. The solvent composition started and was maintained at 100% A for 2 min after the sample injection. The solvent composition was then changed linearly from 100% A to 100% B over 4 min followed by a quick change back to 100% A.
- **Detectors.** UV detector (TSP, UV 100).
- **Column Temperature.** Temperature gradient in the first dimension
- **Sample Concentration.** 5 g/L (2D-LC)
- **Injection Volume.** 100 μL

Preparatory Investigations

The comb-shaped polymer and its precursors (h-PS backbone and d-PS branch) were first characterized by SEC as displayed in Fig. 6.51. The molar mass of the polymer species for each peak can be estimated from the relative intensity of the RI signal to the LS detector signal. In the SEC chromatograms of both branch and

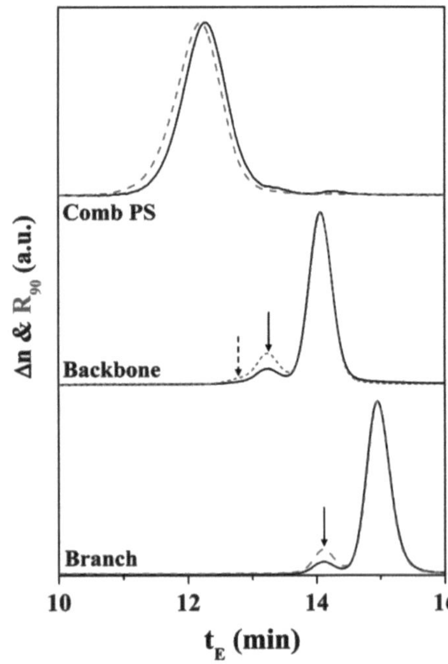

Fig. 6.51 SEC chromatograms of the two PS precursors (backbone and branch) and the PS comb recorded by RI (*solid line*) and light scattering (*dashed line*) detectors, columns: two PL mixed C, mobile phase: THF, column temperature 40 °C (reprinted with permission from [102]. Copyright (2011) American Chemical Society)

backbone PS, a small amount of coupled products (labeled with solid arrows showing double molar mass) is apparent eluting earlier than the narrow peak of the main products. The fronting of the coupled backbone PS peak at $t_E \approx 13$ min (marked with dashed arrow) revealed the possibility of a threefold backbone product. The PS comb contains a trace amount of the PS precursors resolved as separate peaks, but the major product of the PS comb shows a unimodal elution peak. It is apparent that the SEC separation cannot resolve the polymer species having a coupled backbone but yields a broad unimodal peak due to the non-uniform number of branches. From the SEC results alone, it is difficult to obtain more details beyond the average branch numbers determined by the average molar mass.

Separations

In the next step the comb polymer was separated by normal phase TGIC. Figure 6.52 shows an NP-TGIC chromatogram of the comb PS. It separates the comb-shaped polymers according to molar mass, i.e. species with different numbers of branches are resolved. The peak molar mass increases as integral multiples of branch molar mass, but the peaks start to overlap as the number of branches increases due to the finite MMD of the branch and backbone. The contribution of coupled backbone species also needs to be taken into account since a single backbone comb and coupled backbone comb with a fewer number of branches may have similar molar

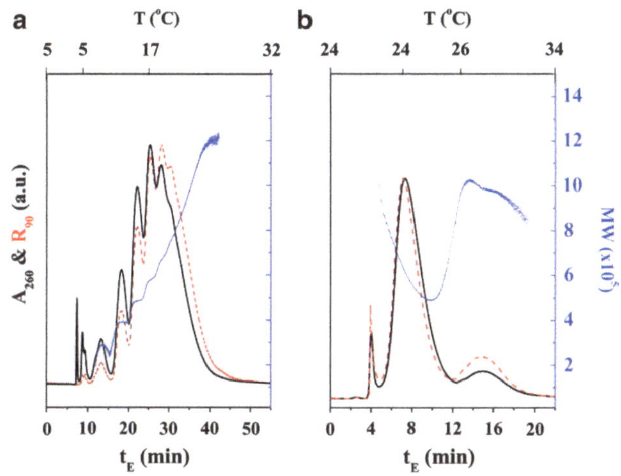

Fig. 6.52 NP-TGIC (**a**) and RP-TGIC (**b**) chromatograms of PS comb recorded by UV (*solid line*) and LS (*dashed line*) detectors, molar mass determined by LS, column temperature is shown in the *top* abscissa, columns: (**a**) Nucleosil, 500 Å, (**b**) Nucleosil, C18, 500 Å, mobile phase: (**a**) iso-octane-THF 52:48 vol.%, (**b**) CH_2Cl_2-CH_3CN 57:43 vol.%, flow rate 0.5 mL/min (reprinted with permission from [102]. Copyright (2011) American Chemical Society)

masses. Figure 6.52b displays the reversed phase TGIC chromatograms of the comb polymer recorded by LS and UV detectors. Unlike the well-resolved NP-TGIC chromatogram, RP-TGIC shows only three peaks indicating a significantly lower selectivity. The peculiar phenomenon in RP-TGIC was found to be due to the chromatographic selectivity regarding isotopes. d-PS interacts with the RP stationary phase less strongly than h-PS, and the unreacted branches elute early at $t_E \approx$ 4 min. The comb polymers containing a single backbone elute later at $t_E \approx$ 6–11 min as the main peak in which the grafted d-PS branches seem to reduce the retention time of the comb-shaped polymer as the number of grafted d-PS branches (thus molar mass) increases. After the main peak, molar mass increases abruptly toward the last peak eluting at $t_E \approx$ 12–18 min which was assigned to the comb species with coupled backbone.

To characterize the comb-shaped PS more rigorously, 2D-LC separation was conducted by combining NP-TGIC (separating the PS comb mainly by molar mass regardless of the isotope content) and the RPLC (separating the comb PS mainly by the isotope composition). Since the NP-TGIC separation of the comb PS exhibits better resolved peaks than the RP-TGIC separation as shown in Fig. 6.52, it is desirable not to lose the high resolution of this separation. Since a fast second dimension separation has to be done at the cost of resolution to some extent, NP-TGIC was employed for the first dimension. For the second RPLC separation, the solvent gradient elution method was employed instead of the temperature gradient elution since the column temperature cannot be changed as rapidly as the solvent composition. As a second point, the solvent compatibility of the mobile phases of the first and second dimensions had to be considered. In the present case, a trapping system was used as described in [106].

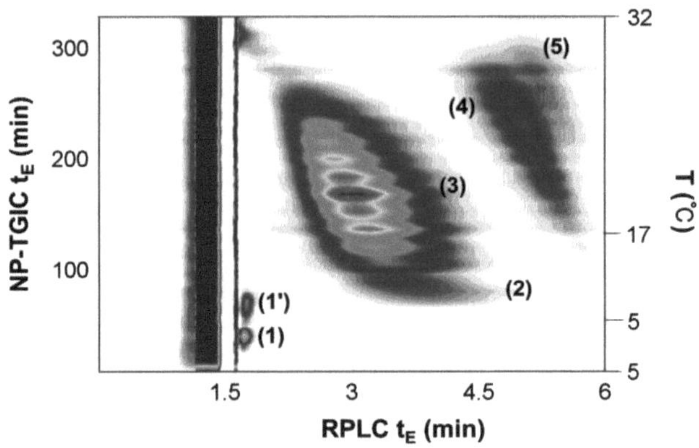

Fig. 6.53 Contour plots of TGIC-RPLC 2D-LC separation of comb PS, first dimension: Nucleosil, 500 Å, iso-octane-THF 52:48 vol.%, flow rate 0.5 mL/min, second dimension: Nucleosil C18, 500 Å, gradient CH_2Cl_2–CH_3CN 57:43–59:41 vol.% in 6 min, flow rate 1.5 mL/min (reprinted with permission from [102]. Copyright (2011) American Chemical Society)

Figure 6.53 displays the TGIC-RPLC 2D-LC separation of the comb-shaped PS. In this 2D mapping, more details of the comb-shaped PS are revealed. Unreacted d-PS branches are further resolved by TGIC into two peaks corresponding to single (1) and coupled branches (1'). The faint peak (2) is the single backbone without any branches. The major peak group (3) corresponds to the comb-shaped PS having different numbers of branches. As the number of branches increases, TGIC retention increases while RPLC retention decreases as already shown in the 1D separation. The peak group (4) corresponds to the comb-shaped PS with a coupled backbone. Since molar mass of the h-PS portion (backbone) is high, it is retained longer in RPLC while its NPLC retention is overlapped with the single backbone combs having similar molar masses (i.e., a larger number of branches) and not resolved in one-dimension TGIC separation. The coupled backbone combs show the same trend as the single backbone combs, i.e., TGIC retention increases while RPLC retention decreases as the number of branches increases.

Another feature to note is the faint peak group (5) eluting after (4) in the RPLC separation. It is apparent that it cannot be resolved either from the other comb groups or from the coupled backbone combs by one-dimensional separation. Nonetheless, it is clearly resolved by virtue of the 2D-LC separation. Judging from its high NP-TGIC retention and the resolved sub-peaks in the group, it is identified as the triply coupled backbone species.

Evaluation

The branch number distribution of the comb-shaped polymer could be obtained from the intensity of the 2D-LC diagram, see Fig. 6.54, by integrating the intensity profile. But in order to obtain the branching distribution more precisely, off-line 2D-LC analysis was performed. The comb-shaped PS was fractionated by RPLC to separate polymers containing different backbone species and subsequently

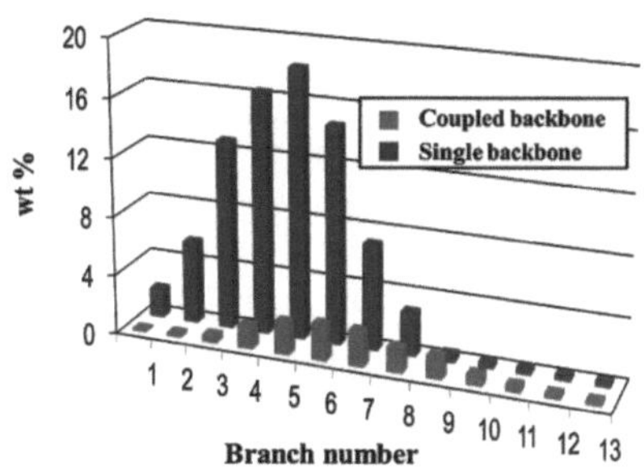

Fig. 6.54 Distribution of branches in the comb-shaped polystyrene (reprinted with permission from [102]. Copyright (2011) American Chemical Society)

subjected to NP-TGIC separation to obtain the branch distribution. The analysis of the polymer species of triple backbone was not attempted since the amount was too small for a reliable analysis.

6.7.2 Analysis of Fatty Alcohol Ethoxylates by LCCC-LAC [107]

Aim

Fatty alcohol ethoxylates (FAEs) are one of the most important classes of functional homopolymers. They are also termed alkyl- or aryloxy-terminated polyethylene oxides (PEOs). These oligomers and polymers have a hydrophilic PEO polymer chain and a hydrophobic fatty alcohol or aryl endgroup. They are industrially used as amphiphilic surfactants, emulsifiers, dispersants etc. FAEs are prepared by anionic polymerization of ethylene oxide in the presence of mixtures of fatty alcohols having different chain lengths, mainly in the range of C_{10} to C_{18}. Thus, macromolecules with different endgroups are formed and the samples exhibit functionality type distributions (FTD). Like all other synthetic polymers, FAE exhibit molar mass distributions, i.e. oligomers with the same endgroups have different chain lengths. Finally, the endgroups may appear as different isomeric structures. To analyze the structure-property correlations of these complex species it is most desirable to have a method that separates FAE according to the endgroups and the oligomer distributions. Such a method would provide the molecular heterogeneity in terms of FTD x MMD.

It has been demonstrated several times that polyalkylene oxides can be separated efficiently by different methods of interaction chromatography [18, 25, 45, 108, 109]. In particular, the functionality type analysis of PEO can be conducted efficiently by LCCC. The molar mass separation of the oligomers can be conducted by SEC. The combination of LCCC and SEC for the analysis of PEO in 2D-LC has been described by Murphy et al. [10] and Kilz [110].

It is known that typical FAE have a degree of polymerization of roughly 5–15. Their molar masses are sufficiently low to obtain oligomer separations in SEC. Under the conditions of 2D-LC, however, this oligomer separation is not achieved (high flow rates, short columns, lower resolution). Alternatively, FAE can be separated

Table 6.17 FAE products that were used to prepare the model blend, *n*: degree of oligomerization

Product	*n*	Endgroup
(1) C_{10}-Oxoalcohol	7	C_{10}
(2) C_{12},C_{14}-Oxoalcohol	7	C_{12}, C_{14}
(3) Nonylphenyl Oxoalcohol	10	Nonylphenyl
(4) C_{13},C_{15}-Oxoalcohol	7	C_{13}, C_{15}
(5) C_{16},C_{18}-Oxoalcohol	6	C_{16}, C_{18}

into oligomers by adsorption chromatography. This has been shown for a number of samples by Trathnigg [111, 112], Rissler and Fuchslueger [113] and others.

The present application describes the combination of two interactive modes of liquid chromatography in 2D-LC. In the first dimension gradient HPLC is used for the functionality type separation while in the second dimension adsorption chromatography is used for oligomer separation. Using this combination, a simultaneous separation according to the endgroups and the degree of oligomerization can be achieved.

Polymers

- A representative very complex FAE was prepared by solution blending of five commercial FAE products of BASF AG (Ludwigshafen, Germany). The products were mixed in equal amounts to produce a blend that contains nine different functionality fractions. The types of endgroups and the average degrees of polymerization as given by the producer are summarized in Table 6.17.

Equipment

- **Chromatographic System.** The separations according to fatty alcohol endgroup (first dimension) were carried out on a Shimadzu HPLC system comprising a DGU-14A degasser, a FCV-10ALVP solvent mixing chamber, a LC-10ADVP pump and a SL 10ACVP auto sampler. Sample fractions from the first dimension were transferred to the second dimension via an eight-port valve system (type EHC8W, VICI Valco instruments, Houston, Texas, USA), attached with two 100 μL loops. The second dimension consisted of a Shimadzu LC-10ATVP pump. A second column, Chromolith C_{18} from Merck (100 × 4.6 mm i.d., C_{18}- grafted monolithic silica) was added before the silica column in the second dimension in order to separate the polymer from methanol coming from the first dimension system. The flow rate was 0.025 mL/min in the first dimension and 1.5 mL/min in the second dimension. Polymer solutions were prepared in methanol-water (80/20% v/v). Data were collected with the software package PSS-WinGPC 7 (Polymer Standards Service GmbH, Mainz, Germany).

- **Columns.** First dimension: Waters (Milford, USA) X-Terra RP-18 (2.5 μm average particle size, 127 Å average pore size, 30 × 4.6 mm i.d.), second dimension: Chromolith Si from Merck (Darmstadt, Germany) (100 × 4.6 mm i.d. monolithic bare silica)
- **Mobile Phase.** First dimension: binary mobile phase gradient starting with methanol/water (80/20% v/v) and going linearly to 100% of methanol in 160 min, second dimension: isopropanol/water (88/12% v/v).
- **Detectors.** ELSD 1000 (Polymer Laboratories, Church Stretton, England)
- **Column Temperature.** 25 °C
- **Sample Concentration.** 20 g/L (2D-LC)
- **Injection Volume.** 25 μL

Preparatory Investigations

In the first step the functionality type separation must be optimized. Since the FAE endgroups were long alkyl or alkylaryl chains, functionality type separation was performed on a reversed phase material. As the non-polar stationary phase C_{18}-modified silica gel was used. This stationary phase was expected to show strong enthalpic interactions with the endgroups but only weak interactions with the polar ethylene oxide chains. The mobile phase composition at the start of the experiment was set to the critical conditions for PEG (methanol/water 80:20 % v/v). These critical conditions were specifically determined for the present system based on several examples of critical conditions for this polymer that are presented in the review article of Macko and Hunkeler [101]. At LCCC conditions, PEG which is always present in technical FAEs eluted without interaction from the stationary phase close to the dead volume of the column. The end-functionalized FAE species showed strong interactions and were retained on the stationary phase. An increase of the mobile phase elution strength (by increasing the methanol content in the mobile phase) was needed to elute the functionality fractions in narrow peaks. A linear increase of the methanol percentage in the mobile phase allowed for a progressive desorption according to the hydrophobicity of the endgroups (i.e. the fatty alcohol chain length). As the ethylene oxide part of the molecules did not interact with the stationary phase, the separation was independent of the number of EO units and based only on the endgroup chain length.

Figure 6.55 presents the separation of the model blend that was achieved under these conditions. As can be seen, all functionality fractions are baseline-separated from each other eluting in the order of increasing hydrophobicity of the endgroups. The baseline separation of the different functionality fractions achieved using this method can be accomplished in a very short period of time: only 8.5 min (equal to an elution volume of 8.5 mL) are required for the separation. Such fast separations were described earlier by Pasch et al. [109]. Peak assignment has already been confirmed by previous LC-NMR experiments [114]. The order of elution confirms a retention mechanism based on hydrophobic interactions.

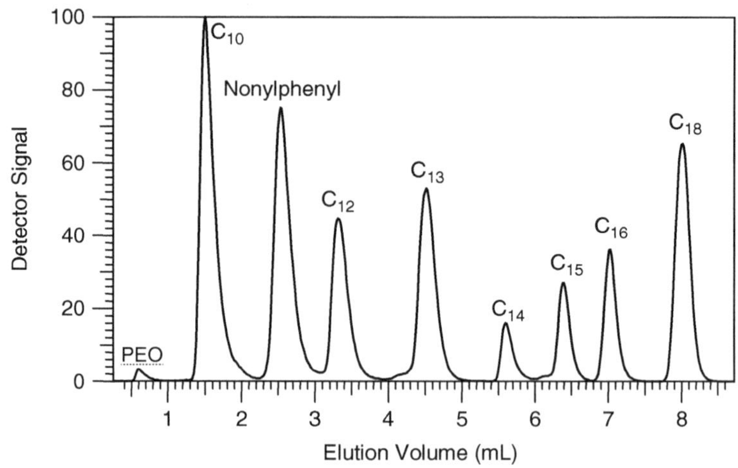

Fig. 6.55 Gradient chromatography of the FAE model blend; stationary phase: X-Terra C18, 30 × 4.6 mm I.D.; mobile phase: linear gradient from methanol-water 80:20 v/v to 100 % of methanol; detection: ELSD (reprinted from [107] with permission of Wiley-VCH)

The separation according to oligomer lengths was performed isocratically since this mode is easier to use as the second dimension for a 2D-LC system. In 2D chromatography, fractions from the first dimension are consecutively injected into the second dimension. Normally, 50–100 such injections are made requiring significant amounts of time. Since these injections directly determine the flow rate value (and the total elution time) of the first dimension, each second dimension experiment must be performed as fast as possible. As standard HPLC pumps have difficulties to properly perform a gradient at very low flow rates (limit generally observed at 0.025 mL/min), a fast separation according to the oligomer lengths (in the second dimension) was required in order to accommodate a reasonable flow rate in the first dimension. To fulfill the two requirements of EO selectivity and fast separation, a short bare silica monolithic column (Chromolith Si, 100 × 4.6 mm I.D.) was selected for the separations. The stationary phase in this case was mesoporic and thus gave lower back pressure than conventional HPLC columns. It allowed for using higher flow rates without being limited by the pump maximum pressure. The mobile phase was a mixture of isopropanol and water (88:12 % v/v). Figure 6.56 shows the oligomer separation obtained for the C_{12},C_{14}-FAE with the fast operating chromatographic system.

The chromatogram obtained for this FAE shows one peak for each oligomer which confirms that separation occurs without interference of the endgroups. Since the separation was accomplished by adsorption chromatography, the shortest oligomers eluted first. Indeed, retention was directly dependent on the number of EO repeating units as each of them could be considered as an adsorbing point. When performing the separation of the complete model blend comprising nine functionality fractions, see Fig. 6.57, the EO oligomer separation was also

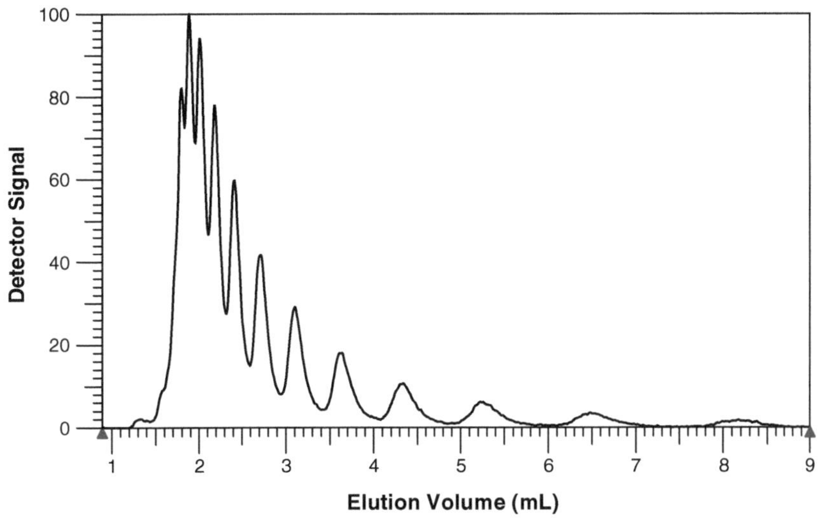

Fig. 6.56 Normal phase separation of the C_{12},C_{14}-FAE according to the EO content; stationary phase: Chromolith Si, 100 × 4.6 mm I.D.; mobile phase: isopropanol-water 88:12 % v/v; detection: ELSD (reprinted from [107] with permission of Wiley-VCH)

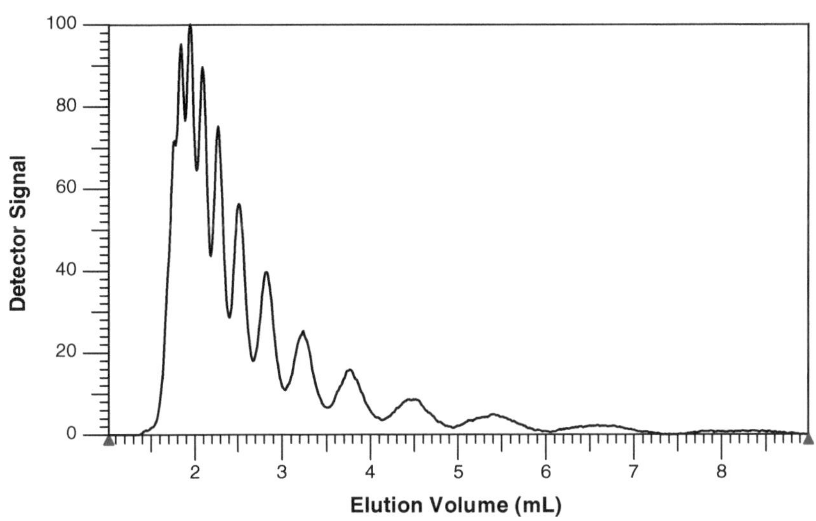

Fig. 6.57 Normal phase separation of the model FAE blend according to the EO content; stationary phase: Chromolith Si, 100 × 4.6 mm I.D.; mobile phase: isopropanol-water 88:12 % v/v; detection: ELSD (reprinted from [107] with permission of Wiley-VCH)

observed, but a slight peak broadening of the later eluting peaks was seen as compared to Fig. 6.56. This different behaviour was most likely caused by the presence of FAE functionalized with longer endgroups (C_{15}, C_{16} and C_{18}) in addition to the previous ones.

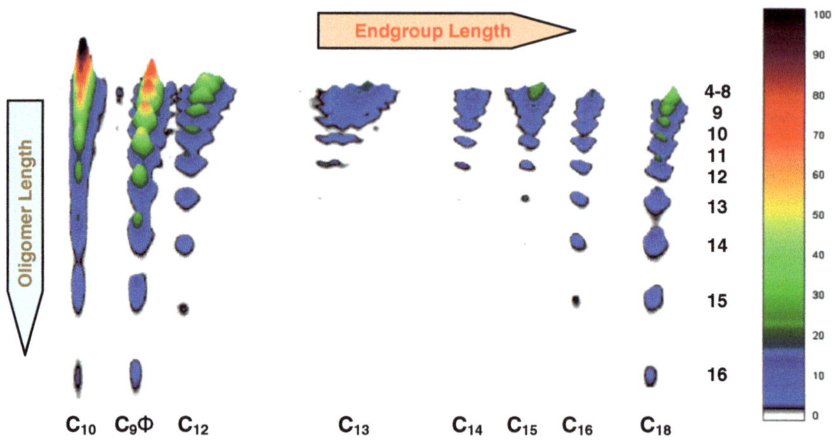

Fig. 6.58 2D plot of the FAE model blend (20 mg/mL in methanol-water 80:20); endgroup separation (first dimension) along the *X*-axis and oligomer separation (second dimension) along the *Y*-axis; experimental conditions are described in the text (reprinted from [107] with permission of Wiley-VCH)

Separations

The ultimate goal of the present method development was the separation of the complex FAE blend simultaneously with regard to the endgroups and the oligomer distributions. As was shown in the preparatory part, selective endgroup separation can be achieved by gradient HPLC while adsorption chromatography is able to provide oligomer separation. The next logical step was now the combination of the two separation protocols in a 2D-LC setup. Figure 6.58 shows the results of the 2D chromatography on the model FAE blend. The directions of the separations are indicated in the plots as arrows.

The 2D plot provides a clear idea of the complexity of the model blend. No doubt, the 2D separation yields selectivities in the different dimensions that are comparable to the one-dimensional separations. The projection of the plot on the *X*-axis would give the chromatogram presented in Fig. 6.55 with the eight baseline separated peaks which is characteristic of the separation according to endgroups. Accordingly, a projection on the *Y*-axis would give a chromatogram equivalent to the one presented in Fig. 6.57. Careful examination of the 2D-plot shows that the series of spots for each functionality fraction represent the corresponding oligomer separations. Thus, in this kind of representation, each spot corresponds exactly to one EO oligomer with a defined endgroup (i.e. one spot for each oligomer). The third dimension is the signal intensity detected by the ELSD coded by colours.

Evaluation

The 2D plot clearly indicates that the separations in the two dimensions are nearly orthogonal as was expected. Only slight curvatures for the oligomer distributions are obtained in the *X*- and *Y*-axis directions. As can be seen, the oligomer distributions are very different for the different functionality fractions. For example, the average degree of oligomerization for sample 2 (C_{12},C_{14}-FAE) is 7, see Table 6.17. Nothing is known about the differences between the C_{12}- and the C_{14}-fractions. The 2D plot indicates significant differences between the two fractions. The highest degree of oligomerization that can be detected for the C_{14}-fraction is $n = 12$. For the C_{12}-fraction, however, oligomers up to $n = 15$ can be detected. A similar behaviour is observed for sample 6 where the higher oligomer part of the C_{18}-fraction is more pronounced as compared to the C_{16}-fraction.

These results show how useful it is to use 2D chromatography for the detailed analysis of complex polymer mixtures. As was shown, direct coupling of information coming from two different kinds of separation enable one to draw a precise map of polymer heterogeneity in only one experiment which is more suitable than average measurements on the bulk samples.

6.8 Further Applications and Outlook

As has been shown, two-dimensional liquid chromatography is one of the most powerful methods for characterizing complex polymers in different coordinates of molecular heterogeneity. Using a chromatographic separation which is selective towards functionality or chemical composition in the first dimension and SEC in the second dimension, truly "orthogonal" separation schemes can be established. Thus, the combination of gradient HPLC versus SEC yields quantitative information on CCD and MMD, while coupling LCCC and SEC is useful for the analysis of functional homopolymers and block copolymers in the coordinates FTD-MMD and CCD-MMD, respectively. Even more complex systems, such as graft copolymers and polymer blends, in which each component may be chemically heterogeneous itself, can be analysed. Further applications of 2D chromatography to complex polymers are summarized in Table 6.18. This Table is not intended to be exhaustive but to give a representative overview.

Although 2D-LC is experimentally more demanding than other chromatographic techniques, the complete characterization yields much more qualitative and quantitative information about the sample and results are presented in an impressively simple way. The contour plot of a 2D separation maps all obtainable information and allows a fast and reliable comparison between two samples. For future developments, the automated comparison of the results of different samples can be considered as an important step to improve process control and quality management. It can be expected, that in addition to LAC and LCCC other separation modes will be combined with SEC or with each other. Extensive summaries on recent advances in on-line multidimensional liquid chromatography were published recently by Malerod et al. [132] and Jandera [134]. A detailed review on experimental conditions for the separation of complex polymers by multidimensional chromatography including information on stationary and mobile phases has been presented by Schubert et al. [138].

Table 6.18 Further applications of 2D chromatography for the analysis of complex polymers

Polymer	Chromatographic System[a]	Remarks	References
PEO-PPO block copolymer	S(1): RP-18 M(1): ACN-water S(2): SDV M(2): THF	(1): Separation of the block copolymer with regard to the PPO block length at LCCC conditions for PEO, PPO is eluted in adsorption mode. (2) SEC mode for PEO and PPO	[54]
PS-PMMA block copolymer	S(1A): RP-18 M(1A): THF-ACN S(1B): silica gel M(1B): MEK-CH S(2): SDV M(2): THF	(1A) separation with regard to the PMMA block at LCCC conditions for PS. (1B) separation with regard to the PS block at LCCC conditions for PMMA. (2) SEC mode for PS and PMMA. Total analysis of diblock copolymers with regard to both blocks and total molar mass	[115]
PS-PB-PBA graft copolymer	S(1): silica gel M(1): THF-CH S(2): SDV M(2): THF	Preparation of a block graft copoly-mer by grafting butyl acrylate onto a PS-PB block copolymer, (1) separation with regard to chemical composition at LCCC conditions for PBA, PS elutes in adsorption mode, PB elutes in SEC mode. (2) SEC separation with regard to total molar mass of the copolymer. Coupling of 2D-LC with FTIR	[116]
PB-g-PMMA graft copolymer	S(1): CN-modified silica gel M(1): chloroform-CH (gradient) S(2): SDV M(2): THF	Preparation of copolymers by grafting methyl methacrylate onto PB, (1) separation with regard to chemical composition by gradient HPLC, (2) SEC separation with regard to total molar mass	[117]
PMMA-grafted natural rubber	S(1): CN-modified silica gel M(1): THF-CH S(2): SDV M(2): THF	Preparation of copolymers by grafting methyl methacrylate onto natural rubber, (1) separation with regard to chemical composition by LCCC, critical conditions for PMMA, (2) SEC separation with regard to total molar mass	[118]
Blends of SBR and BR	S(1): CN-modified polymer M(1): chloroform-CH (gradient) S(2): SDV M(2): THF	Blends are separated into the SBR and BR components, components were analysed by FTIR to obtain chemical composition. (1) gradient separation into SBR and BR, (2) SEC for analysis of the component molar masses	[119]
Fatty alcohol ethoxylates, fatty acid ethoxylates	S(1): RP-18 M(1): MeOH-water S(2): RP-18 M(2): acetone-water	(1) separation with regard to the fatty alcohol endgroups, LCCC conditions for PEO. After (1) fractions are reconcentrated on a full adsorption-desorption column and are then injected sequentially into LEAC. (2) separation into the single oligomers by liquid	[120, 121]

(continued)

Table 6.18 (continued)

Polymer	Chromatographic System[a]	Remarks	References
		exclusion-adsorption chromatography (LEAC)	
Fatty alcohol ethoxylates	S(1): Zorbax silica M(1): ACN-water (gradient) S(2): Pecosphere C_{18} M(2): MeOH-water	(1) separation with regard to oligomer chain length by NPLC, (2) separation according to endgroups by RPLC	[122]
PEO-b-PPO block copolymers	S(1): Prodigy ODS3 M(1): THF-water S(2): Symmetry C_{18} M(2): MeOH-water (gradient)	(1) separation at critical point for PPO, analysis of homopolymer and different functionality fractions, (2) fractions from first dimension separated in LAC according to PPO oligomer distribution	[137]
PEO-b-PPO block copolymers	S(1): Zorbax-SB C18 M(1): ACN-water (gradient) S(2): Polaris NH_2 M(2): EtOH-DCM-water	(1) separation according to PO repeat units in RP mode, (2) separation according to EO repeat units in HILIC mode	[123, 124]
PS-b-PI block copolymers	S(1), LCCC for PS: Nucleosil Si 300 M(1), LCCC for PS: THF-cyclohexane S(1), LCCC for PI: Nucleosil C18 300 M(1), LCCC for PI: MEK-cyclohexane S(2): SDV M(2): THF	Comparison of samples prepared by sequential copolymerization and block coupling, (1) separation at critical conditions of PS and PI, (2) SEC regarding molar mass distributions of all components, comparison of PS and PI calibration	[129]
PS-b-PI block copolymers	S(1): Nucleosil Diol M(1): THF-isooctane S(2): PL Mixed-C M(2): THF	(1) separation in terms of PI block by NP-TGIC, (2) separation according to the PS block length by RPLC	[125]
Stereoregular PEMA	S(1): PL mixed D M(1): THF-cyclohexane S(2): Develosil SG-NH_2 M(2): THF-cyclohexane	(1) Size separation by SEC, (2) tacticity separation by LCCC, fractions were collected offline from SEC and subsequently analysed by LCCC	[126]
Star-shaped PS-b-PB block copolymers	S(1): silica gel M(1): i-octane-THF (gradient) S(2): SDV M(2): THF	(1) gradient HPLC separation which is influenced by chemical composition and molar mass, (2) molar mass separation by SEC, the complementary separation by HPLC and SEC resulted in 16 components as was expected	[16]

(continued)

Table 6.18 (continued)

Polymer	Chromatographic System[a]	Remarks	References
Octylphenoxy-PEO	S(1): Nucleosil RP18 M(1): MeOH-water S(2): SDV M(2): THF	(1) separation according to endgroups by LCCC, (2) molar mass separation by SEC	[54]
PEG-g-PVAc graft copolymers	S(1): Nucleodur C_{18} Pyramid M(1): THF-water S(2): SDV high speed M(2): THF	(1) separation according to the PEG backbone at LCCC for PVAc, (2) molar mass separation for all components by SEC	[128]
P(2EHA)-b-PMA block copolymers	S(1): PLRP-S M(1): THF-MeOH Gradient S(2): PL HTS-C M(2): THF	(1) gradient separation according to chemical composition, not all components are resolved, (2) molar mass separation, due to different molar masses of the components, separation is significantly enhanced	[130]
PS-b-PEO block copolymers	S(1): Macrosphere RP18 M(1): THF-water S(2): Waters Styragel M(2): THF	(1) separation at LCCC conditions for PS, separation of PS homopolymer and block copolymer, (2) molar mass analysis by SEC	[144]
PS-b-PEO block copolymers	S(1), LCCC for PS: Symmetry RP M(1), LCCC for PS: THF-DMF S(1), LCCC for PEO: Nucleosil Si 300 M(1), LCCC for PEO: THF-DMF S(2): PSS Gram M(2): DMF	(1) LCCC separation at critical conditions for PS and PEO, separation and quantification of homopolymer fractions, (2) molar masses by SEC, comparison of SEC calibration for PS and PEO, semiprep fractionations and FTIR analysis to determine block copolymer composition	[131]
PEG-g-MAA graft copolymers	S(1): Nucleosil RP18 M(1) MeOH-water S(2): Suprema linear M(2): water + NH_4Ac	(1) separation regarding the PMAA grafts at LCCC conditions for PEG, (2) molar mass analysis by SEC	[127]
PCL-b-PS-b-PCL	S(1): YMC ODSA M(1): THF-water-acetic acid S(2): SDV M(2): THF	(1) chemical composition separation at LCCC conditions for PS, analysis of block lengths of PCL, (2) molar mass analysis by SEC	[133]
Polysorbates	S(1): Spherisorb Si M(1): acetone-water S(2): Onyx C18 monolith M(2): acetone-water	(1) separation into sorbate mono- and diesters, (2) separation according to PEG oligomer chain length	[135]

(continued)

Table 6.18 (continued)

Polymer	Chromatographic System[a]	Remarks	References
PEO-b-PCL block copolymers	S(1): Spherisorb ODS2 M(1): ACN-water S(2): Synergy Fusion RP M(2): acetone-water (gradient)	(1) critical conditions for PEG, separation according to PCL block in LAC, (2) fractions from first dimension are manually collected and analysed, separation according to EO repeat units	[136]
PMMA-b-PtBMA block copolymers	S(1), LCCC for PtBMA: Nucleosil RP18 M(1), LCCC for PtBMA: THF-ACN S(1), LCCC for PMMA: Nucleosil Si M(1), LCCC for PMMA: THF-hexane S(2): SDV M(2): THF	(1) separations at both LCCC conditions for PtBMA and PMMA, analysis of homopolymers and the blocks that elute in SEC mode, (2) molar mass analysis of all components by SEC	[139]
S-co-MMA random copolymers	S(1): Nucleosil C_{18} M(1): THF-ACN (gradient) S(2): PL Mixed-C M(2): THF	(1) separation according to chemical composition, PMMA elutes first followed by copolymers, PS eluting last, (2) molar mass analysis of all components by SEC	[140]
OH-terminated PMMA	S(1): Hypersil Si M(1): ACN-DCM S(2): PL Oligopore M(2): THF	(1) separation at LCCC conditions for PMMA according to number of endgroups into non-, mono- and difunctional PMMA, (2) molar mass analysis of components by SEC	[141]
PS-PI miktoarm star polymers	S(1): Nucleosil C_{18} M(1): dioxane S(2): PL PolyPore M(2): THF	(1) TGIC separation according to type and number of arms, i.e. PSPI, $(PS)_2PI$, $PS(PI)_2$, $PS(PI)_3$, (2) molar mass analysis of components by SEC	[142]
Polycarbonate degradation	S(1): Hypersil Si M(1): chloroform-diethylether S(2): Waters HSP gel M(2): chloroform	(1) separation at LCCC conditions for PC according to chemical composition before and after degradation, (2) molar mass analysis of components by SEC	[143]
Polyamide 6 and polyamide 6.6	S(1): Supelco Discovery C8 M(1): HFIP-MeOH S(2): SDV M(2): HFIP+NH_4Ac	(1) separation of different functionality fractions at LCCC conditions for the PA, (2) molar mass analysis by SEC	[145]

[a]S(1): stationary phase first dimension, M(1): mobile phase first dimension, S(2): stationary phase second dimension, M(2): mobile phase second dimension

References

1. Grushka E (1970) Anal Chem 42:1142
2. Consden R, Gordon AH, Martin AJP (1944) Biochem J 38:224
3. Grinberg N, Kalász H, Han SM, Armstrong DW (1990) In: Grinberg N (ed) Modern thin-layer chromatography, Chapter 7. New York, Marcel-Dekker Publishing
4. O'Farrell PH (1975) Biol Chem 250:4007
5. Celis JE, Bravo R (1984) Two-dimensional gel electrophoresis of proteins. Academic, New York
6. Anderson NL, Taylor J, Scandora AE, Coulter BC, Anderson NG (1981) Clin Chem 27:1807
7. Efron ML (1959) Biochem J 72:691
8. Gankina E, Belenkii B, Malakhova I, Melenevskaya E, Zgonnik V (1991) J Planar Chromatogr 4:199
9. Litvinova LS, Belenkii BG, Gankina ES (1991) J Planar Chromatogr 4:304
10. Murphy RE, Schure MR, Foley JP (1998) Anal Chem 70:1585
11. Balke ST (1984) Quantitative column liquid chromatography. Journal of Chromatography Library, Vol. 29, Elsevier Publishing, New York
12. Bushey MM, Jorgenson JW (1990) Anal Chem 62:161
13. Cortes HJ (1990) Multidimensional chromatography: techniques and applications. Marcel Dekker, New York
14. Larmann JP, Lemmo AV, Moore AW, Jorgenson JW (1993) Electrophoresis 14:439
15. Liu Z, Sirimanne SR, Patterson DG, Needham LL, Phillips JB (1993) Anal Chem 66:3086
16. Kilz P, Krüger RP, Much H, Schulz G (1995) In: Provder T, Urban MW, Barth HG (Eds.) Chromatographic characterization of polymers: hyphenated and multidimensional techniques. Adv Chem Ser 247. American Chemical Society, Washington, D.C.
17. Venema E, de Leeuw P, Kraak JC, Poppe H, Tijssen RJ (1997) J Chrom A 765:135
18. Glöckner G (1991) Gradient HPLC and chromatographic cross-fractionation. Springer, Berlin, Heidelberg, New York
19. Balke ST, Patel RD (1980) J Polym Sci B Polym Lett 18:453
20. Balke ST (1982) Sep Purif Methods 1:1
21. Balke ST, Patel RD (1983) In: Craver CD (Ed) Polymer characterization. Adv Chem Ser 203, American Chemical Society, Washington, DC
22. Kilz P (1992) Laborpraxis 6:628
23. Kilz P, Krüger RP, Much H, Schulz G (1993) Polym Mater Sci Eng 69:114
24. Pasch H (2004) Characterization of polymer heterogeneity by 2D-LC. In: Striegel AM (ed) Multiple detection in size-exclusion chromatography. ACS Symp Ser 893, American Chemical Society: Washington, DC
25. Pasch H, Trathnigg B (1998) HPLC of polymers. Springer, Berlin-Heidelberg, New York
26. http://www.polymer.de/solutions/copolymer-characterization-and-2d-chromatography.html
27. Ogawa T, Sakai M (1982) J Polym Sci Polym Phys Ed 19:1377
28. Glöckner G, van den Berg JHM, Meijerink NL, Scholte TG (1986) In: Kleintjens I, Lemstra P (eds) Integration of fundamental polymer science and technology. Barking, Elsevier Applied Science
29. Schure MR (1999) Anal Chem 71:1645
30. http://www.pssgpcshop.com/documents/WinGPC_Brochure.pdf
31. Danielewicz M, Kubin M (1981) J Appl Polym Sci 26:951
32. Glöckner G, Koschwitz H, Meissner C (1982) Acta Polym 33:614
33. Sato H, Takeuchi H, Tanaka Y (1986) Macrommolecules 19:2613
34. Mourey TH (1986) J Chromatogr 357:101
35. Mori S (1989) J Appl Polym Sci 38:95
36. Belenkii BG, Gankina ES, Tennikov MB, Vilenchik LZ (1976) Dokl Akad Nauk USSR 231:1147
37. Tennikov MB, Nefedov PP, Lazareva MA, Frenkel SJ (1977) Vysokomol Soedin A19:657

38. Skvortsov AM, Belenkii BG, Gankina ES, Tennikov MB (1978) Vysokomol Soedin A20:678
39. Entelis SG, Evreinov VV, Gorshkov AV (1986) Adv Polym Sci 76:129
40. Belenkii BG, Gankina ES (1977) J Chromatogr 141:13
41. Skvortsov AM, Gorbunov AA (1990) J Chromatogr 507:487
42. Gorshkov AV, Prudskova TN, Guryakova VV, Evreinov VV (1986) Polym Bull 15:465
43. Gorshkov AV, Verenich SS, Evreinov VV, Entelis SG (1988) Chromatographia 26:338
44. Gorshkov AV, Much H, Becker H, Pasch H, Evreinov VV, Entelis SG (1990) J Chromatogr 523:91
45. Pasch H, Zammert I (1994) J Liq Chromatogr 17:3091
46. Pasch H, Brinkmann C, Much H, Just U (1992) J Chromatogr 623:315
47. Pasch H, Augenstein M (1993) Makromol Chem 194:2533
48. Pasch H, Gallot Y, Trathnigg B (1993) Polymer 34:4986
49. Pasch H, Augenstein M, Trathnigg B (1994) Makromol Chem 195:743
50. Pasch H, Deffieux A, Henze I, Schappacher M, Rique-Lurbet L (1996) Macromolecules 29:8776
51. Pasch H (1993) Polymer 34:4095
52. Pasch H, Rode K (1996) Macromol Chem Phys 197:2691
53. Pasch H, Rode K, Chaumien N (1996) Polymer 37:4079
54. Adrian J, Pasch H, Braun D (1998) LC-GC Int 11:32
55. Just U, Much H (1994) Proc Int GPC Symp, 861, Milford
56. Glöckner G, Stickler M, Wunderlich W (1989) J Appl Polym Sci 37:3147
57. Augenstein M, Stickler M (1990) Makromol Chem 191:415
58. Mori S (1990) J Chromatogr 503:411
59. Mori S (1988) Anal Chem 60:1125
60. Mori S (1981) Anal Chem 53:1813
61. Kilz P, Krüger RP, Much H, Schulz G (1995) ACS Adv Chem 247:223
62. Siewing A, Schierholz J, Braun D, Hellmann G, Pasch H (2001) Macromol Chem Phys 202:2890
63. Siewing A (2002) PhD Thesis, University of Technology, Darmstadt, Germany
64. Gerber J, Radke W (2005) e-polymers No. 45
65. Gerber J, Radke W (2005) Polymer 46:9224
66. Gorbunov AA, Vakhrushev AV (2009) Polymer 50:2727
67. Krüger RP, Much H, Schulz G (1994) J Liquid Chromatogr 17:3069
68. Vakhtina IA, Okunieva AG, Techritz R, Tarakanov OG (1976) Vysokomol Soedin A18:471
69. Filatova NN, Rovina DY, Evreinov VV, Entelis SG (1978) Vysokomol Soedin A20:2367
70. Filatova NN, Gorshkov AV, Evreinov VV, Entelis SG (1988) Vysokomol Soedin A30:953
71. Adrian J, Braun D, Rode K, Pasch H (1999) Angew Makromol Chem 267:73
72. Adrian J, Braun D, Pasch H (1999) Angew Makromol Chem 267:82
73. Biela T, Duda A, Penczek S, Rode K, Pasch H (2002) J Polym Sci Part A: Polym Chem 40:2884
74. Biela T, Duda A, Rode K, Pasch H (2002) Polymer 44:1851
75. Barrows TH (1990) Synthetic bioabsorbable polymers. In: Szycher M (ed) High performance biomaterials. Technomic Publishing Co. Inc., Lancaster, Basel, p 243
76. Kharas GB, Sanchez-Riera F, Severson DK (1994) Polymers of lactic acid. In: Plastics from microbes, Hanser Publishers, Munich, New York, p. 93
77. Lewis DH (1990) In: Chasin M, Langer R (eds) Drugs and the Pharmaceutical Sciences. Vol 45: Biodegradable Polymers as Drug Release Systems. Marcel-Dekker Inc., New York, p 1
78. Bhardwaj R, Blanchard J (1998) Int J Pharm 170:109
79. Langer R (2000) Acc Chem Res 33:94
80. Sinclar RG (1996) J Macromol Sci-Pure Appl Chem A33:585
81. Doi Y, Steinbüchel A (eds) (2002) Polyesters III—applications and commercial products (Biopolymers, Vol 4). Weinheim, Wiley-VCH

82. Zhang X, Goosen MFA, Wyss UP, Pichora D (1993) J Macromol Sci- Rev Macromol Chem Phys C33:81
83. Winet H, Bao J (1998) Biomed Mater Res 40:567
84. Kowalski A, Libiszowski J, Duda A, Penczek S (2000) Macromolecules 33:1964
85. Kowalski A, Duda A, Penczek S (2000) Macromolecules 33:7359
86. Kowalski A, Duda A, Penczek S (1998) Macromol Rapid Commun 19:567
87. Krüger RP, Much H, Schulz G (1996) Macromol Symp 110:155
88. Malik MI, Sinha P, Bayley GM, Mallon PE, Pasch H (2011) Macromol Chem Phys 212:1221
89. Malik MI, Harding GW, Pasch H (2012) Anal Bioanal Chem 403:601
90. Somasundaran P, Mehta SC, Purohit P (2006) Adv Colloid Interface Sci 128–130:103
91. Krulevitch P, Benett W, Hamilton J, Maghribi M, Rose K (2002) Biomed Microdevices 4:301
92. LeBaron PC, Wang Z, Pinnavaia TJ (1999) Appl Clay Sci 15:11
93. Mark JE (1996) J Macromol Sci, Pure Appl Chem A 33:2005
94. Chen X, Gardella JA, Kumler PL (1992) Macromolecules 25:6631
95. Chen X, Gardella JA, Kumler PL (1993) Macromolecules 26:3778
96. Chen X, Gardella JA (1998) Macromolecules 31:9328
97. Miller PJ, Matyjaszewski K (1999) Macromolecules 32:8760
98. Shinoda H, Miller PJ, Matyjaszewski K (2001) Macromolecules 34:3186
99. Shinoda H, Matyjaszewski K, Okrasa L, Mierzwa M, Pakula T (2003) Macromolecules 36:4772
100. Matyjaszewski K, Xia J (2001) Chem Rev 101:2921
101. Macko T, Hunkeler D (2003) Adv Polym Sci 163:61
102. Ahn S, Im K, Chang T, Chambon P, Fernyhough CM (2011) Anal Chem 83:4237
103. Graessley WW (2008) Polymeric liquids and networks: dynamics and rheology. Garland Science, New York
104. McLeish TCB (2002) Adv Phys 51:1379
105. Chambon P, Fernyhough CM, Im K, Chang T, Das C, Embery J, McLeish TCB, Read DJ (2008) Macromolecules 41:5869
106. Im K, Park HW, Kim Y, Chung B, Ree M, Chang T (2007) Anal Chem 79:1067
107. Raust J, Bruell A, Sinha P, Hiller W, Pasch H (2010) J Sep Sci 33:1373–1381
108. Keil C, Esser E, Pasch H (2001) Macromol Mat Eng 286:161
109. Pasch H, Bruell A, Cabrera K (2005) e-polymers, no. 20
110. Kilz P (2004) Chromatographia 59:3
111. Trathnigg B, Thamer D, Yan X, Kinugasa S (1993) J Liq Chromatogr 16:2439
112. Trathnigg B, Thamer D, Yan X, Maier B, Holzbauer HR, Much H (1994) J Chromatogr A 665:47
113. Rissler K, Fuchslueger U, Grether HJ (1994) J Liq Chromatogr 17:3109
114. Hiller W, Brüll A, Argyropoulos D, Hoffmann E, Pasch H (2005) Magn Reson Chem 43:729
115. Pasch H, Mequanint K, Adrian J (2002) e-polymers No. 005
116. Adrian J, Esser E, Hellmann G, Pasch H (2000) Polymer 41:2439
117. Siewing A, Lahn B, Braun D, Pasch H (2003) J Polym Sci Polym Chem 41:3143
118. Graef SM, van Zyl AJP, Sanderson RD, Klumperman B, Pasch H (2003) J Appl Polym Sci 88:2530
119. Heinz LC, Siewing A, Pasch H (2003) e-polymers No. 065
120. Trathnigg B, Rappel C (2002) J Chromatogr A 952:149
121. Trathnigg B, Rappel C, Raml R, Gorbunov A (2002) J Chromatogr A 953:89
122. Murphy RE, Schure MR, Foley JP (1998) Anal Chem 70:4353
123. Jandera P, Fischer J, Lahovska H, Novotna K, Cesla P, Kolarova L (2006) J Chromatogr A 1119:3
124. Jandera P, Holcapek M, Kolarova L (2001) Int J Polym Anal Charact 6:261
125. Park S, Cho D, Ryu J, Kwon K, Lee W, Chang T (2002) Macromolecules 35:5974
126. Janco M, Hirano T, Kitayama T, Hatada K, Berek D (2000) Macromolecules 33:1710
127. Pasch H, Adler M, Rittig F, Becker S (2005) Macromol Rapid Commun 26:438

128. Knecht D, Rittig F, Lange R, Pasch H (2006) J Chromatogr A 1130:43
129. Mass V, Bellas V, Pasch H (2008) Macromol Chem Phys 209:2026
130. Raust JA, Houillot L, Charleux B, Moire C, Farcet C, Pasch H (2010) Macromolecules 43:8755
131. Malik MI, Harding GW, Grabowsky ME, Pasch H (2012) J Chromatogr A 1244:77
132. Malerod H, Lundanes E, Greibokk T (2010) Anal Methods 2:110
133. Schmid C, Weidner S, Falkenhagen J, Barner-Kowollik C (2012) Macromolecules 45:87
134. Jandera P (2012) Cent Eur J Chem 10:844
135. Trathnigg B, Abrar S (2010) Proc Chem 2:130
136. Ahmed H, Trathnigg B (2009) J Sep Sci 32:1390
137. Malik MI, Trathnigg B, Saf R (2009) J Chromatogr A 1216:6627
138. Baumgaertel A, Altunas E, Schubert US (2012) J Chromatogr A 1240:1
139. Falkenhagen J, Much H, Stauf W, Müller AHE (2000) Macromolecules 33:3687
140. van der Horst A, Schoenmakers PJ (2003) J Chromatogr A 1000:693
141. Jiang X, van der Horst A, Lima V, Schoenmakers PJ (2005) J Chromatogr A 1076:51
142. Im K, Park HW, Lee S, Chang T (2009) J Chromatogr A 1216:4606
143. Coulier L, Kaal ER, Hankemeier T (2005) J Chromatogr A 1070:79
144. Beaudoin E, Dufils PE, Gigmes D, Marque S, Petit C, Tordo P, Bertin D (2006) Polymer 47:98
145. Rittig F, Pasch H (2008) In: Cohen S, Schure M (eds) Multidimensional liquid chromatography: theory and applications in industrial chemistry and life sciences. Wiley, Hoboken, NJ

Hyphenation of Polymer Chromatography with Information-Rich Detectors

7

The determination of compositional changes across the molar mass distribution (MMD) of a polymer or the detection of a specific component in a complex polymer mixture is of considerable interest. This information allows to predict physical properties and ultimately the performance of the polymer. Several analytical techniques are of use in determining these properties. Mass spectrometry, NMR, and infrared spectroscopy can be used to provide data about the compositional details of the sample.

When only spectroscopic methods are used, they are able to identify polymer components with respect to their chemical nature. However, in many cases they are unable to answer the question as to whether two chemical structures are combined to yield a copolymer or a blend or both. For example, when analyzing a rubber mixture by a single spectroscopic method like FTIR or NMR one is able to identify styrene and butadiene as the monomer units. However, it is impossible to decide if the sample is a mixture of polystyrene (PS) and polybutadiene (PB), or a copolymer of styrene and butadiene, or a blend of a styrene-butadiene copolymer and PB. In the case of the latter, even the copolymer composition cannot be determined just by running a FTIR or NMR spectrum.

For the precise determination of the complex polymer composition including the chemical composition and MMD of the components in most cases a separation step is required. Only when fractions are obtained that comprise the different polymer components, an analysis with regard to chemical composition and MMD can be conducted. The present chapter discusses different options to use liquid chromatography (SEC, HPLC, LCCC) in conjunction with FTIR, mass spectrometry and NMR for the separation and analysis of complex polymers.

7.1 Coupling with FTIR Spectroscopy

When analysing a complex polymer, very frequently the first step must be the determination of the bulk composition. Only when the chemical structures of the polymer components (monomers) are known, sophisticated separation techniques

H. Pasch and B. Trathnigg, *Multidimensional HPLC of Polymers*, Springer Laboratory,
DOI 10.1007/978-3-642-36080-0_7,

such as gradient high performance liquid chromatography (HPLC) or liquid chromatography at critical conditions (LCCC) can be optimized for a specific analysis.

The most frequently used techniques for a "flash" analysis are infrared spectroscopy and size exclusion chromatography (SEC). Infrared (IR) spectroscopy provides information on the chemical substructures present in the sample, while SEC gives a first indication of the molar mass range. Information on both molar mass and composition is obtained when SEC or a comparable chromatographic method is combined with an IR detector. In the past, numerous workers have tried to use IR detection of the SEC column effluent in liquid flow cells. The problems encountered relate to obtaining a sufficient signal-to-noise (S/N) ratio even with FTIR instruments, flow-through cells with minimum path lengths and mobile phases with sufficient spectral windows. A major limitation with all flow-through cells is the limited selection of solvents/mobile phases that exhibit sufficiently large spectral windows for high sensitivity measurements.

One of the few very fortunate cases is the SEC-FTIR analysis of polyolefins. In this case 1,2,4-trichlorobenzene (TCB) is used as the mobile phase which is sufficiently transparent in the range of 2,700–3,000 cm^{-1} that is used for polyolefin detection. As has been shown by DesLauriers and others, the compositional heterogeneity (short chain branching, SCB) in polyolefins can be analyzed sensitively by on-flow SEC-FTIR [1–7]. Chromatograms are generated from ratio-recorded transmittance spectra where the spectrum of the pure mobile phase is used as background. Typical sample concentrations are 1–3 mg/mL and rather large injection volumes of 400–1,000 μL are used for sufficient signal-to-noise ratio. In the case of low density materials branching is determined as the levels of methyl (2,958 cm^{-1}) and methylene endgroups (2,928 cm^{-1}) [1, 2, 4]. For high density materials with low degrees of branching multivariable statistical techniques are preferred [6].

A rather broad applicability of FTIR as a detector in liquid chromatography can be achieved when the mobile phase is removed from the sample prior to detection. In this case the sample fractions are measured in pure state without interference from solvents. Experimental interfaces to eliminate volatile mobile phases from HPLC effluents have been tried with some success [8–10] but the breakthrough towards a powerful FTIR detector was achieved only by Gagel and Biemann, who formed an aerosol from the effluent and sprayed it on a rotating aluminum mirror. The mirror was then deposited in a FTIR spectrometer and spectra were recorded at each position in the reflexion mode [11–13].

This principle is used in an interface that originally was developed by Lab Connections Inc. introducing the LC-Transform [14–16] and modified further several times [17]. The design concept of the interface is shown in Fig. 7.1. The system is composed of two independent modules, the sample collection module and the optics module. The effluent of the liquid chromatography column is split with a fraction (frequently 10 % of the total effluent) going into the heated nebulizer nozzle located above a rotating sample collection disc. The nozzle rapidly evaporates the mobile phase while depositing a tightly focused track of the solute. After collecting a chromatogram on the sample collection disc, the disc is transferred to the optics module in the FTIR for analysis of the deposited sample track. A control module defines the sample collection disc position and rotation rate in order

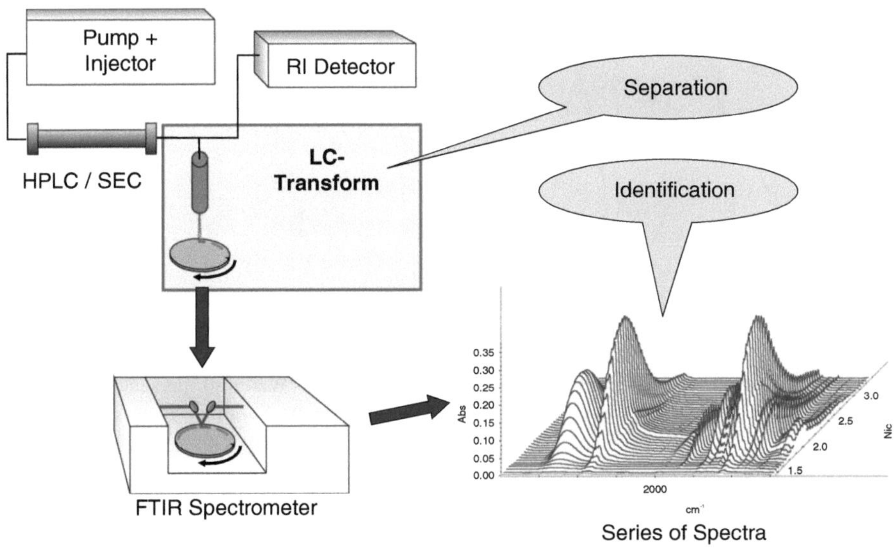

Fig. 7.1 Schematic representation of the principle of coupled LC-FTIR using the LC-Transform Interface (reprinted from [18] with permission of Elsevier)

to be compatible with the run time and peak resolution of the chromatographic separation. Data collection is readily accomplished with suitable software packages. The sample collection disc is made from Germanium (Ge) which is optically transparent in the range 6,000–450 cm^{-1}. The lower surface of the disc is covered with a reflecting aluminum layer.

As a result of the investigation a complete FTIR spectrum for each position on the disc and, hence, for each sample fraction is obtained. This spectrum bears information on the chemical composition of each sample fraction. The set of all spectra can be arranged along the elution time axis and yields a three-dimensional plot in the co-ordinates elution time-FTIR frequency-absorbance.

Modifications of this concept have been introduced in recent years, including the interfaces SECurity LC600r and LC600xy [17]. The former uses the rotating Ge disk design, while the latter uses a x-y stage.

One of the benefits of coupled SEC-FTIR is the ability to identify directly the individual components separated by chromatography. A typical SEC separation of a polymer blend is shown in Fig. 7.2a [20]. Two separate elution peaks 1 and 2 were obtained, indicating that the blend contained at least two components of significantly different molar masses. A quantification of the components with respect to concentration and molar mass, however, could not be carried out as long as the chemical structure of the components is unknown.

The analysis of the chemical composition of the sample was conducted by coupled SEC-FTIR using the LC Transform interface. The sample was fractionated with respect to molecular size, the fractions were deposited on the Ge disc and FTIR spectra were recorded continuously along the sample track. In total, a set of about 80 spectra was obtained which was presented in a three-dimensional plot, see

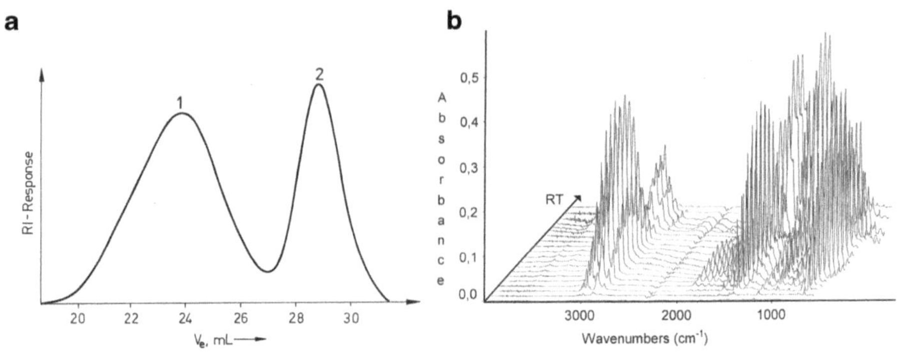

Fig. 7.2 SEC separation (**a**) and FTIR analysis of fractions in a 'Waterfall' presentation (**b**) of a binary polymer blend (reprinted from [19] with permission of Springer Science + Business Media)

Fig. 7.2b. The projection of the 3D plot on the retention time-IR frequency coordinate system yielded a two-dimensional representation where the intensities of the absorption peaks were given by a colour code. Such a "contour plot" readily provides information on the chemical composition of each chromatographic fraction, see Fig. 7.3. It was obvious that the chromatographic peaks 1 and 2 had different chemical structures. By comparison with reference spectra which are accessible from corresponding data bases, component 1 could be identified as polystyrene (PS), while component 2 was polyphenylene oxide. With this knowledge, appropriate calibration curves could be used for quantifying the composition and the component molar masses of the blend.

Coupled SEC-FTIR becomes an inevitable tool when blends comprising copolymers have to be analysed. Very frequently components of similar molar masses are used in polymer blends. In these cases resolution of SEC is not sufficient to resolve all component peaks, see Fig. 7.4a for a model binary blend containing an additive. The elution peaks of the polymer components 1 and 2 overlapped and, thus, the molar masses could not be determined directly. Only the additive peak 3 at the low molar mass end of the chromatogram was well separated and could be quantified.

A first indication of the composition of the present sample could be obtained from the contour plot in Fig. 7.4b. Component 3 showed typical absorption peaks of a phenyl benzotriazole and could be identified as a UV stabilizer of the Tinuvin type. Component 2 exhibited absorption peaks that were characteristic for nitrile groups (2,237 cm^{-1}) and styrene units (760, 699 cm^{-1}), while component 1 showed a strong ester carbonyl peak around 1,740 cm^{-1} and peaks of styrene units. In agreement with the peak pattern of spectra reported in literature, component 2 was identified as a styrene-acrylonitrile copolymer (SAN). Component 1 could have been a mixture of PS and polymethyl methacrylate (PMMA) or a styrene-methyl methacrylate copolymer. Since the FTIR spectra were uniform over the entire elution peak, it was more likely that component 1 is a copolymer.

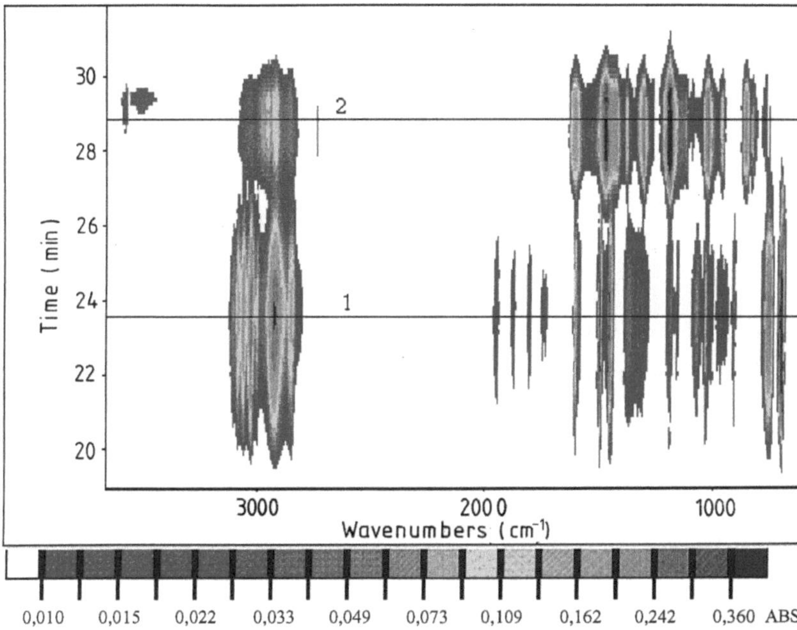

Fig. 7.3 SEC-FTIR analysis of a binary polymer blend as 'contour plot' representation (reprinted from [19] with permission of Springer Science + Business Media)

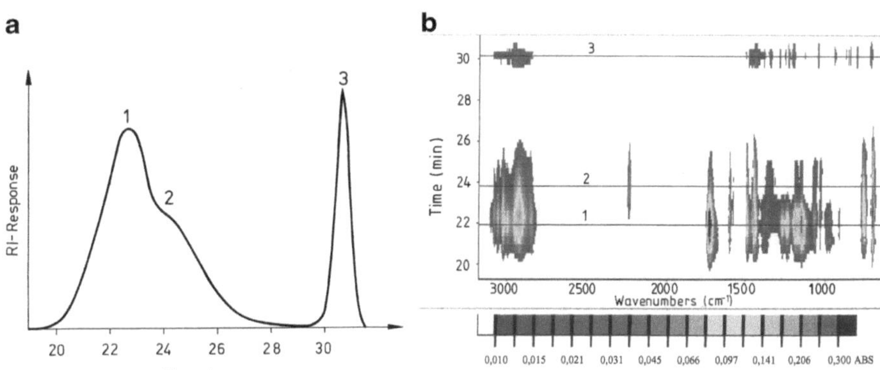

Fig. 7.4 SEC separation of a blend of two copolymers and an additive (**a**) and contour plot of the SEC-FTIR analysis of the blend (**b**), colour code represents relative absorbance (reprinted from [19] with permission of Springer Science + Business Media)

One important feature of the SEC-FTIR software is that from the contour plot specific elugrams at one absorption frequency can be obtained. Taking the elugram at 2,230 cm^{-1} which is specific for the nitrile group, the elution peak of the SAN copolymer could be presented individually. For the presentation of component 1 the

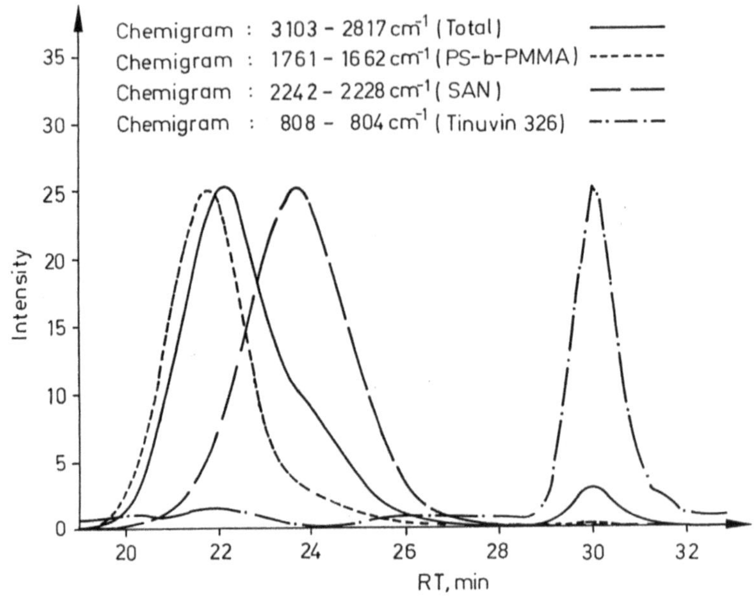

Fig. 7.5 Chemigrams taken from the contour plot in Fig. 7.4b (reprinted from [19] with permission of Springer Science + Business Media)

elugram at the carbonyl absorption frequency was drawn. Thus, via the "chemigram" presentation the elution peak of each component is obtained, see Fig. 7.5. The total concentration profile could be obtained from the chemigram at the frequency of the C–H valence vibrations (2,800–3,100 cm^{-1}). The specific chemigrams which were characteristic for each component represented the elution profile of this component. Accordingly, the chemigrams could be used for the calculation of the molar masses of the components.

The most advanced application of this approach is to use it as a detector in 2D-LC. Adrian et al. presented the deformulation of a complex polymer that was prepared by grafting butyl acrylate onto a poly(styrene-b-butadiene) backbone [21]. The separation of the graft copolymer was conducted by comprehensive LCCC × SEC.

In a relatively short period of time the LC Transform system found its way into a large number of laboratories. Applications of the technique have been discussed in various fields. Willis and Wheeler demonstrated the determination of the vinyl acetate distribution in ethylene-vinyl acetate copolymers, the analysis of branching in high-density polyethylene, and the analysis of the chemical composition of a jet oil lubricant [22]. Provder et al. [23] showed that in powder coatings all additives were positively identified by SEC-FTIR through comparison with known spectra. Even biocides could be analyzed in commercial house paints. The comparison of a PS-PMMA blend with a corresponding copolymer gave information on the chemical drift. The analysis of a modified vinyl polymer sample by SEC/FTIR showed

Table 7.1 Typical characteristics of SEC/FTIR online flow cells and off-line solvent elimination interfaces (adapted from [26] with permission from Elsevier)

Condition	Flow cell interface	Solvent elimination interface
Gradient separations	No	Yes
Qualitative information	Limited, depends on mobile phase	Yes
Quantitative information	Excellent	Limited
Sensitivity	Moderate	Excellent
Limit of detection	Low, depends on mobile phase	High
Spectral S/N ratio	Moderate, spectra collection on the fly	High, post-run scanning possible
Peak asymmetry	Not affected	Not affected
Ease of operation	User friendly	Time consuming optimization
Application area	SEC	SEC, gradient HPLC

that some of the components of the binder could be identified readily as vinyl chloride, ethyl methacrylate and acrylonitrile, and an epoxidized drying oil additive was detected [23].

The quality of the results from SEC-FTIR strongly depends on the surface quality of the deposited sample fractions. Cheung et al. demonstrated that the surface wetting properties of the substrate dominate the deposit morphology [24]. The spectra fidelity, film quality, resolution and polymer recovery were considered by Balke et al. [25]. For different interface designs it was found that the morphology of the deposited polymer film was a key parameter for quantitative measurements.

A quite useful comparison of the flow-through cell and the LC Transform system was given by Kok et al. [26], see Table 7.1.

In a number of more recent applications Esser et al. [27, 28] and Pasch et al. [29–32] addressed the analysis of complex rubber formulations and styrene-acrylate copolymers. The LC Transform interface cannot only be used for low boiling point mobile phases. A modified version is able to evaporate high boiling point solvents like trichlorobenzene by applying high vacuum. This system was coupled to high-temperature SEC and HPLC for the analysis of complex polyolefins [33–37].

7.1.1 Analysis of Cross-linked Styrene-Butadiene Rubber by SEC and FTIR Spectroscopy [31]

Aim

Commercial styrene-butadiene rubbers (SBR) are roughly classified into three types of copolymers according to the sequence distribution of the monomer units. Depending on the preparation procedures, random copolymers, styrene-butadiene (SB) or styrene-butadiene-styrene (SBS) block copolymers, or partially blocked products are obtained. The significance of the chemical composition and sequence distribution in SBR has long been recognized as a dominant factor in characterizing mechanical and thermal properties. SBS type triblock copolymers are typical thermoplastic elastomers which exhibit unique structure–property relationships. Their properties are significantly influenced by the numbers and chain lengths of the S and B blocks. SB and SBS block copolymers were successfully analyzed by LCCC where both S and B block lengths were determined with good accuracy [38]. In addition, residual PS and PB homopolymers were separated from the block copolymers. The identification of the different fractions was carried out by on-line FTIR spectroscopy.

SBRs and SBR-butadiene rubber (BR) blends are of extraordinary importance for the production of automobile tyres. Depending on the molar mass and chemical composition of the SBR and BR, and the SBR-to-BR ratio, the performance of the materials can vary in a wide range. It is, therefore, important to determine the following parameters: molar mass and S/B ratio of the SBR, molar mass of the BR, SBR/BR ratio, exact chemical microstructure of the butadiene units in SBR and BR (cis- vs. trans-, 1,2- vs. 1,4-units). The separation of blends of SBR and BR has been accomplished by LCCC [27]. The exact chemical structure of the blend components was analysed by coupling the chromatographic separation to FTIR detection. The FTIR spectra of the components revealed information on the styrene and butadiene content and the conformation of the butadiene units (1,2-, 1,4-cis, 1,4-trans units). Poorly soluble high molar mass samples were separated by combining critical separation with a gradient elution technique.

Typically, SBR rubbers are cross-linked in order to obtain their final properties. While there is a variety of techniques to separate and analyse additives and oils in cross-linked rubbers, there are no efficient techniques to separate the polymer itself. Once the rubber is cross-linked, it is insoluble and cannot be analyzed by chromatography. In previous work [39, 40] ozonolysis has been used to degrade SBR and to obtain soluble degradation products. These soluble products have been analyzed by HPLC. Due to severe degradation, however, only the styrene fragments could be analyzed properly. The present application is aimed at using a soft pyrolysis technique to degrade the rubbers only slightly. The soluble (if possible) high molar mass part shall be separated by SEC and analyzed by FTIR spectroscopy. The present work focuses on the polymeric part of the samples, while other components like processing oils are just identified but not analyzed in detail.

Materials

- **Polymers**. The BR and SBR samples were technical products of BAYER AG, Leverkusen, Germany, and Dunlop GmbH, Hanau, Germany. Their analytical data as provided by the manufacturers are summarized in Table 7.2. The cross-linking of the rubbers was conducted at 170 °C for 4 h under nitrogen. Since the rubbers were technical formulations, they contained the necessary cross-linking agents and, therefore, could be cross-linked just by heating. The pyrolysis was

Table 7.2 Chemical composition of rubber samples as given by the suppliers

Polymer	Sample	Composition[a]
SBR	Krylene 1500	23.5 % Styrene
	SBR 1712	23.5 % Styrene + 37.5 phr processing oil
	SBR 1721	40 % Styrene + 37.5 phr processing oil
	SBR 1	100 phr Krylene 1500 + 50 phr carbon black + 10 phr processing oil + 5 phr other additives
BR	BR 1203	+ 37.5 phr processing oil
	Ubepol 150 L	+ 37.5 phr processing oil
SBR-BR blend	SBR-BR	69 phr SBR 1712 + 50 phr BR 1203 + 70 phr carbon black + 19 phr processing oil + 13 phr other additives

[a]*phr* parts per hundred

carried out at 220 °C under nitrogen atmosphere in a programmable oven. The extraction was conducted in the following way: 750 mg of the pyrolyzed sample were swollen in 20 mL THF for 18 h at 40 °C. The extractables were filtered off, THF was evaporated and the residue was dried in vacuum at 40 °C.

Equipment

- **Chromatographic System**. The separations were carried out on a modular chromatographic apparatus, comprising a Waters model 510 pump, a Rheodyne six-port injection valve and a Waters column oven. The experiments of coupled SEC-FTIR were carried out using the LC Transform® Model 100 of Lab Connections, Marlborough, USA. The system was composed of two independent modules, the sample collection module and the optics module. The effluent of the SEC column was split with a fraction (10% of the total effluent) going into the heated nebulizer nozzle while 90% of the total effluent going to the concentration detector. When a chromatogram had been collected on the sample collection disc, the disc was transferred to the optics module in the FTIR for analysis of the deposited sample track. A control module defined the sample collection disc position and rotation rate in order to be compatible with the run time and peak resolution of the chromatographic separation. As a result, a complete FTIR spectrum for each position on the disk and, hence, for each sample fraction was obtained.
- **Column**. SDV linear column, 300 × 8 mm i.d. (PSS GmbH, Mainz, Germany)
- **Mobile Phase**. THF, HPLC grade.
- **Detectors**. PL 1000 evaporative light scattering detector, Waters 486 tunable UV detector.
- **Column Temperature**. 25 °C
- **Sample Concentration**. 1–3 mg/mL. All samples are dissolved in THF
- **Injection Volume**. 100 μL

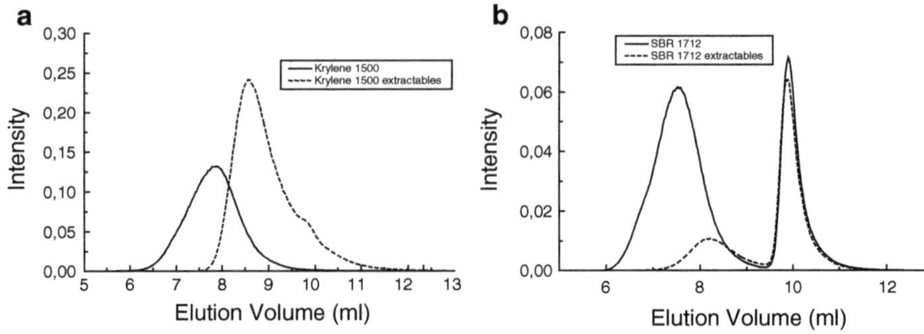

Fig. 7.6 SEC chromatograms of Krylene 1500 (**a**) and SBR 1712 (**b**) and the extractables after pyrolysis, stationary phase: SDV linear, mobile phase: THF, detector: ELSD (reprinted from [31] with permission of Wiley-VCH)

Preparatory Investigations

The protocol that shall be followed includes the analysis of the non-crosslinked precursor rubbers, the thermal cross-linking of the rubbers, the solvent extraction of the pyrolysates, and the analysis of the extracts. As representative examples, the SEC chromatograms of two rubbers are given in Fig. 7.6a, b. As can be seen clearly, the molar masses of the extractables are significantly lower than the molar masses of the original samples. For Krylene 1500 only one elution peak is obtained before and after the pyrolysis because this rubber does not contain processing oil. For SBR 1712 in Fig. 7.6b the elution peak at roughly 10 mL can be assigned to the processing oil, while the peak at lower elution volume corresponds to the polymer. Table 7.3 summarizes the amounts of extractables, the molar masses of the precursor rubbers and the molar masses of the extractables.

As is shown in Table 7.3, the SBR and BR rubbers exhibit significantly different pyrolysis behaviours. For the BR rubbers, extractables in the range of 6–10 wt% are obtained, while the amount of extractables for SBR rubbers are significantly higher. Despite of the fact that the amounts of extractables in some cases are low, they have relatively high molar masses for all samples that indicate an average of 10–100 chain scissions per molecule. For example, the initial molar mass of SBR 1721 is 323 kg/mol while the molar mass of the extractables is 3.6 kg/mol. Accordingly, an average of 80 chain scissions should have taken place. Considering the molar mass and the styrene content (23.5 wt%), the precursor rubber was composed of roughly 700 styrene units and 4,500 butadiene units. Since chain scission is likely to take place at the butadiene units, only about 2–3 % of the initial butadiene units will change their chemical structure in the pyrolysis step. Accordingly, it can be assumed that the original chain structure changes only marginally during the pyrolysis.

Table 7.3 Amounts and molar masses of the extractables after pyrolysis of the cross-linked precursor rubbers

Sample	M_w (kg/mol)	Amount of extractables[a] (wt%)
SBR 1712	460.9	
SBR 1712 extractables	6.9	41
SBR 1721	323.3	
SBR 1721 extractables	3.6	36
SBR 1	135.7	
SBR 1 extractables	7.0	29
Krylene 1500	262.9	
Krylene 1500 extractables	25.6	12
BR 1203	242.5	
BR 1203 extractables	7.0	6
Ubepol	181.8	
Ubepol extractables	8.9	10
SBR-BR	330.0	
SBR-BR extractables	24.1	47

[a]related to the total polymer amount

Separations

For the analysis of the chemical composition of the original rubbers and the extractables after pyrolysis, the SEC separation was coupled to FTIR spectroscopy. Using the LC Transform interface, the chemical composition at each point of the elution curve was determined. The results of the SEC-FTIR analyses are presented as Gram-Schmidt plots and chemigrams. While the Gram-Schmidt is a presentation of the total concentration across the elution profile, the chemigrams are measured at specific wavenumbers and present the concentration profiles of specific groups, e.g. styrene or butadiene. Both presentations are shown in Fig. 7.7a for sample Krylene 1500. The concentration profiles of the different structural units are presented as chemigrams at the following wavenumbers: 1,482–1,500 cm^{-1} for styrene, 910 cm^{-1} for 1,2-vinyl-butadiene, 966 cm^{-1} for 1,4-trans-butadiene, and 738 cm^{-1} for 1,4-cis-butadiene. The uniform distribution of the styrene and butadiene units across the elution peak indicates the uniform composition of the SBR rubber.

Using the same approach, the extractables of the vulcanized Krylene 1500 after pyrolysis were analyzed, see Fig. 7.7b. As compared to the analysis of the original sample the only significant difference is the shift of the elution peak towards higher elution volumes indicating a decrease in molar mass. The distribution of the substructures, as indicated by the different chemigrams, is as homogeneous as in the precursor rubber sample, except for a small peak at roughly 9 mL that indicates a very low molar mass styrene-enriched fraction. For a more precise comparison of the original sample and the extractables of the degraded sample, the compositions of both were determined quantitatively.

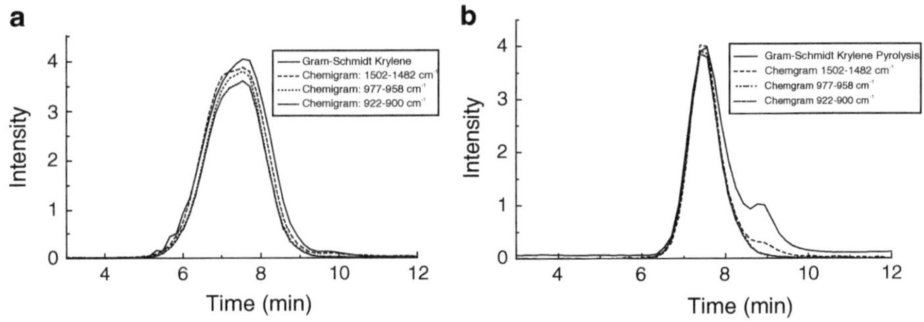

Fig. 7.7 Gram-Schmidt plot and chemigrams of Krylene 1500 (**a**) and the extractables of Krylene 1500 (**b**), chromatogram is given in Fig. 7.6, stationary phase: SDV linear, mobile phase: THF, detector: FTIR (reprinted from [31] with permission of Wiley-VCH)

Based on Lambert-Beer's law, the relative concentrations of the substructures can be determined from the relative intensities of the corresponding absorption peaks. The following compositions for Krylene 1500 and the corresponding extractables have been calculated, see also Table 7.4:

	Krylene 1500	Krylene 1500 extractables
Styrene	24 wt%	22 wt%
1,4-Trans-butadiene	59 wt%	62 wt%
1,2-Vinyl-butadiene	17 wt%	16 wt%
1,4-Cis-butadiene	Not found	Not found

The results indicate that the chemical composition of the original rubber is also present in the extractables from the pyrolysate. Although the extractables constitute only 12 wt% of the original sample, the chemical composition of this part is representative for the total sample.

A similar result was obtained for the more complex sample SBR 1. The results of the SEC-FTIR analysis of the precursor rubber and the corresponding extractables are shown in Fig. 7.8a, b. As was expected, peak 1 is due to the SBR rubber while peak 2 with significantly increased aromatic absorptions is due to the low molar mass additives, mainly the processing oil. This is also confirmed by the corresponding FTIR spectra at the peak maxima. Unfortunately, the spray technique that is used in the LC Transform interface does not permit complete separation of the polymer peak from the additive peak. This is due to the significant spreading of the low viscosity oil over the Ge disc. The quantification of the polymer peak shows that the chemical composition of the extractables after pyrolysis is in excellent agreement with the chemical composition of the precursor.

Table 7.4 Chemical compositions of the initial rubbers and the extractables from the vulcanizates as determined by coupled SEC-FTIR

Sample	Styrene [wt%]	1,4-trans-butadiene [wt%]	1,2-vinyl-butadiene [wt%]	1,4-cis-butadiene [wt%]
SBR 1712	26	58	16	–
SBR 1712 extractables	48	43	9	–
SBR 1721	38	51	11	–
SBR 1721 extractables	29	58	13	–
Krylene 1500	24	59	17	–
Krylene 1500 extractables	22	62	16	–
BR 1203	–	4	1	95
BR 1203 extractables	–	8	1	91
Ubepol	–	3	1	96
Ubepol extractables	–	7	2	91
SBR 1	32	52	16	–
SBR 1 extractables	29	57	14	–
SBR-BR	10	27	7	56
SBR-BR extractables	18	37	9	36

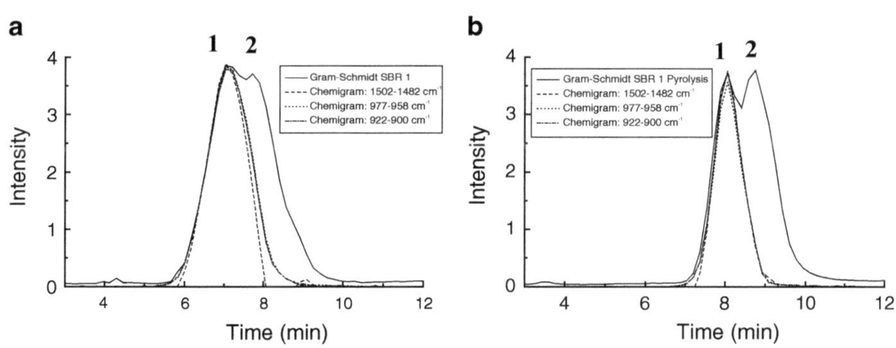

Fig. 7.8 Gram-Schmidt plot and chemigrams of SBR 1 (**a**) and the extractables of SBR 1 (**b**), stationary phase: SDV linear, mobile phase: THF, detector: FTIR (reprinted from [31] with permission of Wiley-VCH)

	SBR 1	SBR 1 extractables
Styrene	32 wt%	30 wt%
1,4-Trans-butadiene	52 wt%	55 wt%
1,2-Vinyl-butadiene	16 wt%	15 wt%
1,4-Cis-butadiene	Not found	Not found

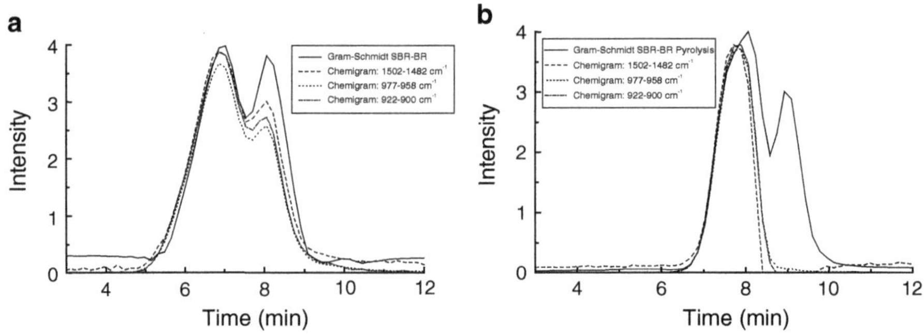

Fig. 7.9 Gram-Schmidt plot and chemigrams of SBR-BR blend (**a**) and the extractables of SBR-BR (**b**), stationary phase: SDV linear, mobile phase: THF, detector: FTIR (reprinted from [31] with permission of Wiley-VCH)

Finally, a blend of SBR and BR was investigated. This blend was a 1:1:1 mixture of SBR 1712, BR 1203, and carbon black. In addition, the blend contained different types of additives and processing oil.

The blend was vulcanized and degraded under the above mentioned conditions and yielded 47 wt% extractables with a molar mass of 24.0 kg/mol. The SEC-FTIR analyses of the original blend and the extractables are given in Fig. 7.9. The quantification of the polymeric part shows that there is a significant difference between the chemical composition of the extractables after pyrolysis and the chemical composition of the precursor.

	SBR-BR	SBR-BR extractables
Styrene	10 wt%	18 wt%
1,4-Trans-butadiene	27 wt%	37 wt%
1,2-Vinyl-butadiene	7 wt%	9 wt%
1,4-Cis-butadiene	56 wt%	36 wt%

While the concentration of the SBR-related structural units is higher, the concentration of the BR-related structural units is lower in the extractables as compared to the original sample. This result is straightforward since the pyrolysis of the blend yields mainly soluble degradation products from the SBR component. Therefore, this part is overestimated in the blend. SEC did not separate SBR and BR from each other and, accordingly, FTIR could not quantify them separately.

Evaluation

As has been shown cross-linked SBR and BR can efficiently be analyzed by liquid chromatography and FTIR spectroscopy. By pyrolyzing the vulcanizate and extracting the soluble part, rubber fragments are obtained that can be analyzed by SEC and FTIR spectroscopy. By quantitative analysis of the FTIR spectra the concentrations of the different structural units, including styrene, 1,4-trans-butadiene, 1,2-vinyl-butadiene and 1,4-cis-butadiene in the polymeric part are determined.

It has been shown that the chemical composition of the original non-crosslinked rubbers and the chemical composition of the extractables are rather identical. Deviations from the original composition are obtained for the SBR-BR blend. Obviously, both components exhibit different pyrolysis behaviours. For the accurate analysis of such samples a chromatographic technique must be used that separates the sample by chemical composition into the SBR and the BR parts, such as HPLC or 2D-LC, see for example Heinz et al. [41]. The quantitative compositions that were obtained by SEC-FTIR are summarized in Table 7.4.

7.1.2 Fast Chemical Composition Analysis of Random Styrene-Butylacrylate Copolymers by HPLC-FTIR [32]

Aim

Combinatorial methods are high-efficiency methods to create large composition libraries of materials, e.g. with continuous composition variations. These compositions are tested systematically in parallel for specific properties of interest, in contrast to the time-consuming one-composition-at-a-time approach. Combinatorial methods are of particular interest for the development of new or better polymeric materials. Given the huge variety in monomers, catalysts, and polymerization techniques leading to the possible preparation of unlimited numbers of new formulations it is reasonable, if not obvious, to use combinatorial techniques. The ability to map out structure–property relationships in a relatively short period of time is crucial for the discovery of polymers with new or improved properties. The preparation of large numbers of target compounds, however, is only the first step in a combinatorial set-up. As important as the fast synthesis is the fast analysis of the prepared sample sets. Synthetic polymers are highly complex materials that very frequently exhibit a chemical or functional heterogeneity in addition to the molar mass distribution. In order to evaluate a new material properly, the different distributions have to be determined quantitatively.

It has been shown that the throughput in SEC can be increased by 5–10 times when new stationary phases and column shapes are developed that are designed for high-throughput experimentation [42–44]. Using these new materials the time per SEC run can be decreased to about 2 min as compared to more traditional analysis times of 30–60 min. When new column technology is applied to HPLC of polymers, functionality type separations can be conducted in less than 5 min [44, 45]. However, while the chromatographic separation has been conducted in a very short period of time, the spectroscopic (e.g., FTIR) analysis of chromatographic fractions has taken much longer.

The present application makes use of the LC Transform system for the high-throughput analysis of random styrene-butylacrylate copolymers (SBA) with regard to molar mass and chemical composition distribution.

Table 7.5 Bulk chemical compositions and molar masses of the SBA copolymers analysed by high-throughput SEC with RI detection[a]

Sample	BA/S content (mol%)	M_n (kg/mol)	M_w (kg/mol)
1	0/100	57	120
2	20/80	64	149
3	30/70	65	154
4	50/50	72	182
5	70/30	75	199
6	80/20	82	227
7	100/0	44	75
8	20/80	88	223
9	30/70	89	230
10	40/60	93	256
11	50/50	91	247
12	60/40	96	276
13	70/30	101	299
14	80/20	86	223

[a]Samples 1–7: low conversion polymers with narrow CCD, samples 8–14: high conversion polymers with broad CCD

Materials

- **Polymers**. Thirteen random copolymers were prepared using a parallel experimental set-up. Polymerizations were carried out in a Polymer Laboratories PL-SP 260 high-temperature sample preparation instrument. Screw-neck vials of 4 mL volume were filled with destabilized styrene and *n*-butyl acrylate, recrystallised benzoyl peroxide and distilled toluene. The vials were agitated and heated at a temperature of 70 °C. For the low-conversion samples the polymerisation time was 1 h, for the high-conversion samples the reaction time was 20 h. In this case the amount of toluene was optimized in order to avoid cross-linking. The copolymer solutions were precipitated in methanol and dried in vacuum at 40 °C. The copolymer blends were prepared by mixing solutions of the copolymers in toluene. Their bulk chemical composition and the molar masses obtained by high-throughput SEC are summarized in Table 7.5.

Equipment

- **Chromatographic System**. Agilent 1100 Series HPLC modular system (Agilent Technologies) comprising a quadruple pump, an auto-sampler and a column oven was used. For data collection and processing the software package 'WinGPC-Software' (Polymer Standards Service GmbH, Mainz, Germany) was used. For the HPLC-FTIR experiments the LC Transform interface LCT-1 was used. Experimental conditions were as follows: evaporation temperature 110 °C, rotation of the Ge disc with variable speed at 45°/min. The FTIR spectra were taken with a Nicolet Protegé 460 Magna-IR Technology spectrometer, one sample and background scan, 8 cm^{-1} spectral resolution, 600–2,000 cm^{-1} spectral range. The flow rate for SEC was 5 mL/min while the HPLC separations were conducted at a flow rate of 2 mL/min.

- **Columns**. SEC: high-speed linear SDV (PSS GmbH, Mainz, Germany), average particle size 10 μm, column size 50 × 20 mm i.d.; HPLC: Luna silica gel (Phenomenex), average particle size 3 μm, column size 30 × 4.6 mm i.d.
- **Mobile Phase**. SEC: THF; HPLC: hexane-toluene-MEK
- **Detectors**. PL evaporative light scattering detector ELS 1000 (gas flow: 1.5, nebulizer temperature: 90 °C, evaporation temperature: 120 °C), Agilent 1100 Series refractive index detector
- **Column Temperature**. 40 °C
- **Sample Concentration**. Samples were dissolved in a concentration of 0.5 mg/mL (SEC) or 6 mg/mL (HPLC) in the mobile phase and filtered through a 0.45 μm filter prior to analysis.
- **Injection Volume**. 100 μL (SEC) and 20 μL (HPLC)

Preparatory Investigations

The most useful LC technique for random copolymer separation is gradient HPLC. Under favourable conditions gradient HPLC separates copolymers strictly with regard to chemical composition rather irrespective of molar mass. For styrene-ethyl acrylate copolymers it was shown by Krämer et al. that they can be separated according to chemical composition using a RP-18 stationary phase and a linear gradient of acetonitrile-THF [46]. At a flow rate of 1 mL/min the time requirement per analysis was roughly 20 min. For the analysis of the present samples this method was not suitable. In this case a gradient of hexane-toluene-MEK on a silica gel stationary phase was found to separate the copolymers according to chemical composition. As it was the aim of the present study to run fast HPLC separations, a very short column (Luna of Phenomenex) with an average particle size of 3 μm was used. The gradient is given in Table 7.6.

Figure 7.10 shows the chromatograms of the different low-conversion SBA copolymers. As can be seen, very narrow and uniform elution peaks are obtained. Since gradient HPLC separates according to chemical composition, the narrow and uniform elution peaks are a clear confirmation of the fact that these samples have a very narrow chemical composition distribution.

The separations were conducted at a flow rate of 2 mL/min requiring less than 1.5 min per analysis. The total analysis time for the three blends including column washing and re-establishment of the gradient conditions took less than 8 min. Accordingly, this technique can be used for the fast separation of large numbers of samples with regard to chemical composition. As compared to conventional gradient HPLC separation experiments time savings of more than 90% were achieved.

For the analysis of the copolymers with regard to chemical composition distribution the HPLC system must be calibrated, i.e. a correlation of the elution volume

Table 7.6 Gradient for the HPLC separation of SBA copolymers

Time (min)	Hexane (%)	Toluene (%)	MEK (%)
0	60	40	0
0.15	33	60	7
0.60	45	40	15
1.5	20	40	40
1.75	100	0	0
2.25	100	0	0

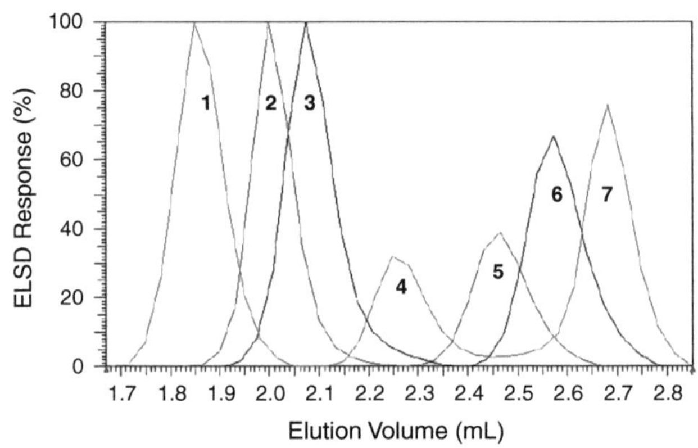

Fig. 7.10 Gradient HPLC of blends of PS, PBA and low-conversion SBA copolymers, stationary phase: Luna, mobile phase: hexane-toluene-MEK, detector: ELSD, blend 1: samples 1 + 4 + 7, blend 2: samples 2 + 5, blend 3: samples 3 + 6 (reprinted from [32] with permission of European Polymer Federation)

and the chemical composition must be established. In the present case, the calibration was conducted by measuring blends of PS and PBA in different well-known proportions. From the FTIR spectra of the blends the ratios of the absorption peaks at 700 cm^{-1} for PS (I_{PS}) and 1,727 cm^{-1} for PBA (I_{PBA}) were determined and plotted against the PBA content of the blend, see Fig. 7.11. The calibration curve was subsequently used to calculate the BA contents of the copolymers from the FTIR peak areas.

Separations

The fast compositional analysis of the chromatographic peaks was conducted using FTIR as a direct detector in HPLC. As a result of the measurements chemigrams are obtained presenting the intensity profile of a certain absorption band across the elution profile. The deposited copolymer fractions that were eluted from HPLC and sprayed onto the Ge disc are shown in Fig. 7.12. As can be seen the whole disc can be loaded with HPLC fractions that are automatically scanned in the FTIR spectrometer. After optimization of the FTIR settings, the measurement of the background and of seven copolymer peaks can be carried out in less than 6 min. The

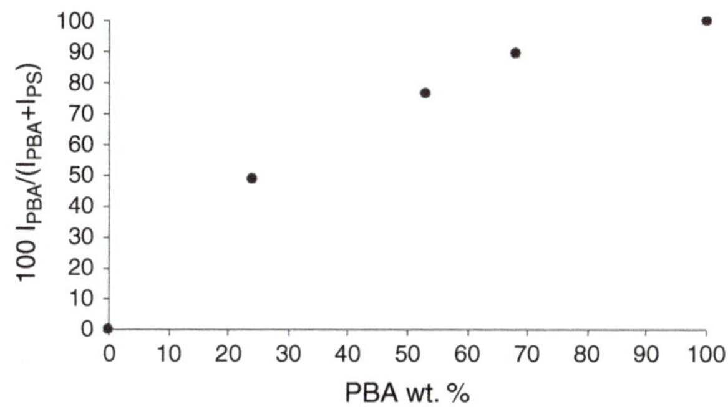

Fig. 7.11 Calibration curve for the analysis of SBA copolymers with regard to chemical composition (reprinted from [32] with permission of European Polymer Federation)

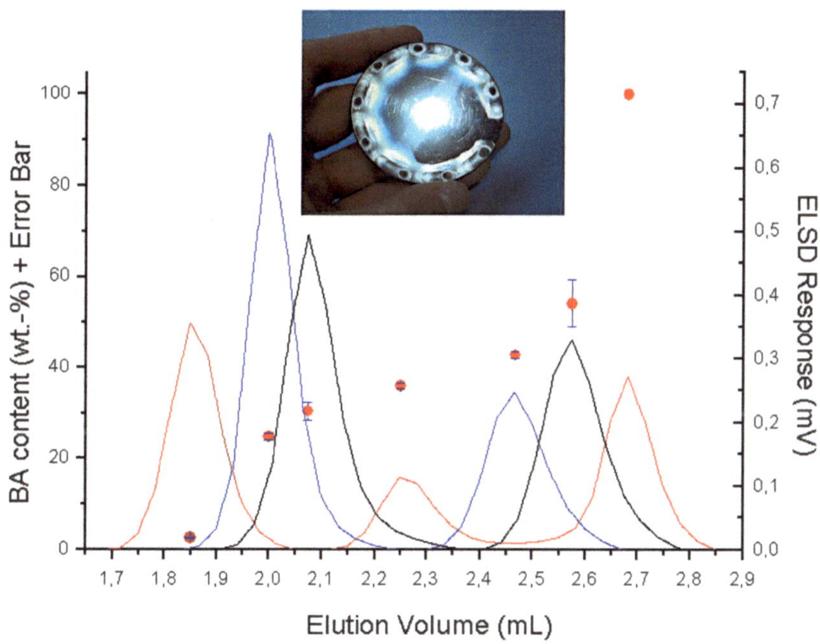

Fig. 7.12 Multiple HPLC separations of blends of SBA copolymers and analyses of the BA content by FTIR, experimental conditions and sample assignment see Fig. 7.10, errors are indicated by corresponding error bars, inset shows the Ge disc after polymer deposition (reprinted from [32] with permission of European Polymer Federation)

low-conversion copolymers were measured within 18 min, whereas with conventional measurements this would have taken more than 3 h. Figure 7.12 shows the HPLC-FTIR results of the copolymer blend separations.

As has been shown in Fig. 7.10, one blend with three polymer components and two blends with two copolymer components were analyzed. The calculated BA

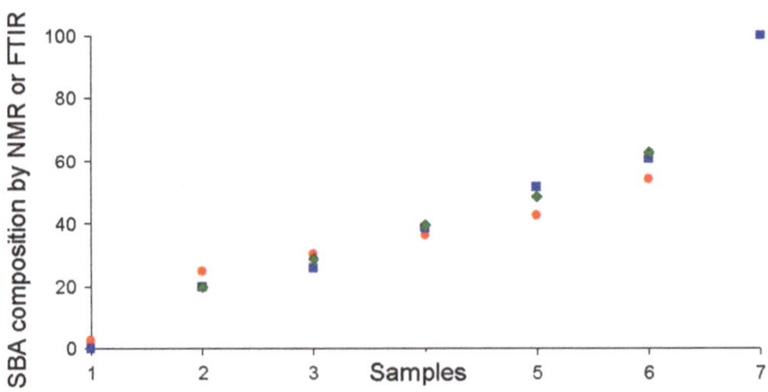

Fig. 7.13 Comparison of SBA chemical composition as determined by ^{1}H NMR (*squares*), off-line FTIR (*diamonds*), and coupled HPLC-FTIR (*circles*) (reprinted from [32] with permission of European Polymer Federation)

contents of the copolymer fractions are plotted as a function of the elution volume. To check for reproducibility, every copolymer blend was separated and analyzed consecutively three times. The accuracy of these measurements is presented as an error plot. Considering the errors in the calibration of FTIR, the evaluation of sample spectra and the experimental peculiarities of the LC Transform system, a maximum error of about 10% can be regarded as a good result.

To check the accuracy of the FTIR analyses, the data of ^{1}H NMR and FTIR bulk analyses were compared with the results of the coupling experiments. The ^{1}H NMR measurements were conducted in deuterated chloroform. For the determination of the copolymer compositions the signals of the aromatic protons (styrene) and the $-CH_2O-$ protons (ethyl acrylate) were used. As can be seen in Fig. 7.13, a very good agreement between the different techniques was achieved. Maximum error was less than 9 %.

For the analysis of the high-conversion SBA copolymers the same gradient conditions were used. Different from the low-conversion copolymers these copolymers exhibit very broad elution peaks, see Fig. 7.14. This is a clear indication of the high chemical heterogeneity of these copolymers.

Evaluation

Using the BA content versus elution volume dependence given in Fig. 7.12, the elution curves of the high-conversion SBA copolymers in Fig. 7.14 can be re-calculated to chemical composition distribution curves, see Fig. 7.15. Considering the limited accuracy of the concentration axis, these curves give a very clear presentation of the chemical heterogeneity of the copolymers. Depending on the composition of the monomer feed, distribution curves of different shape and broadness are obtained. For example, the copolymer with a nominal content of 50% BA contains copolymer fractions with BA contents of 28–97%.

To summarize, the fast and reproducible chromatographic analysis of SBA copolymers can be achieved when existing techniques are modified with regard to

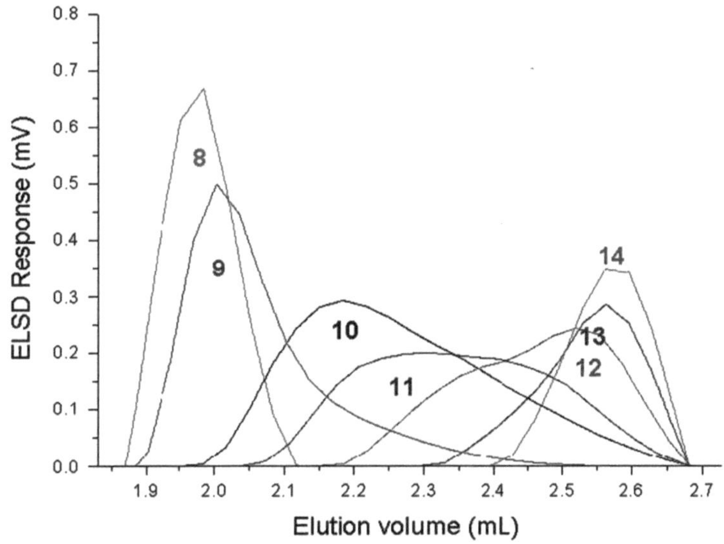

Fig. 7.14 Gradient HPLC analyses of high-conversion SBA copolymers, experimental conditions see Fig. 7.10, numbers indicate sample numbers (reprinted from [32] with permission of European Polymer Federation)

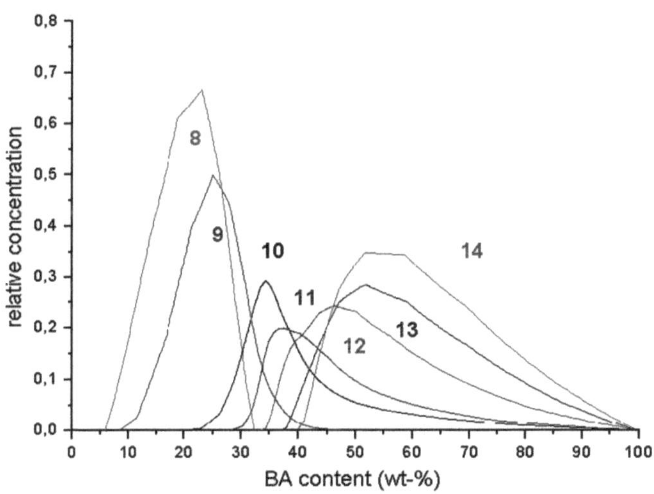

Fig. 7.15 Chemical composition distribution (CCD) of high-conversion SBA copolymers, numbers indicate sample numbers (reprinted from [32] with permission of European Polymer Federation)

the stationary and mobile phases, and the flow rate. Fast and high-throughput SEC analyses can be conducted within less than 3 min per sample. The time requirements for gradient HPLC separations can be decreased to less than 2 min per sample when very short and highly efficient columns and high flow rates are used. The calibration of the HPLC separation can be conducted by coupling the HPLC system with FTIR through the LC Transform interface.

7.2 Coupling with Mass Spectroscopy

From the early days of modern mass spectrometry (MS), the value of its combination with chromatography was appreciated. The coupling of GC with MS was a natural evolution since they are both vapour phase techniques, and very quickly GC-MS was accepted as a standard component of the organic analytical laboratory. It has taken considerably longer to develop suitable modes of HPLC-MS coupling. The difficulties with HPLC-MS were associated with the fact that vaporization of typically 1 mL/min from the HPLC translates into a vapour flow rate of approx. 500–1,000 mL/min. Other difficulties were related to the eluent composition as result of the frequent use of non-volatile modifiers, and the ionization of non-volatile and thermally labile analytes. However, during the past two decades commercial interfaces have been developed which have led to a broad applicability of HPLC-MS [47–49].

The techniques necessary for the successful introduction of a liquid stream into a mass spectrometer are based on the following principles: electrospray ionization [50], atmospheric pressure chemical ionization [51], thermospray ionization [52], and particle beam ionization [53]. In a thermospray interface [54–56] a jet of vapour and small droplets is generated out of a heated vaporizer tube. Nebulization takes place as a result of the disruption of the liquid by the expanding vapour that is formed at the tube wall upon evaporation of part of the liquid in the tube. Prior to the onset of the partial evaporation inside the tube a considerable amount of heat is transferred to the solvent. This heat later assists in the desolvation of the droplets in the low-pressure region. By applying efficient pumping by means of a high-throughput mechanical pump attached directly to the ion source up to 2 mL/min of aqueous solvents can be introduced into the MS vacuum system. The ionization of the analytes takes place by means of solvent-mediated chemical ionization reactions and ion evaporation processes. In a particle beam interface [53, 57, 58] the column effluent is nebulized either pneumatically or by thermospray nebulization, into a near atmospheric-pressure desolvation chamber, which is connected to a momentum separator where the high-mass analytes are preferentially transferred to the MS ion source while the low-mass solvent molecules are efficiently pumped away. The analyte molecules are transferred as small particles to a conventional ion source, where they disintegrate upon collisions at the heated source walls. The released gaseous molecules are ionized by electron impact or chemical ionization.

Two different sample-introduction approaches are used in combination with atmospheric pressure ionization (API) devices. They primarily differ in the nebulization principle and in the application range they cover. In a heated nebulizer or APCI interface [59], the column effluent is pneumatically nebulized in a heated tube, where the solvent evaporation is almost completed. Atmospheric-pressure chemical ionization (APCI), initiated by electrons from a corona discharge needle, is achieved in the same region. Subsequently, the ions generated are sampled into the high vacuum of the mass spectrometer for mass analysis. In an electrospray interface [60, 61], the column effluent is nebulized into the atmospheric-pressure region as a result of the action of a high electric field resulting from a 3 kV potential difference between the narrow-bore spray capillary and a surrounding counter electrode. The solvent emerging from the capillary breaks into fine threads which subsequently disintegrate in small droplets. In some designs, the electrospray nebulization is assisted by pneumatic nebulization. Such an approach is called an ionspray interface [62].

From the point of view of polymer analysis, a mass spectrometric detector would be a most interesting alternative to the conventional detectors because this detector could provide absolute molar masses of polymer components [63, 64]. Provided that fragmentation does not occur, intact molecular ions could be measured. The measured mass of a particular component could then be correlated with chemical composition or chain length. However, the major drawback of most conventional HPLC-MS techniques is the limited mass range, preventing higher oligomers (molar mass above 2,000–3,000 g/mol) from being ionized without fragmentation [65–67].

The use of MS for detailed polymer analysis has been becoming more established due to the introduction of soft ionization techniques that afford intact oligomer or polymer ions with less fragmentation [68–71]. One of these techniques, electrospray ionization MS (ESI-MS), has been widely applied in biopolymer analysis. Proteins and biopolymers are typically ionized through acid–base equilibria. When a protein solution (the effluent from a HPLC separation) is exposed to an electrical potential it ionizes and disperses into charged droplets. Solvent evaporation upon heat transfer leads to the shrinking of the droplets and the formation of analyte ions. Larger molecules aquire more than one single charge, and, typically, a mixture of differently charged ions is obtained.

Unfortunately, ESI-MS has had limited applications in polymer analysis [72, 73]. Unlike biopolymers, most synthetic polymers have no acidic or basic functional groups that can be used for ion formation. Moreover, each molecule gives rise to a charge distribution envelope, thus further complicating the spectrum. Therefore, synthetic polymers that can typically contain a distribution of chain lengths and a variety in chemical composition or functionality furnish complicated mass spectra, making interpretation nearly impossible.

To overcome the difficulties of ESI-MS, salts were added to the mobile phase to facilitate ionization [74, 75]. To simplify the resulting ESI spectra, the number of components entering the ion source was reduced. Combining SEC with electrospray detection, a number of polymeric systems have been analysed including PEO [76], aliphatic polyesters [77], phenolic resins [78], methyl methacrylate macromonomers [79] and polysulfides [80]. The detectable mass range for different species, however, was well below 5,000 g/mol, indicating that the technique is not really suited for polymer analysis.

Matrix-assisted laser desorption/ionization time-of-flight MS (MALDI-TOF-MS) is the most powerful soft ionization technique that allows desorption and ionization of very large molecules even in complex mixtures. In polymer analysis, the great potential of MALDI-TOF-MS is to perform the direct identification of mass-resolved polymer chains, including intact oligomers within a molar mass distribution, and the simultaneous determination of structures and endgroups in polymer samples. This most promising method for the ionization of large molecules and the analysis according to their molar mass and functionality has been introduced by Tanaka et al. [80] and Karas and Hillenkamp [81–83]. Compared to other mass spectrometric techniques, the accessible mass range has been extended considerably. In principle, the sample to be investigated and a matrix solution are mixed in such a ratio that matrix separation of the sample molecules is achieved. After drying, a laser pulse is directed onto the solid matrix to photo-excite the matrix material. This excitation causes the matrix to explode, resulting in the expulsion and soft ionization of the sample molecules. Once the analyte is ionized, it is accelerated and analysed in a TOF mass spectrometer. As a result, the analyte is separated according to the molar mass of its components, and in the case of heterogeneous polymers additional information on chemical composition may be obtained. In a number of papers it was shown that polymers may be analyzed up to relative molar masses of about 500,000 Da [84–88]. It was shown in a number of applications that functionally heterogeneous polymers can be analysed with respect to the degree of polymerization and the type of functional groups [89–92].

The on-line combination of liquid chromatography and MALDI-TOF-MS would be of great value for polymer analysis. In particular, for chemically or functionally heterogeneous polymers liquid chromatography could provide separation with respect to chemical composition while MALDI-TOF would analyse the fractions with respect to oligomer distribution or molar mass. Unfortunately, MALDI-TOF is based on desorption of molecules from a solid surface layer and, therefore, a priori not compatible with liquid chromatography. In an attempt to take advantage of the MALDI-TOF capabilities, off-line LC separations are conducted and the resulting fractions subjected to MALDI-TOF measurements. This can be done manually or using an interface and it has the advantage that virtually any type of chromatographic separation can be combined with MALDI-TOF.

The different options for online coupling have been summarized by Weidner and Falkenhagen [93], including (1) spray (aerosol) methods and (2) continuous flow methods. In the first case, the solvent is evaporated by spraying into a heated tube. The aerosol that is formed by nebulization with nitrogen enters the mass spectrometer and is ionized by the laser, see Fig. 7.16a.

The matrix is added before the nebulization via a T-piece. The obtained MALDI spectra, however, were of rather low quality [94–98]. In the second case, the dissolved sample is introduced into the spectrometer and the laser is focussed at the tip of the inlet system, see Fig. 7.16b. In order to maintain the vacuum the flow rates are typically 1–5 μL/min. Later, a number of modifications have been presented that describe an atmospheric pressure approach [99–102].

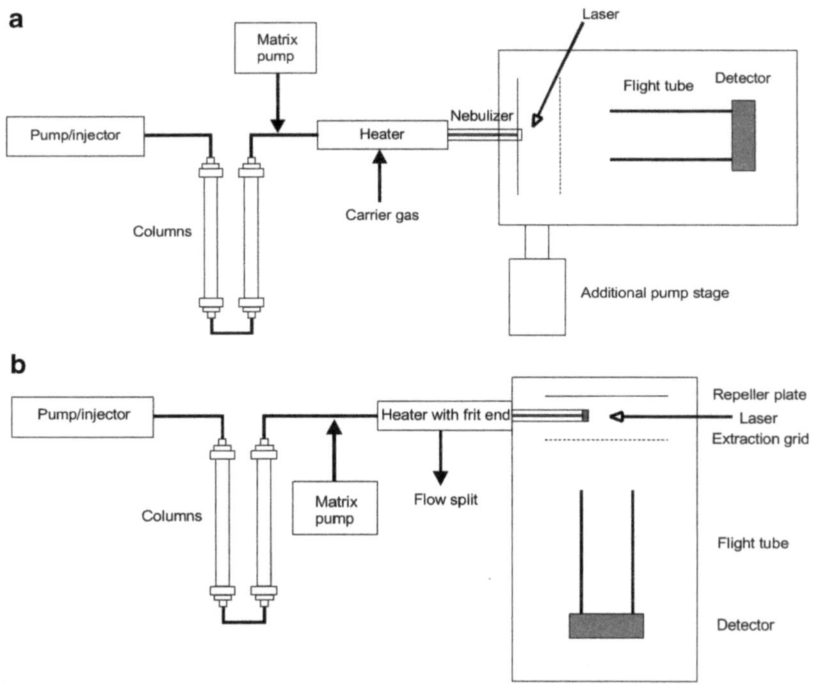

Fig. 7.16 Principles of online coupling of LC and MALDI-TOF, (**a**) spray method, (**b**) continuous flow method (reprinted from [93] with permission of John Wiley & Sons)

While direct coupling is a problem, it has been shown that at least the LC sample collection and subsequent preparation of the samples for MALDI analysis can be automated. Different interfaces have been developed that can be coupled directly to the LC system. In these interfaces the eluate stream is collected in vials or deposited on the MALDI target via a spray or drip process. The matrix required for the MALDI process is either co-added to the eluate stream or matrix-precoated MALDI targets are used. Typical examples for interfaces are: Probot™ Microfraction Collector, LC Packings; DiNa Map MALDI Spotter, Kromatek, UK; MALDILC™ System, Gilson; SunChrom, Germany [103–106]. Eluates can be collected in vials and, after finishing fractionation, small volumes can be premixed with matrix solution. Afterwards, a small amount of this mixture (normally only a few microliters) will be deposited on the MALDI—TOF target plate [107–109]. Out of the spray interfaces, the LC Transform system is the most popular one. This system is a modified version of the device that is used for LC-FTIR interfacing [110]. A more recent development is the heated droplet interface [111, 112].

Over the years a significant number of applications for the coupling of LC and MALDI-TOF have been published. Comprehensive overviews can be found in [113–115]. One option is to use MALDI-TOF for the calibration of SEC. In SEC of low molar mass samples the separation into individual oligomers and the quantitative determination of the MMD via an oligomer calibration can be

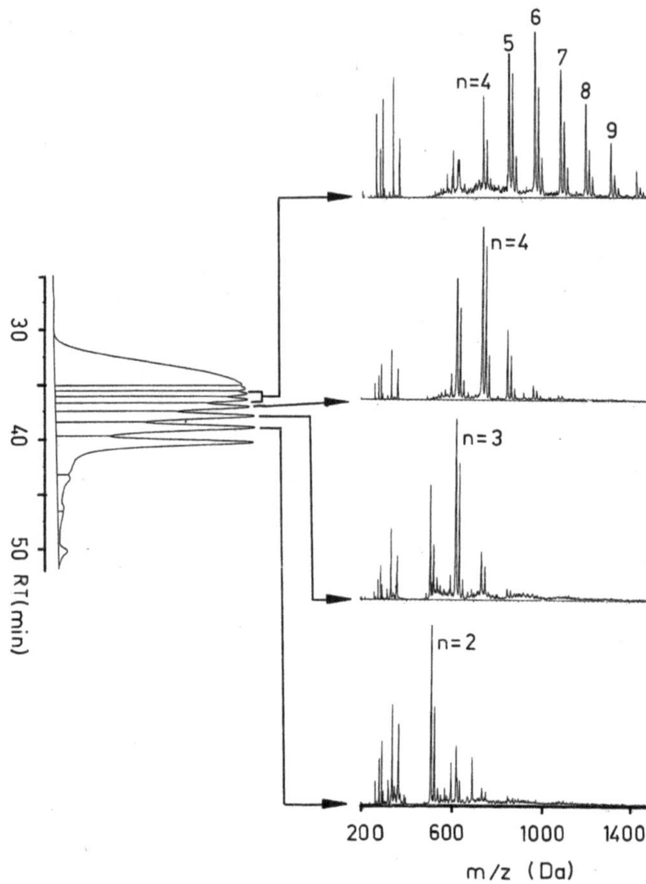

Fig. 7.17 SEC of oligo(caprolactone) and offline analysis of fractions by MALDI-TOF, peak assignment indicates degree of polymerisation (n) (reprinted from [115] with permission of Elsevier)

achieved, see Fig. 7.17 for oligo(caprolactone). The lower oligomers appeared as well separated peaks at the high retention time end of the chromatogram. For the analysis of the peaks, i.e. the assignment of a certain degree of polymerization (n) to each peak, MALDI-TOF MS was used. The SEC separation was conducted at the usual analytical scale and the oligomer fractions were collected, resulting in amounts of 5–20 ng substance per fraction in THF solution. The solutions were directly mixed with the matrix solution, placed on the sample slide and subjected to the MALDI experiments. For the lower oligomers the spectra consisted of a number of peaks of high intensity having a peak-to-peak mass increment of 114 Da, which equals the mass of the caprolactone repeating unit. These peaks represented the $M + Na^+$ molecular ions, whereas the peaks of lower intensity in their vicinity were due to the formation of $M + K^+$ molecular ions. $M + Na^+$ and $M + K^+$ molecular ions were formed due to the presence of small amounts of Na^+ and K^+ ions in the samples and/or the matrix. Further peaks of low intensity indicated a functional heterogeneity in the samples. From the masses of the M + Na^+ peaks the degree of polymerization of the corresponding oligomer was calculated. By this procedure, the first peak in the chromatogram was assigned to $n = 1$, the second peak to $n = 2$, and so on. From the elution time and the degree of polymerization of each oligomer peak an oligomer calibration curve log molar mass versus elution time was constructed. The conventional calibration curve based on PS standards differed remarkably from the oligomer calibration curve.

Other examples of successful combinations of liquid chromatography and MALDI-TOF were given by Krüger et al., separating linear and cyclic fractions of polylactides by LC-CC [116]. Just and Krüger were able to separate cyclic siloxanes from linear silanols and to characterize their chemical composition [117]. The calibration of a SEC system by MALDI-TOF was discussed by Montaudo [118]. Polydimethyl siloxane (PDMS) was fractionated by SEC into different molar mass fractions. These fractions were subjected to MALDI-TOF for MMD. The resulting peak maximum molar masses were combined with the elution volumes of the fractions from SEC to give a PDMS calibration curve log M versus V_e. The calibration of SEC by MALDI-TOF-MS for PMMA, polyvinyl acetate and vinyl acetate copolymers has been discussed by Danis et al. In addition to obtaining proper calibration curves, band broadening of the SEC system was detected [119]. The analysis of random copolyesters has been described by Montaudo [120]. Falkenhagen et al. [121] and Pasch and Rode [115] described the characterization of EO-PO copolymers. The chromatographic fractionation was conducted using LCCC. Lee et al. [122] addressed the analysis of L-lactide-ethylene oxide block copolymers while Montaudo and Montaudo described the separation and analysis of polyacrylate [123] and styrene-maleic anhydride copolymers [124].

At least for the lower mass range the combination of different LC separation modes with ESI-TOF MS is a useful option. Recently, Falkenhagen and Weidner reported on the coupling of ultrahigh pressure liquid chromatography (UPLC) and ESI-TOF MS for the analysis of complex copolymers [125]. They showed that critical conditions for PEO and PPO can be determined easily without running a large number of molar mass standards, see Fig. 7.18.

Experimental parameters for SEC, LCCC and liquid adsorption chromatography (LAC) can be adjusted very quickly by running mass spectra at various UPLC elution times. The LCCC mode can be clearly identified by the fact that for different elution times the same oligomer distributions (number of EO units) are obtained. In contrast, in the SEC mode increasing elution times correspond to lower numbers of EO units. This approach was then used to analyze a variety of copolymers.

Over the last few years MALDI-TOF became a popular tool for the 2D mapping of copolymers, see for example Weidner et al. [126]. A dedicated software tool 'MassChrom2D' has been developed to process the data from HPLC separation and

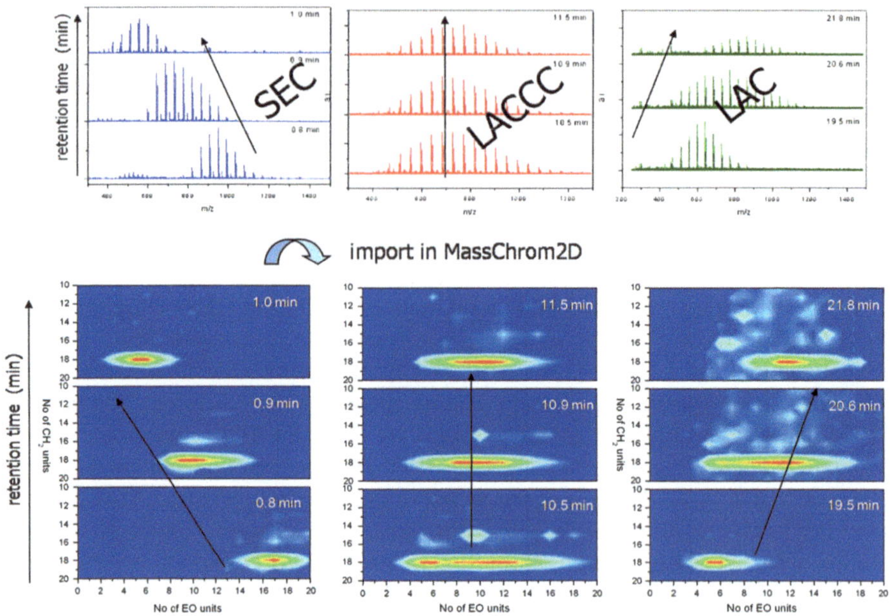

Fig. 7.18 Determination of different modes of liquid chromatography (SEC, LACCC, LAC) for PEO by UPLC/ESI-TOF MS (reprinted with permission from [125]. Copyright (2009) American Chemical Society)

MALDI-TOF analysis of fractions to produce 2D plots similar to those from 2D-LC [127]. Latest state-of-the-art reports on the diversity of MALDI-TOF applications for complex polymers have been given by Weidner and Trimpin [128], Hart-Smith and Barner-Kowollik [129] and Weidner and Falkenhagen [93].

In the following sections representative examples for the analysis of complex polymers by LC-MALDI-TOF are given. These examples were taken from a more extensive compilation given in [113].

7.2.1 Molar Mass Distribution of Polyester Copolymers [130, 131]

Aim

Polyesters are polycondensation products with a broad MMD. The determination of MMD by SEC is complicated due to the fact that calibration standards with narrow polydispersity are not commercially available. The direct MMD determination by MALDI-TOF MS is not possible because the expected polydispersity index of $M_w/M_n \sim 2$ is too high for a direct measurement. Even more complicated is the situation for polyester copolymers where different diols and diacids are reacted with each other. In this case an accurate MMD by SEC is not possible unless molar mass sensitive detectors (light scattering, viscometer detectors) are coupled to SEC.

By pre-fractionating different polyester samples, fractions of low polydispersity can be obtained which subsequently can be analyzed by MALDI-TOF MS. These analyses yield calibration curves for polyesters of different composition which can be used for computing molar mass averages for homo- and copolymers.

Materials

- Laboratory products of different polyesters have been prepared by melt polycondensation starting from stoichiometric amounts of dimethyl esters and 1,4-butane diol. Dimethyl esters of the following acids were used: adipic acid, succinic acid, and sebacic acid.

Equipment

- **Chromatographic System**. Waters Model 600A apparatus and a Waters Model 401 differential refractometer. 30 drops of each fraction were collected, fractionation was carried out several times to accumulate sufficient amounts for further analyses.
- **Columns**. Five Ultrastyragel connected in series, average particle size 10 μm, column size 300 × 7.8 mm, pore sizes 10^5, 10^3, 500, 10^4, 100 Å
- **Mobile Phase**. THF or chloroform
- **Sample Amount**. 60 μL of a 15 g/L solution in THF or chloroform
- **MALDI-TOF System**. Bruker Reflex MALDI-TOF with linear and reflectron detectors, acceleration voltage of positive 30 kV.
- **MALDI-TOF Sample Preparation**. 0.1 mL of each collected fraction was mixed with 2 mL HABA solution (0.1 M 2-(4-hydroxyphenyl azo)benzoic acid in THF/$CHCl_3$ 1:1). 0.2 μL of this volume were dropped on the MALDI sample target, dried and analyzed.

Preparatory Investigations

As the first step of these investigations, the homopolymers polybutylene adipate (PBA), polybutylene succinate (PBSu) and polybutylene sebacate (PBSe) are fractionated by SEC. Typically, injecting about 0.5–1 mg of polymer into the SEC and collected 25–100 fractions yields sufficient quantities for MALDI-TOF analysis. Selected nearly monodisperse fractions are analyzed by MALDI-TOF and the data are used for constructing the corresponding log M versus elution volume curves. Figure 7.19 shows the calibration curves obtained for PBA, PBSu and PBSe.

As can clearly be seen from these plots, the calibration curves of the polyesters are different in terms of position and slope. Apparently, the hydrodynamic volumes are in the order PBSe > PBA > PBSu, showing a correlation with their chemical structure.

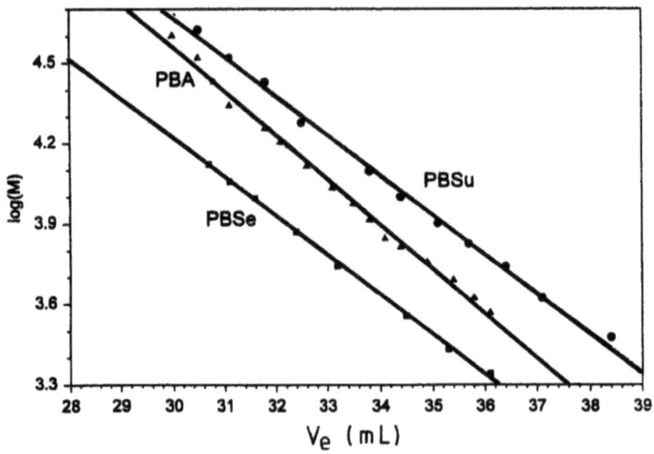

Fig. 7.19 Plots of log molar mass versus elution volume for PBSe (*filled square*), PBA (*filled triangle*), and PBSu (*filled circle*) (reprinted with permission from [131]. Copyright (1998) of American Chemical Society)

MALDI-TOF Analysis

Copolyesters which were prepared from 50:50 mixtures of different dimethyl esters are fractionated in a similar way. Figure 7.20 shows a typical SEC trace of polybutylene adipate/sebacate (PBA-Se) together with the MALDI-TOF spectra of fractions taken at different elution volumes. The corresponding calibration curve for the copolymer together with the calibration curves for the homopolymers is given in Fig. 7.21a. It is clear from this plot that the copolymer deviates over its entire molar mass range from the additivity of hydrodynamic volumes, which is normally assumed in conventional SEC experiments. The same behaviour is obtained for the copolymer polybutylene adipate/succinate (PBA-Su). In this case, the calibration curve for the copolymer is even lower than the calibration lines of the corresponding homopolymers, see Fig. 7.21b.

Comparing the molar mass data obtained by conventional SEC with the data from SEC-MALDI-TOF, significant deviations are obtained for PBSe-Su and PBA-Se-Su. This is a further strong indication that the hydrodynamic properties of the copolymers are more complex than it can be extrapolated from the behaviour of the homopolymers.

7.2.2 Molar Mass and Chemical Composition Analysis of PnBMA-b-PMMA Block Copolymers [132]

Aim

Block copolymers are frequently prepared by sequential polymerization using different techniques including anionic, group transfer (GTP) or atom transfer radical polymerisation (ATRP). The polymerisation is started by homopolymerising the first

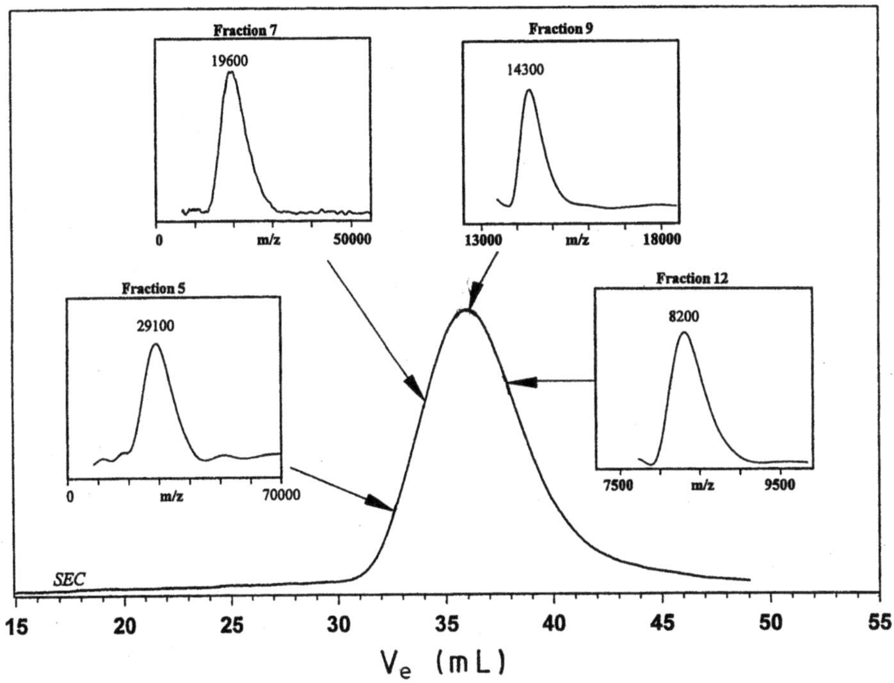

Fig. 7.20 SEC chromatogram for PBA-Se and MALDI-TOF spectra of fractions 5, 7, 9, and 12, solvent THF (reprinted with permission from [131]. Copyright (1998) of American Chemical Society)

monomer to form the first block of the block copolymer. When the first monomer is consumed, the second monomer is added to the reaction mixture to form the second block. Due to impurities in the reaction mixture, very frequently after the first polymerization step a number of chain ends are terminated and a homopolymer fraction of the first monomer is formed.

As has been discussed before, accurate molar mass determination of copolymers is a problem due to the fact that the SEC system cannot be calibrated by proper calibration standards. Therefore, a calibration shall be conducted through SEC-MALDI-TOF. In addition, information shall be obtained on the presence of homopolymers and the chemical composition of the copolymer fractions.

Materials

- Diblock copolymer of *n*-butyl methacrylate (nBMA) and methyl methacrylate (MMA). The sample under investigation has been prepared by group transfer polymerization starting with *n*-butyl methacrylate. The average molar mass of the sample is about 4,000 g/mol.

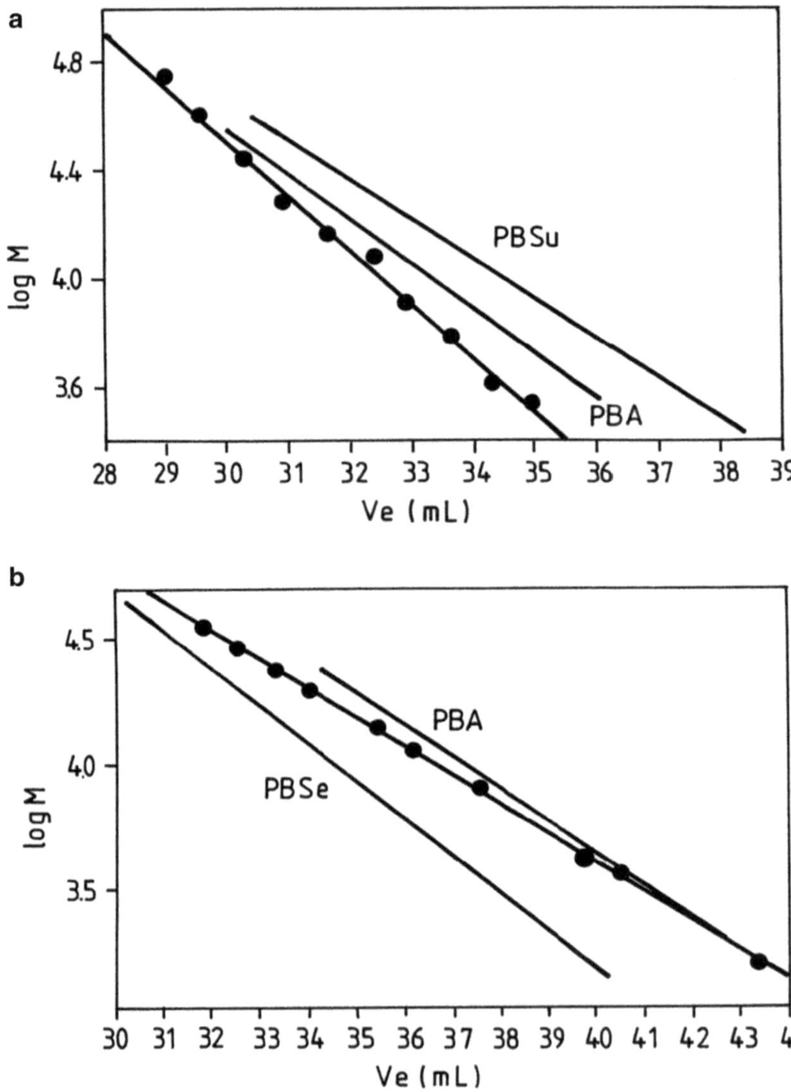

Fig. 7.21 Plots of log molar mass versus elution volume for PBSe, PBA, and PBA-Se (**a**) and for PBA, PBSu, and PBA-Su (**b**) (reprinted with permission from [131]. Copyright (1998) of American Chemical Society)

Equipment

- **Chromatographic System**. Modular HPLC system comprising a Waters model 510 pump, a Rheodyne six-port injection valve and a Waters column oven. For automatic fraction collection and deposition on the MALDI targets the LC-Transform Model 500 (Lab Connections, Marlborough, MA, U.S.A.) is used.

- **Columns**. Set of four PLgel columns 10^5 Å + Mixed-D + Mixed-E + 50 Å average pore diameter. Column size was 300 × 7.5 mm i.d.
- **Mobile Phase**. THF of HPLC grade.
- **Detector**. Waters differential refractometer R 410.
- **Sample Amount**. 100 μL of a 0.2 mg/mL solution in the mobile phase
- **MALDI-TOF System**. Kratos Kompact MALDI 4 with linear and reflectron detectors, acceleration voltage of positive 20 kV.
- **MALDI-TOF Sample Preparation**. 10 % of the SEC effluent were introduced into the MALDI interface via a capillary. The effluent was sprayed via a heated capillary nozzle continuously on a slowly moving Kratos MALDI target pre-coated with the matrix 1,8,9-trihydroxy anthracene (dithranol). The matrix was manually deposited on the MALDI target from a THF solution. For the enhancement of ion production, sodium trifluoroacetate was added to the matrix.

Preparatory Investigations

The experimental setup for the SEC-MALDI-TOF analysis is schematically presented in Fig. 7.22. As already explained, the SEC separation is carried out in the usual way with typical SEC flow rates and concentrations. 10% of the effluent is split off, directed into the interface and deposited on the MALDI targets. The SEC chromatogram of the sample recorded with the refractive index detector does not indicate any peculiarities or by-products, see Fig. 7.23.

MALDI-TOF Analysis

After the SEC separation the fractions are automatically deposited on the MALDI-TOF sample target. Prior to fraction deposition the target was pre-coated with the matrix dithranol and a small amount of sodium trifluoro acetate (NaTFA) to enhance the formation of $[M + Na]^+$ molecular ions. Since the fraction deposition is carried out through a heated capillary nozzle, a solid fraction/matrix film is obtained on the MALDI-TOF target. The spray-deposition procedure must be optimized very carefully to ensure that a uniform sample/matrix track is formed. If the nozzle temperature is too low, the aerosol stream is too wet, resulting in partial desolving and blowing away of the matrix. If the nozzle temperature is too high, the aerosol stream is too dry, resulting in a bilayer film where the matrix and the sample molecules are not mixed at all. In the present experiments where solely THF was used as the eluent, a spray nozzle temperature of 70 °C has been found to be the optimum. In the same manner, the distance between the spray nozzle and the MALDI-TOF target has to be optimized.

The MALDI-TOF target (Kratos) has a length of 70 mm and is scanned continuously with 3,500 laser pulses. Every 50 pulses are summarized to give a complete MALDI-TOF spectrum. With SEC as the pre-separation technique, low positions on the target correspond to high molar masses, while high positions are equivalent to low molar masses. Selected spectra from different positions of the polymer/matrix

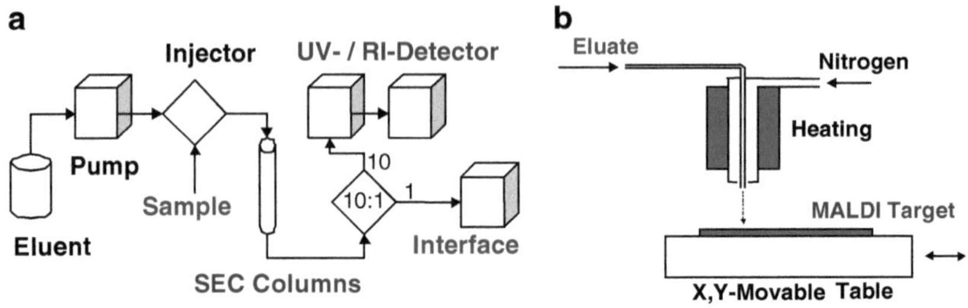

Fig. 7.22 Schematic representation of the experimental setup for the SEC to MALDI-TOF coupling (**a**) and deposition of the fractions onto the MALDI target (**b**) (reprinted from [113] with permission of Springer Science + Business Media)

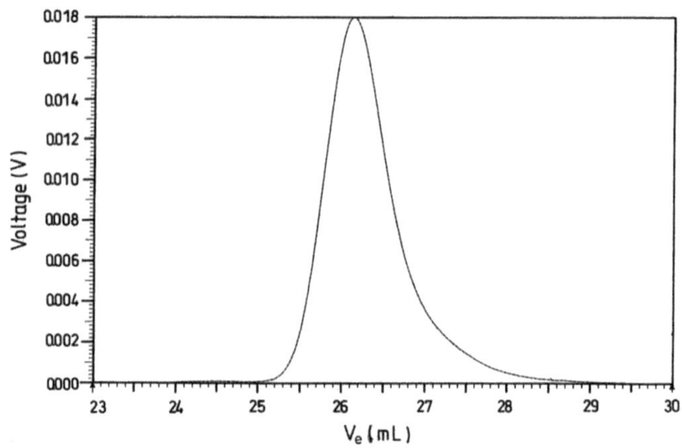

Fig. 7.23 SEC chromatogram of the PnBMA-PMMA block copolymer; stationary phase: PLgel 10^5 Å + Mixed-D + Mixed-E + 50 Å, mobile phase: THF, detector: RI (reprinted from [113] with permission of Springer Science + Business Media)

track of the sample are given in Fig. 7.24. The higher molar mass fractions in (a)-(c) are characteristic for copolymer structures exhibiting typical mass increments of 100 Da for the MMA repeat unit and 142 Da for the nBMA repeat unit. Even these narrow disperse fractions exhibit a multitude of different mass peaks (usually more than 100) indicating the high complexity of the fractions. Different from a-c, the lower molar mass fraction in (d) is very uniform with respect to composition. For this fraction, only peak-to-peak mass increments of 142 Da are observed which are typically for PnBMA. Accordingly, this fraction can be assigned to an unwanted by-product, namely PnBMA.

The total spectrum of the sample is obtained when the individual spectra of all pulses are summed, see Fig. 7.25a. This spectrum clearly indicates the presence of PnBMA having an average molar mass that is about 50% of the total molar mass of 4,200 g/mol. This result is in very good agreement with the expectations. The monomer ratio of the sample under investigation is about 1:1 resulting in average

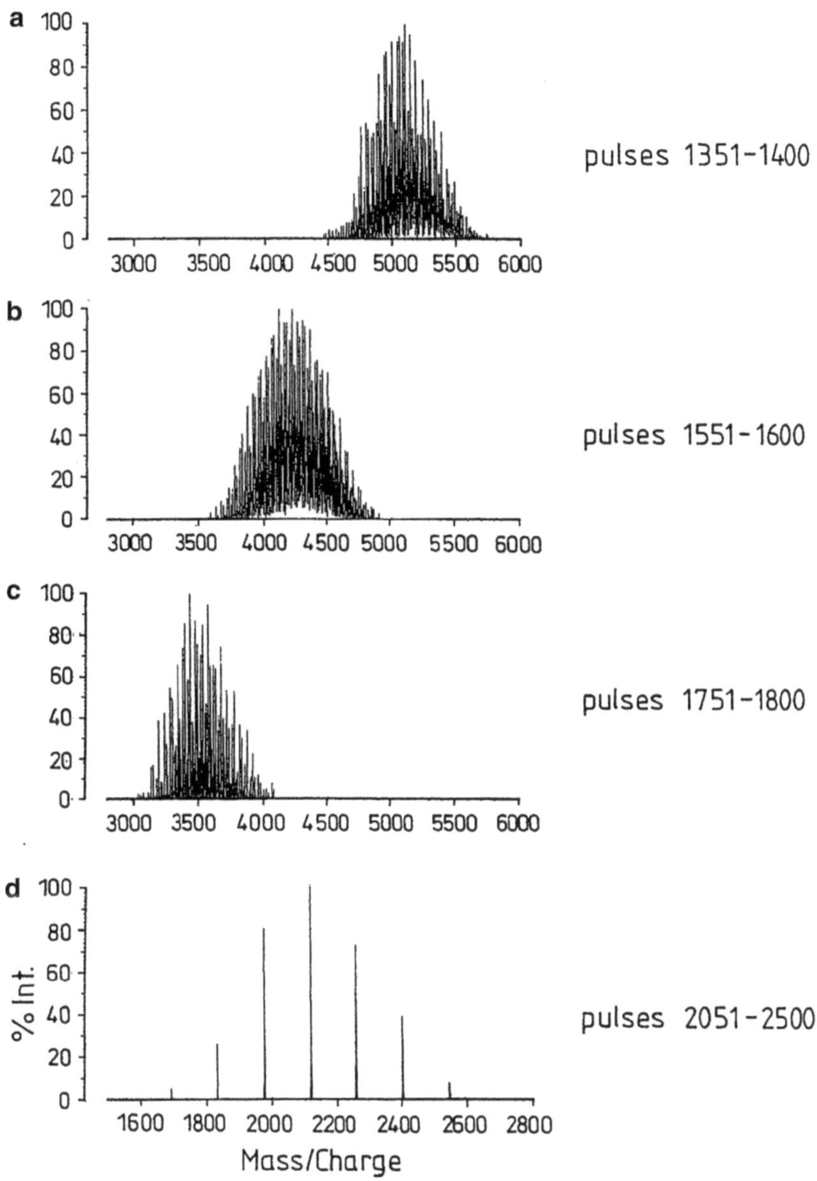

Fig. 7.24 MALDI-TOF spectra of fractions from SEC separation of PnBMA-b-PMMA, on-line SEC-MALDI-TOF analysis, matrix: dithranol, NaTFA (reprinted from [132] with permission of Elsevier)

block lengths of 2,100 g/mol for the PnBMA and the PMMA blocks. Since the polymerization was started with *n*-butyl methacrylate, the formation of a small amount of PnBMA has to be expected. The molar mass of this homopolymer must

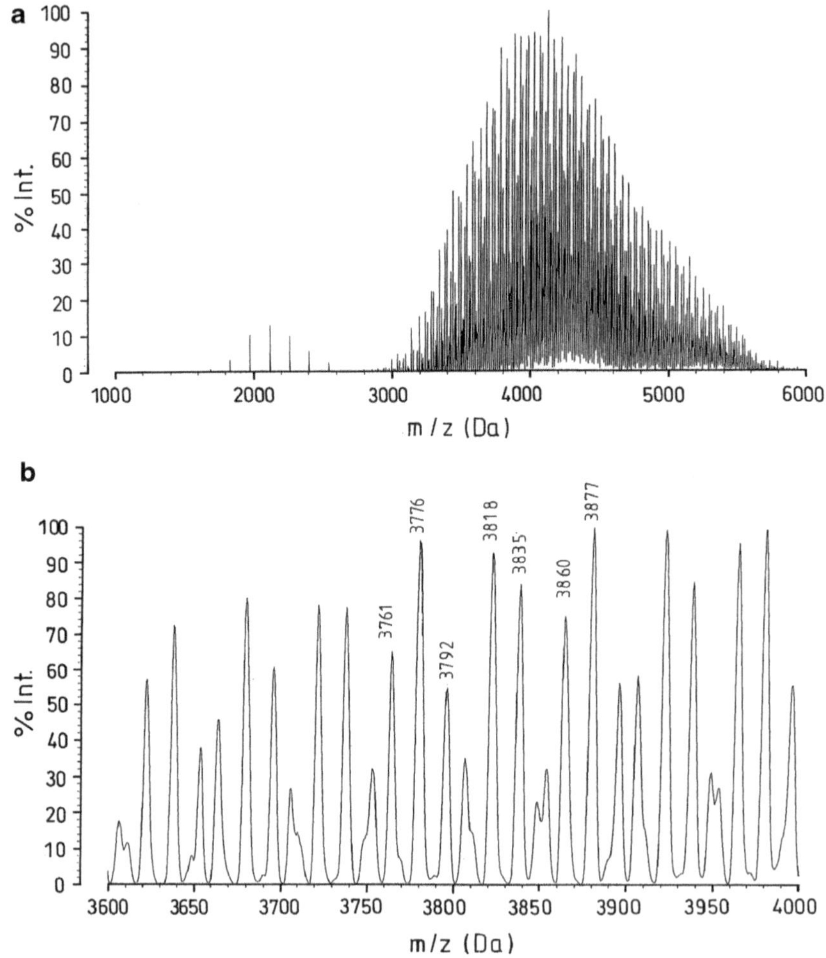

Fig. 7.25 Calculated MALDI-TOF spectrum from on-line SEC-MALDI-TOF analysis of PnBMA-b-PMMA (**a**) and zoomed part of the spectrum (**b**) (reprinted from [132] with permission of Elsevier)

be of the same magnitude as the PnBMA block in the copolymer. This is indeed the case as is shown by the MALDI-TOF measurement.

The chemical composition of the block copolymer can be studied in detail by analysing the different mass peaks, see zoomed part of the spectrum in Fig. 7.25b. Each peak in the spectrum can be assigned to one individual oligomer composition $(nBMA)_X(MMA)_Y$. For example, the mass peak at 3,761 Da corresponds to an oligomer with 15 nBMA and 16 MMA units. Its overall structure is H-$(nBMA)_{15}$-$(MMA)_{16}$-H. The mass peak at 3,835 Da is due to the oligomer H-$(nBMA)_{12}$-$(MMA)_{21}$-H. The calculated and observed molar masses for selected oligomers are summarized in Table 7.7.

Table 7.7 Calculated and observed molar masses for individual oligomers of sample PnBMA-b-PMMA

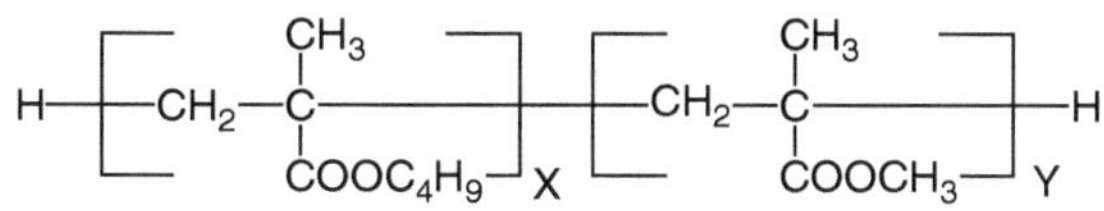

X	*Y*	[M + Na]$^+$ (calculated)	[M + Na]$^+$ (observed)
15	16	3,760	3,761
13	19	3,776	3,776
11	22	3,792	3,792
16	15	3,802	3,803
14	18	3,818	3,818
12	21	3,834	3,835
10	24	3,850	3,850
15	17	3,860	3,860
13	20	3,876	3,877

7.3 Coupling with ^{1}H-NMR

Nuclear magnetic resonance (NMR) spectroscopy is by far the most powerful spectroscopic technique for obtaining structural information about organic compounds in solution. Its particular strength lies in its ability to differentiate between most structural, conformational and optical isomers. NMR spectroscopy can provide all necessary information to unambiguously identify a completely unknown compound. The NMR detection technique is quantitative with individual areas in spectra being proportional to the number of contributing nuclei. The major disadvantage of NMR is the relatively low sensitivity in comparison to MS, the other is the fact that structure elucidation of mixtures of unknown compounds with overlapping NMR signals is difficult and may be nearly impossible in cases with overcrowding signals in a small chemical shift region of the NMR spectrum. Therefore, in many cases it is useful that a separation is performed prior to the use of NMR. For more efficient procedures, a direct coupling of separation with NMR detection would be the method of choice [133].

The direct coupling of liquid chromatography with proton NMR has been dealt with over a period of more than 25 years. Early experiments of coupled HPLC-^{1}H-NMR were conducted in stop-flow mode or with very low flow rates [134–136]. This was necessary to accumulate a sufficient number of spectra per sample volume in order to improve the signal-to-noise ratio. Other problems associated with the implementation of on-line HPLC-NMR have included the need for deuterated solvents. However, with the exception of deuterium oxide the use of deuterated eluents is too expensive for routine analysis. Therefore, proton containing solvents such as acetonitrile or methanol must be used. To get rid of the solvent signals in the spectra, the proton NMR signals of the solvents have to be suppressed.

Advances in HPLC-NMR over the last decade provide evidence that many of the major technical obstacles have been overcome [137, 138]. With the development of more powerful NMR spectrometers combined with new NMR techniques for solvent suppression it became much easier to obtain well resolved spectra in an on-flow mode. In particular, very efficient solvent suppression techniques significantly improved the spectra during the HPLC-NMR run [139, 140]. Even the direct coupling of supercritical fluid chromatography with ^{1}H-NMR [141–143] together with the monitoring of supercritical fluid extraction [144] as well as the coupling of capillary electrophoresis and ^{1}H-NMR [145–147] have been reported. An overview on the applications of online HPLC-^{1}H-NMR in organic chemistry was given by Albert [133].

The first steps of polymer analysis into coupled LC-^{1}H-NMR were performed by Hatada et al. [148]. They linked a size exclusion chromatograph to a 500 MHz proton NMR spectrometer and investigated isotactic PMMA. Using deuterated chloroform as the eluent and running the chromatography at a rather low flow rate of 0.2 mL/min they were able to accumulate well resolved proton spectra. From the intensities of the proton signals of the endgroups and the monomer units they determined the number-average molar mass across the elution curve. In further investigations they developed an absolute calibration method for direct determination of molar masses and molar mass distributions by online SEC-^{1}H-NMR [149]. Ute reported on the chemical composition analysis of EPDM copolymers as a function of molar mass, and the monitoring of stereocomplex formation for isotactic and syndiotactic PMMA [150].

Blends of isotactic and syndiotactic PMMA were also studied regarding the stereocomplexation in non-deuterated solvents [151]. Krämer et al. reported about SEC–NMR for the chemical composition analysis of poly(styrene-co-butyl acrylate) copolymers [152]. Further studies on coupled HPLC–NMR have shown the power of liquid adsorption chromatography (LAC) for the analysis of polymers regarding the chemical composition [153–158]. It was shown that polyethylene oxides can be analyzed with regard to functionality type distribution by identifying the different endgroups [153]. This investigation has been conducted under conditions which are common for HPLC separations, i.e. sufficiently high flow rate, moderate sample concentration, and on-flow detection. Using an octadecyl-modified silica gel as the stationary phase and an eluent of acetonitrile-deuterium oxide 50:50 (v/v) the sample was separated into different functionality fractions that were then analyzed by on-flow ^{1}H-NMR. It has been shown that LCCC coupled to NMR allows for the full assignment of the chemical structure and the degree of polymerization of all oligomer species in complex mixtures [155, 156]. The critical point of adsorption was also used for the analysis of tacticity of poly(ethyl methacrylate)s by HPLC–NMR [157]. In further studies on copolymers, gradient HPLC–NMR was used for the analysis of the CCD of random poly(styrene-co-ethyl acrylate) copolymers [158]. In a number of very recent publications it was shown that LC-NMR was used successfully for the analysis of polymer blends and block copolymers regarding chemical composition and microstructure [159, 160]. The most advanced development, however, was the development of a technology for on-flow high temperature SEC-NMR for the analysis of polyolefins pioneered by Hiller et al. [161].

The online coupling of SEC/HPLC and ^{1}H-NMR requires an HPLC instrument and an NMR spectrometer with a high magnetic field. Typically, a field strength of 400 MHz or higher is necessary to provide sufficient sensitivity. Instead of a conventional NMR probe, a continuous-flow probe is used that is connected to the HPLC system via a capillary. The schematic set-up for online LC-NMR and the design of a typical flow probe are presented in Fig. 7.26. The active volume of the flow probe is typically between 30 and 120 μL.

In an on-flow experiment ^{1}H-NMR data are collected over the entire chromatographic peak(s) and are stored as consecutive series of co-added scans. More details on the experimental part of LC-NMR are available from [162]. Most of the early experiments used deuterated mobile phases to avoid strong proton signals of the solvents that would overlap with the signals of the analyte protons. Chloroform-d was frequently used and proton signals due to impurities were simply eliminated by baseline substraction. A typical example for the analysis of a copolymer by SEC-NMR is shown in Fig. 7.27. In this application the molar mass dependence of the chemical composition has been investigated for a block copolymer PMMA-b-PnBMA. The cross sections at 3.59 ppm (OCH_3 of the MMA units) and 3.95 ppm (OCH_2 of the nBMA units) provided the changes in the content of MMA and nBMA, respectively, as a function of molar mass. It can be seen clearly that the higher molar mass part contains more nBMA units [163].

In daily routine use of a standard size HPLC instrument it will be too expensive to work with deuterated solvents. In this case protonated solvents are regularly used which, however, produce very strong proton signals. Considering the fact that the analyte concentration in the eluate is very low, very effective solvent supression techniques must be used.

With the development of more powerful NMR spectrometers combined with new NMR techniques for solvent suppression, nowadays it has become much easier to obtain well-resolved spectra in the on-flow mode. In particular, the solvent suppression technique developed by Smallcombe et al. [139] significantly improves the spectra during the LC-NMR run. This experiment, which is based on the WET solvent suppression technique of Ogg et al. [140] combines shaped rf pulses, pulsed-field gradients (PFG), and selective ^{13}C decoupling. It allows acquisition of high-quality spectra at on-flow conditions even with steep HPLC gradients.

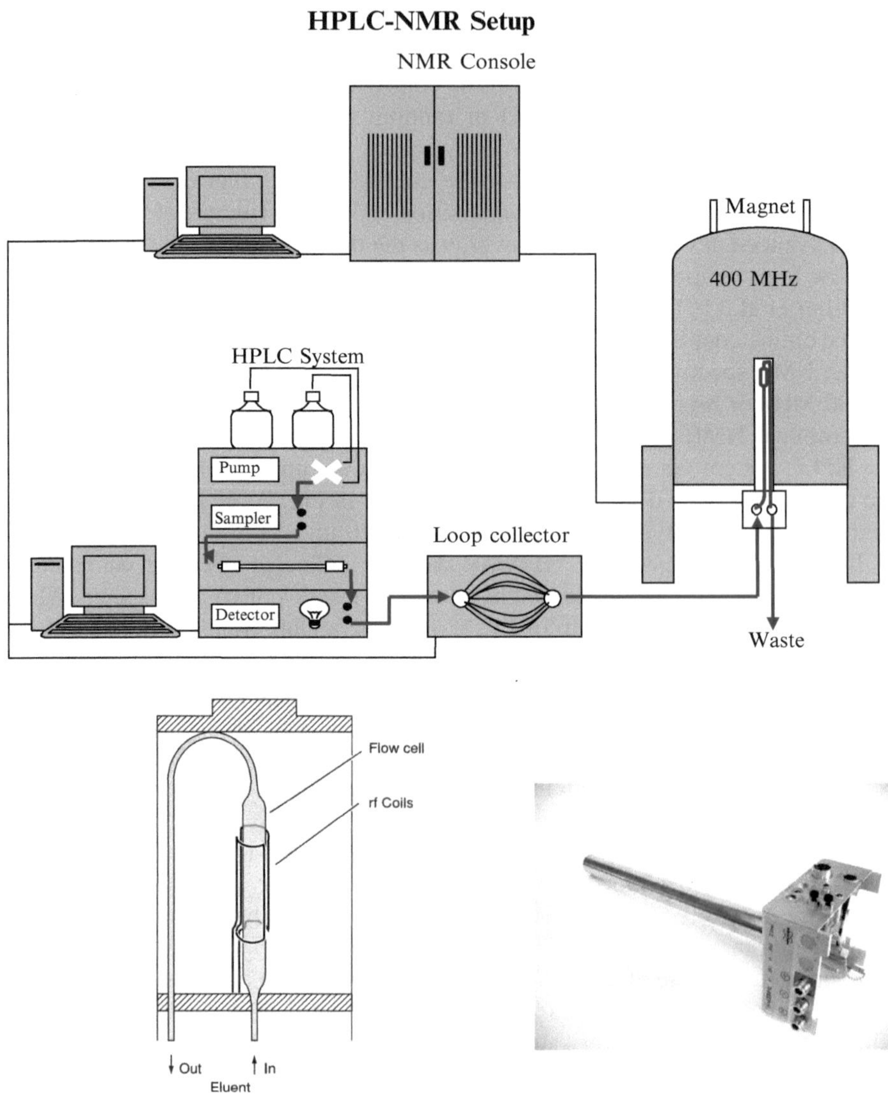

Fig. 7.26 Schematic set-up of LC-NMR for on-line experiments and the design of a flow probe

In a typical experiment the following pulse sequence is used, see Fig. 7.28: WET sequence consisting of four 20 ms selective SEDUCE pulses (98.2, 80.0, 75.0, and 152.2° for the $B1$-insensitive WET), four gradient pulses (duration 1 ms) with amplitudes of 24, 12, 6, and 3 G/cm, respectively, followed by an additional 3 ms delay and a composite 90° read pulse. Carbon decoupling is applied during the selective proton pulses using Waltz-16 decoupling [139].

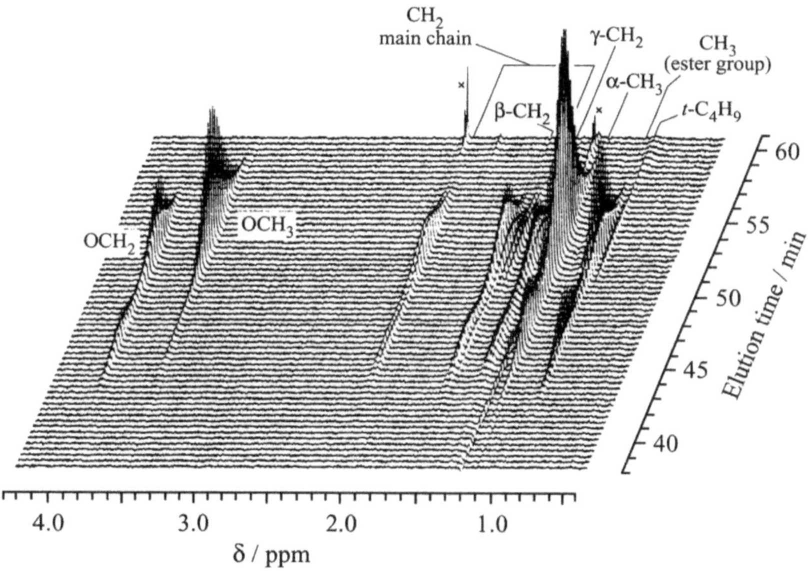

Fig. 7.27 500 MHz online SEC-NMR analysis of PMMA-b-PnBMA block copolymer, impurities are labelled 'x', mobile phase: chloroform-d, flow rate 0.2 mL/min, amount of sample 1 mg, pulse repetition time 1.0 s (reprinted from [162] with permission of Springer Science + Business Media)

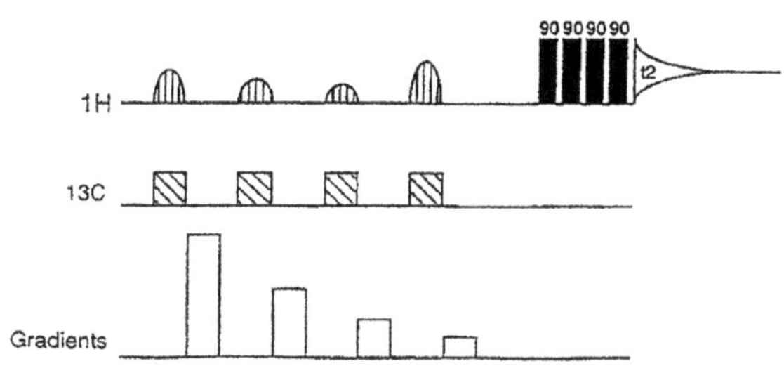

Fig. 7.28 Typical pulse sequence used for HPLC-NMR experiments

In the following sections representative examples for the analysis of complex polymers will be discussed. In those examples LC-NMR is not presented as 'just another' hyphenated technique but it will be shown that LC-NMR is essentially the only choice to obtain the required structural information.

7.3.1 Analysis of Octylphenyl-Terminated PEO [153]

Aim

The analysis of fatty alcohol ethoxylates (FAEs) has been addressed before, see e.g. Sect. 6.7.2. One problem that cannot be addressed by selective fractionation alone is the analysis of the structure of the fatty alcohol endgroups. MALDI-TOF MS provides an 'endgroup mass', however, it is not always possible to correlate this mass with a specific structure. When the endgroups have different topologies but similar masses then standard MALDI-TOF experiments do not provide any useful information. In such cases NMR is the method of choice. The present example is a technical octylphenyl-PEO that by MALDI-TOF has been found to contain various oligomer series. The assignment of these series to specific endgroups, however, was not possible.

Materials

- The FAE under investigation was a technical product of BASF AG, Ludwigshafen, Germany

Equipment

- **Chromatographic System**. Varian modular HPLC system, comprising a Varian 9012 pump and a Valco injection valve. The flow rate was 1 mL/min.
- **Columns**. Varian Res Elut C18 90A, 150 × 4.6 mm i.d.
- **Mobile Phase**. Acetonitrile-deuterium oxide 50:50 % by volume.
- **Detector**. Varian 9050 UV detector at a wavelength of 260 nm
- **Sample Amount**. 100 μL of a 15 mg/mL solution in acetonitrile
- **NMR System**. Varian 500 MHz NMR spectrometer UNITY*plus*. The HPLC-NMR probe containing a 60 μL flow cell was an indirect detection probe with PFG. All measurements were carried out at room temperature. The signal-to-noise ratio of the LC-NMR probe is given by the following specifications: 22:1 for the anomeric proton of sucrose (41 μg/60 μL) at 500 MHz. In the case of on-line measurements, this corresponds to a detection limit of 10 μg per compound in the flow cell or 300 ng in the case of stop-flow measurements. The line width is measured using a solution of 1 % of chloroform in acetone-d and specified as follows: 10 and 20 Hz at 0.55 and 0.11 %, respectively, of the total peak height.

Preparatory Investigations

A separation with respect to the endgroups of the FAE can be obtained by adsorption chromatography or LCCC. The functionality type separation of the present sample is shown in Fig. 7.29. Using an octadecyl-modified silica gel as the stationary phase, the sample is separated into functionality fractions at an eluent composition of ACN-D_2O of 50:50% by volume. The first elution peaks appear between 1.2 and 1.7 min in the region of the injection peak. Further elution peaks are obtained at retention times of 5.14 and 7.80 min. The major fraction of the sample elutes between 14 and 25 min, and for this fraction additionally a partial oligomer separation is obtained. The identification of the fractions by UV detection

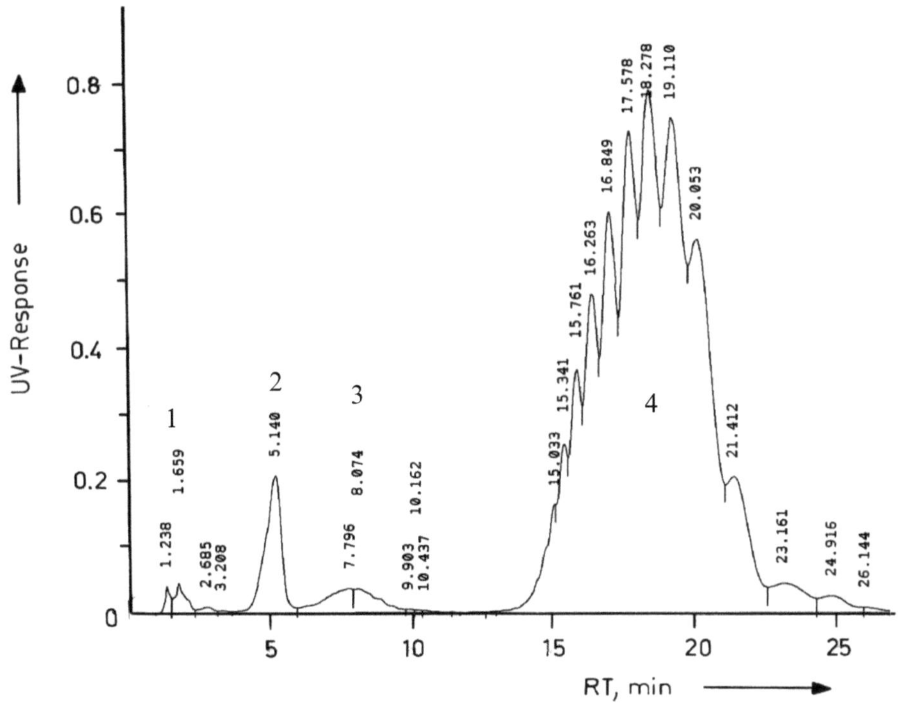

Fig. 7.29 HPLC chromatogram of a technical PEO (reprinted with permission from [153]. Copyright (1996) of American Chemical Society)

is not possible but the fractions at 5.14 and 7.80 min and the major fraction exhibit a significant UV response at 260 nm. This is a first indication for the presence of aromatic moieties. Since the PEO polymer chain is aliphatic, the endgroups are assumed to be aromatic.

For structural identification of the fractions, HPLC was coupled on-line to the NMR spectrometer. The injection of the sample into the HPLC system was automatically initiated by the NMR console via a trigger pulse when starting acquiring NMR data. The pulse sequence given in Fig. 7.28 was used. With this sequence and applying shifted laminar pulses both solvent resonances (ACN at 2.4 ppm and water at 4.4 ppm) could be suppressed simultaneously using only one rf channel and maintaining phase coherence with the transmitter.

NMR Analysis

After leaving the UV detector, the eluate is directly introduced into the NMR cell via capillary tubing. Since a series of free induction decays (FID) was collected, a Fourier transformation via the acquisition times and a combination of the spectra

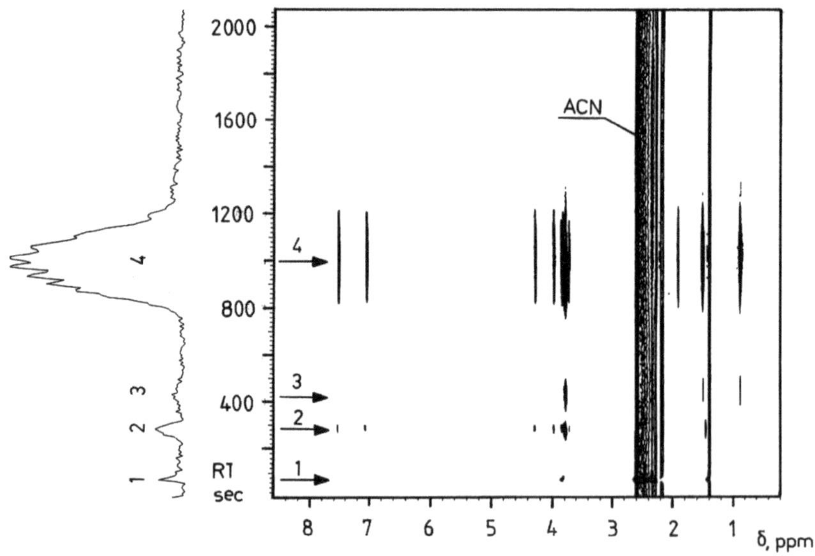

Fig. 7.30 Contour diagram of ^{1}H chemical shift versus retention time and chemigram of the on-flow HPLC-NMR analysis of a technical PEO (reprinted with permission from [153]. Copyright (1996) of American Chemical Society)

could be carried out with the two-dimensional NMR software. As a result of the online HPLC-NMR experiment, a contour plot of ^{1}H chemical shift vs retention time was generated, see Fig. 7.30.

In the present case, due to efficient solvent suppression, the obtainable structural information relates to the entire chemical shift region. Residual signals of the eluent are obtained at 2.1–2.6 and 1.4 ppm due to ACN and its impurities. From the contour plot, four different elution peaks can be identified. The peaks exhibit NMR signals at about 3.8 ppm, which can be attributed to the protons of $-CH_2O-$ groups. Accordingly, it can be assumed that all four elution peaks are due to ethylene oxide oligomers (repeat unit $-CH_2CH_2O-$). The remarkable feature of the present investigation is that even the low-concentration components in peaks 1–3 can clearly be identified in the contour plot. Different from HPLC of low molar mass organic compounds where usually very sharp elution peaks are obtained, in polymer chemistry broad elution peaks appear due to the effect of the MMD. As a result, peak intensity at each point of the chromatogram is rather low, making it difficult to detect these peaks. However, as can be seen from Fig. 7.30, the present NMR detection method is sufficiently sensitive to detect even very small peaks.

Detailed structural information could be obtained from the individual NMR spectra of the fractions at the peak maximum, see Fig. 7.31.

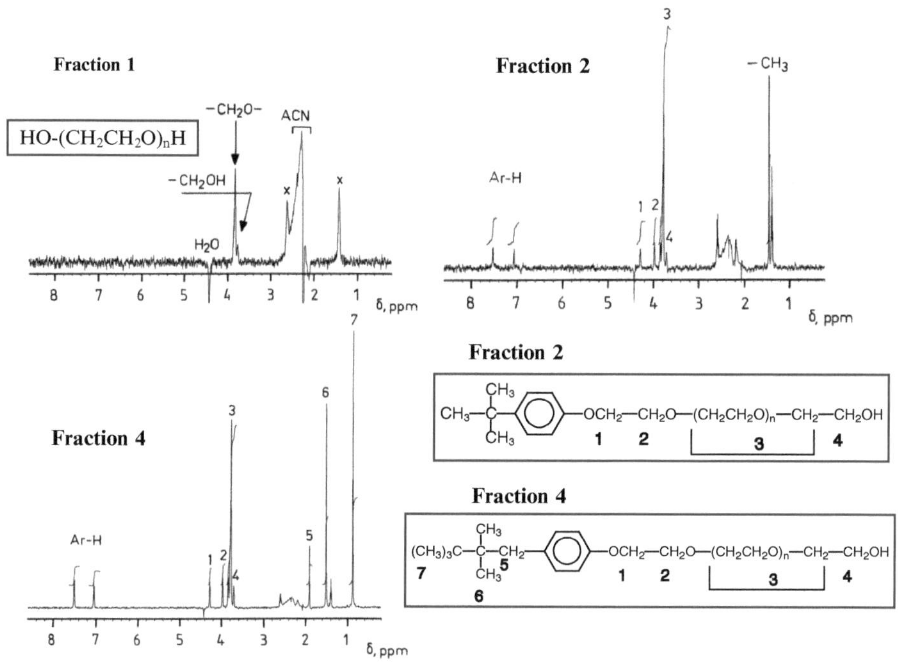

Fig. 7.31 Individual fraction spectra taken from Fig. 7.30 (reprinted with permission from [153]. Copyright (1996) of American Chemical Society)

The first peak was identified as being polyethylene glycol while the other fractions were alkylphenoxy polyethylene oxides. From the intensities of the endgroups and the ethylene oxide repeat units the average degree of polymerization for each fraction was calculated (fraction 2: $n = 6$, fraction 4: $n = 5.5$). Based on the total intensity distribution, a calculated chromatogram (or chemigram) was generated from the NMR contour diagram. On comparing the real chromatogram (Fig. 7.29) with the chemigram (Fig. 7.30) an excellent agreement was obtained even recalling the oligomer separation pattern of the major fraction.

7.3.2 Analysis of Microstructure of Polystyrene [154]

Aim

NMR is of exceptional value in the study of the stereochemistry of polymers. Depending on the NMR field strength, stereochemically different sequences of five or more monomer units can be distinguished. However, in PS due to the large number of possible conformations and long range coupling interactions very complex ^{1}H as well as ^{13}C spectra are obtained, making it very difficult to interpret them completely. The ortho-aromatic proton resonances of PS contain stereosequence information but these are not adequately understood at present. The methine proton resonances occur in two general areas, and it has been shown in studies on partially epimerized isotactic PS that the lower field methine proton resonance area is due to mm stereosequences [164]. Fine structure due to pentad or higher stereosequences has been observed in the methine proton resonances of partially deuterated PS [165, 166]. The methylene and quaternary aromatic carbon resonances are sensitive to hexad and heptad stereosequence effects [167, 168].

A different approach was reported by Sato and co-workers in a number of papers [169–171]. Using SEC, they separated anionically polymerized PS into single oligomers and then subjected the oligomers to multiple HPLC separations to obtain individual steric isomers. In some cases more than 50 cycles of separation were carried out, making the procedure an extensively time- and labour consuming technique. The isolated isomers were then subjected to ^{1}H- and ^{13}C-NMR analysis.

In the present application, oligomeric PS shall be separated regarding the oligomer distribution by LAC and the stereostructure of the oligomers analysed by on-flow ^{1}H-NMR. Information shall be obtained on the stereoregularity of each oligomer.

Materials

- The oligostyrenes were SEC calibration standards prepared by anionic polymerization using sec-butyllithium as the initiator. They were provided by PSS GmbH Mainz, Germany.

Equipment

- **Chromatographic System**. Varian modular HPLC system, comprising a Varian 9012 pump and a Valco injection valve. The flow rate was changed according to the following program (time/flow rate): 0/1.00, 8/1.00 linear to 20/2.00, 25/2.00 linear to 30/1.00 min/(mL/min)
- **Columns**. Macherey & Nagel Nucleosil RP-18, 5 μm average particle size, 100 Å average pore size, 250 × 4 mm i.d.
- **Mobile Phase**. Acetonitrile HPLC grade.
- **Detector**. Varian 9050 UV detector at a wavelength of 260 nm
- **Sample Amount**. 100 μL of a 15 mg/mL solution in acetonitrile
- **NMR System**. Varian 500 MHz NMR spectrometer UNITY*plus*. The HPLC-NMR probe containing a 60 μL flow cell was an indirect detection probe with PFG. All measurements were carried out at room temperature. The signal-to-noise ratio of the LC-NMR probe is given by the following specifications: 22:1 for the anomeric proton of sucrose (41 μg/60 μL) at 500 MHz. In the case of online measurements, this corresponds to a detection limit of 10 μg per compound in the flow cell or 300 ng in the case of stop-flow measurements. The line

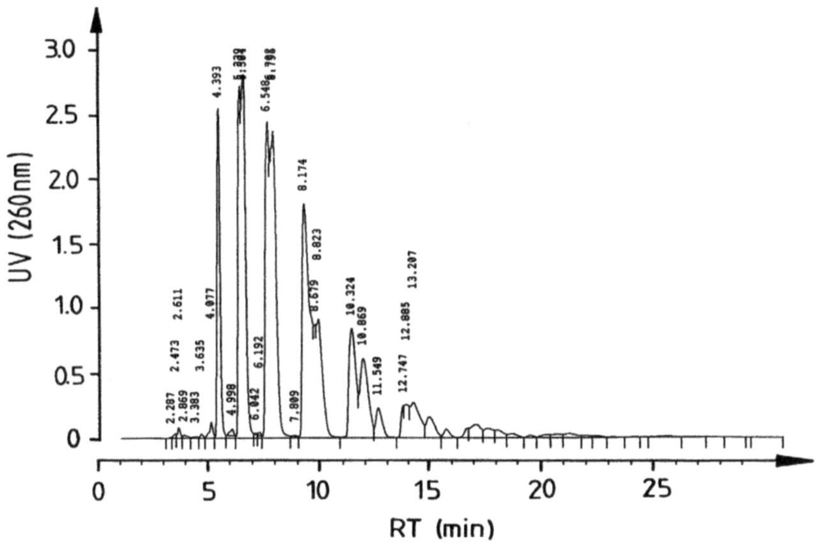

Fig. 7.32 HPLC chromatogram of an oligostyrene, molar mass 530 g/mol (reprinted from [154] with permission of Elsevier)

width is measured using a solution of 1% of chloroform in acetone-d and specified as follows: 10 and 20 Hz at 0.55 and 0.11%, respectively, of the total peak height.

Preparatory Investigations

A rather simple separation scheme for oligostyrenes was used that was adapted from a procedure published by Eisenbeiss et al. [172]. Isocratic elution using ACN as the mobile phase was used. The separation was optimised by applying a flow rate gradient. The chromatogram of one of the oligostyrenes is shown in Fig. 7.32. The first oligomer peak was identified as being the dimer ($n = 2$), the next peak was identified as the trimer ($n = 3$) and, accordingly, the following peaks could be assigned to the tetramer, pentamer etc. The dimer peak appeared uniform, whereas for the following oligomers a splitting of the peaks was obtained. For $n = 3$ and $n = 4$ a splitting into two peaks was observed. For $n = 5$ and further a splitting into three or more peaks occurred, which could be attributed to the presence of different tactic isomers.

For structural identification of the isomerism of the oligomers, the HPLC separation is coupled online to the NMR spectrometer. The major problem in studying styrene oligomers by HPLC-NMR is that for the HPLC separation only organic solvents are used. NMR prefers deuterated solvents which are very expensive and it is shown here that high quality HPLC-NMR experiments can also be performed without adding any deuterated component. In other words, no deuterium

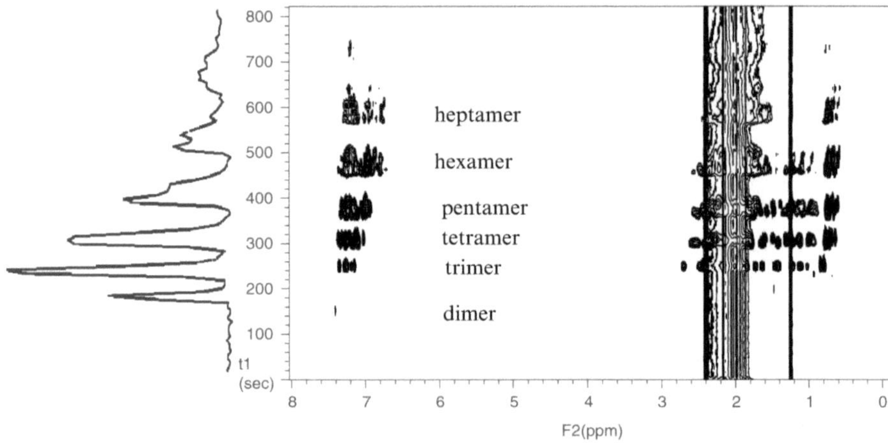

Fig. 7.33 Contour diagram of chemical shift versus retention time and chemigram of the on-flow HPLC-NMR analysis of oligostyrene (reprinted from [154] with permission of Elsevier)

lock is used for the on-flow measurements. All measurements were conducted in normal HPLC grade ACN. These conditions require high stability of the NMR instrument and an efficient solvent suppression technique.

NMR Analysis

After leaving the UV detector, the eluate is directly introduced into the NMR cell via capillary tubing. Since a series of free induction decays (FID) was collected, a Fourier transformation via the acquisition times and a combination of the spectra could be carried out with the two-dimensional NMR software. The obtainable structural information relates to the entire chemical shift region, however residual signals of the eluent are obtained at 1.8–2.4 ppm and 1.3 ppm due to acetonitrile and its impurities, see Fig. 7.33.

The contour plot clearly reveals two signal regions, which can be used for analysis. These are the region of the methyl protons of the sec-butyl endgroup at 0.6–0.8 ppm and the aromatic proton region of the styrene units at 6.5–8.0 ppm. Most informative is the analysis of the methyl protons. The contour plot for the aromatic region shows that systematic changes of the signals occur with changes in the degree of polymerization. For the dimer the aromatic proton signals are located in the narrow range of 7.1–7.4 ppm whereas for the pentamer the signals are distributed over the range 6.6–7.1 ppm. The projection of the aromatic signals on the retention time axis results in a chromatogram presentation, which is fairly similar to the initial chromatogram shown in Fig. 7.32.

For the generation of the contour plot, every 8 s a complete spectrum is produced by co-adding eight scans. Accordingly, for the structural analysis 128 spectra are available over the entire retention time range. For the analysis of a separated

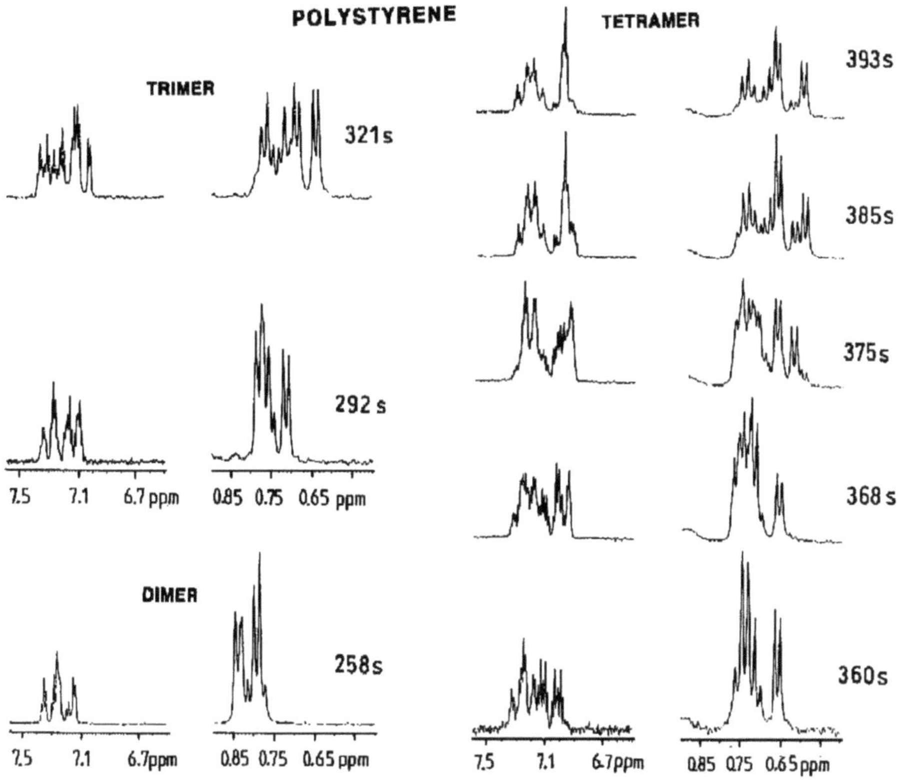

Fig. 7.34 Aromatic and methyl proton regions of different oligomer peaks of PS 530, obtained from coupled HPLC-NMR (reprinted from [154] with permission of Elsevier)

oligomer, a minimum of four spectra can be used. These spectra bear selective information on the tacticity, even without completely separating the tactic isomers chromatographically.

The spectral characteristics of the aromatic and methyl proton regions of the different oligomers are summarized in Fig. 7.34. For the dimer, four similar spectra were obtained, indicating that the dimer is present only in one isomeric form. However, assuming one isomeric form, for the methyl protons the appearance of a doublet and a triplet must be expected, since the chemical structure of the endgroup is CH_3–CH_2–$CH(CH_3)$–. Instead, two doublets and two triplets are obtained, which clearly show that both isomeric forms are present. The resolving power of liquid chromatography is in this case not sufficient for a separation. The trimer gave four spectra, two of them revealing different isomeric structures. The other two spectra turned out to result from overlapping of the isomer spectra. Again, each of these spectra show two doublets and two triplets, thus relating to two different isomeric structures. Accordingly, for the trimer the expected four different

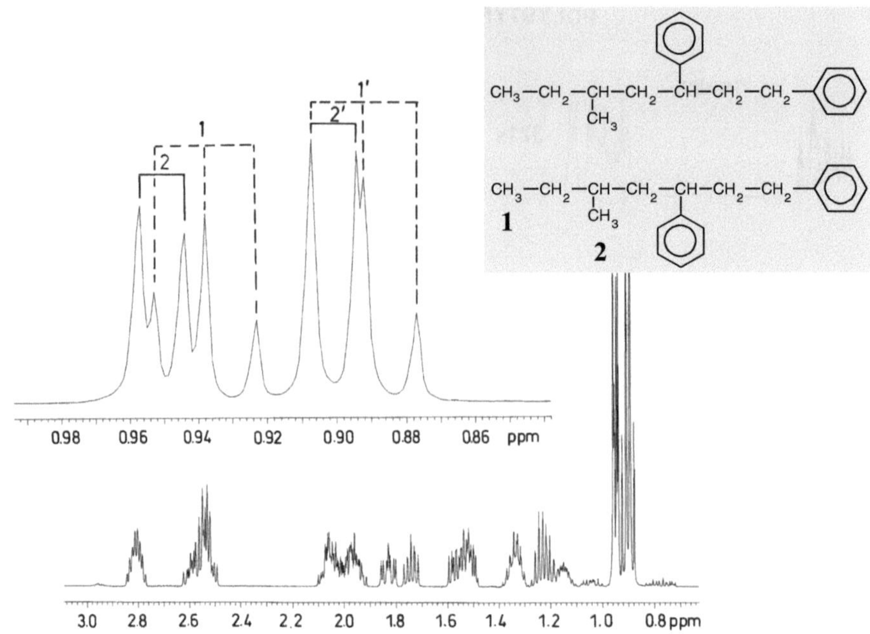

Fig. 7.35 Aliphatic region of the ^{1}H-NMR spectrum of the styrene dimer (reprinted from [154] with permission of Elsevier)

isomeric forms are present. For the tetramer five different spectra are obtained, and here it is very difficult to judge which of the spectra are overlaps. Nevertheless, the different spectra appear due to the iso-, syndio- and atactic configurations. The pentamer gave 11 different spectra, but because of their complexity an identification of isomers seems to be impossible.

Since the spectral information of the HPLC-NMR experiment is quite complex and the interpretation of the spectra is not straightforward, additional offline experiments are conducted on individual oligomers which were obtained by distillation of a PS oligomer and isolation of the monomer, dimer and trimer. The offline ^{1}H-NMR spectrum of the dimer is shown in Fig. 7.35. For the methylene and methine protons complex multiplets are obtained which shall not be analysed in detail. An interesting pattern is obtained for the protons of the methyl groups 1 and 2. In this region nine signals appear which are identical to the signals of the dimer peak in Fig. 7.34 obtained by the online HPLC-NMR experiment. These signals can be grouped into two doublets and two triplets as is shown in the zoomed part of Fig. 7.35. The appearance of two doublets for methyl group 2 and two triplets for methyl group 1 is clear evidence for the presence of two isomeric structures.

In a similar way, the styrene trimer is analysed. In this case four different isomeric structures can be identified.

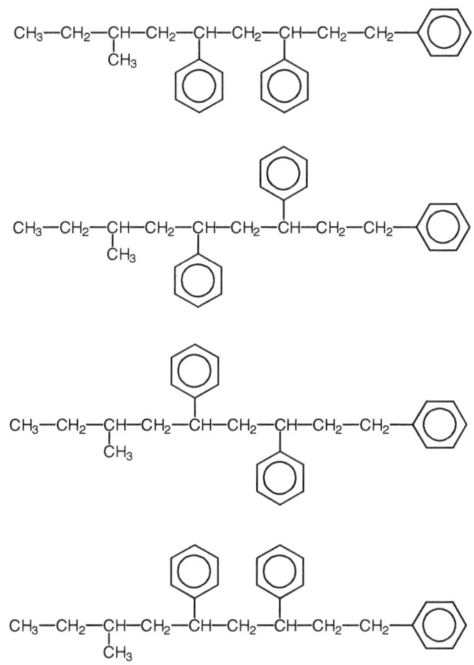

For the trimer it can be concluded that by HPLC-NMR one peak is identified in the chromatographic separation giving two distinctively different ^{1}H-NMR spectra that can be assigned to four different isomers. This result shows convincingly that a coupled HPLC-NMR experiment is by far superior over separate HPLC and NMR experiments.

7.3.3 Microstructure Analysis of PI-b-PMMA Block Copolymers [173]

Aim

The different methods of interaction chromatography have been shown to be powerful tools for the separation and analysis of block copolymers, see e.g. Chap. 6. In particular LCCC has been used numerous times for the analysis of the different blocks in di-and triblock copolymers. Operating at LCCC conditions of one block, the other block can be analysed regarding its chain length distribution. Information on chemical composition can then be obtained by LC-FTIR or LC-MALDI-TOF. These coupled methods cannot, however, provide information about the microstructure (tacticity) of the building blocks of segmented copolymers.

In the present application the power of LCCC-NMR for the analysis of block copolymers shall be demonstrated. As an example, PI-b-PMMA diblock copolymers shall be investigated. These diblock copolymers consist of a polyisoprene (PI) and a PMMA block connected by a covalent linkage. The samples are expected to contain the respective homopolymers as by-products. In addition, both blocks exhibit distinct microstructural features that shall be elucidated.

Materials

- The PI-b-PMMA block copolymers are laboratory products. Their synthesis and purification is described in detail in [173]. The chemical structure of the block copolymers is as follows:

1,2 PI-b-PMMA

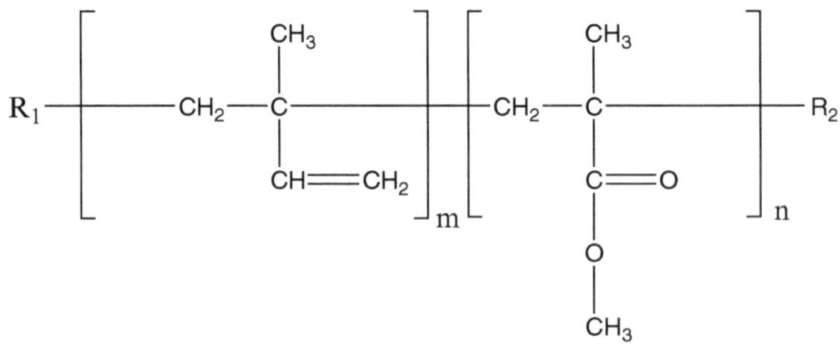

1,4 PI-b-PMMA

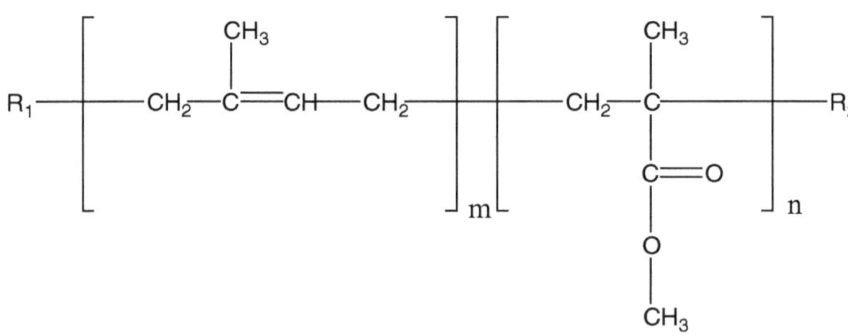

3,4 PI-b-PMMA

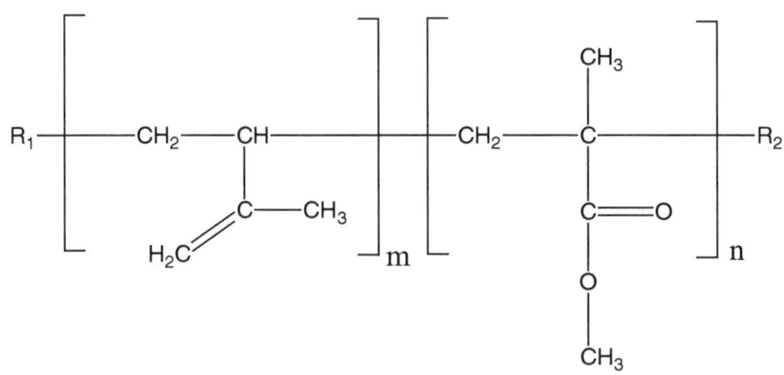

Table 7.8 PI-b-PMMA copolymers of different block lengths estimated by SEC and the average chemical composition of PI and PMMA determined by ^{1}H NMR

	PI-b-PMMA		
Sample	M_w (kg/mol)	M_n (kg/mol)	PI/PMMA (mol%)
1	18.8	17.7	91.4/8.6
2	19.3	18.3	50.7/49.3
3	90.2	85.7	15.9/84.1
4	102.6	99.8	25.4/74.6
5	52.0	45.0	55.6/44.4

The molar masses and chemical compositions of the block copolymers are summarized in Table 7.8.

Equipment

- **Chromatographic System**. Agilent 1100 HPLC system. The flow rate is 0.5 mL/min
- **Columns**. LCCC PMMA: Nucleosil Si 300–5 + Si 1000–7 (Macherey-Nagel, Dueren, Germany) with column sizes of 200 × 4.6 mm i.d.; LCCC PI: Nucleosil C_{18} 300–5 + C_{18} 1000-7 (Macherey-Nagel, Dueren, Germany) with column sizes of 250 × 4 mm i.d.
- **Mobile Phase**. Ethyl acetate (LCCC PMMA), dioxane (LCCC PI)
- **Detector**. PL ELS detector 1000, Agilent UV
- **Sample Amount**. 50 μL of a solution of 10 mg/mL (homopolymers) and 30 mg/mL (block copolymers) in the mobile phase
- **NMR System**. The HPLC-NMR experiments were performed with an AVANCE-400 NMR spectrometer (Bruker BioSpin GmbH) attached to the HPLC system via a loop collector interface. The NMR experiments were carried out with an inverse triple resonance flow probe equipped with a pulsed field gradient coil. The flow cell has an active volume of 60 μL. The ^{1}H 90° pulse was 4.7 μs. On-flow HPLC-NMR experiments were carried out by using protonated HPLC solvents. WET solvent suppression was applied to the HPLC solvents. Eight scans per FID with 16 kB data and 1.1 s repetition delay were acquired. The series of FIDs were Fourier transformed via one time domain and plotted as 2D contour plots of retention times versus chemical shifts.

Preparatory Investigations

One of the major challenges in the analysis of copolymers by LC-NMR is the complex composition of the mobile phase that consists of at least two different solvents (giving different proton signals in the NMR spectrum) with varying compositions. A major step forward is the fractionation of copolymers by HPLC using a single solvent as mobile phase. Suitable conditions for fractionation are then adjusted by the column temperature which does not affect the NMR detection.

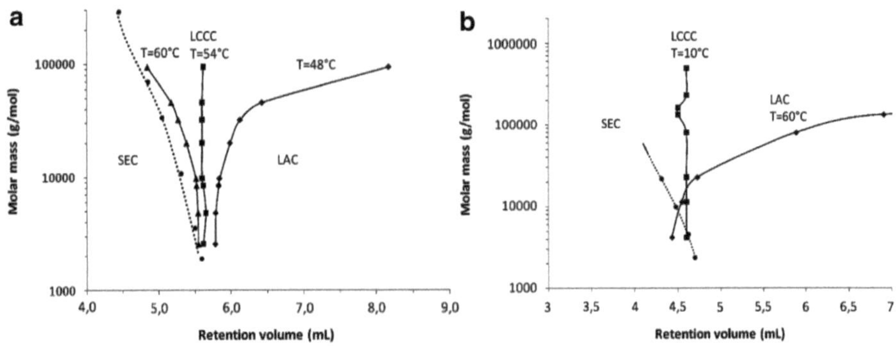

Fig. 7.36 (**a**) Critical diagram of molar mass versus retention volume for PI, detector: ELSD, column temperature: (*filled triangle*) $T = 60$ °C, (*filled square*) $T = 54$ °C, (*filled diamond*) $T = 48$ °C; (*filled circle*) and *dotted line* = PMMA molar mass calibration with LCCC-NMR at critical conditions of PI, (**b**) critical diagram for PMMA, detector: ELSD; column temperature: (*filled square*) $T = 10$ °C and (*filled diamond*) $T = 60$ °C; (*filled circle*) and dotted line = PI molar mass calibration with LCCC-NMR at critical conditions of PMMA (reprinted with permission from [173]. Copyright (2010) of American Chemical Society)

Figure 7.36 shows the adjustment of the critical conditions for PI and PMMA using variations in the column temperature. Critical conditions of PI are established using 1,4-dioxane at a column temperature of 54 °C. Using ethyl acetate as the mobile phase on silica gel the critical conditions of PMMA are established at a column temperature of 10 °C. At the critical conditions of PI, PMMA elutes in SEC mode whereas at the critical conditions of PMMA, PI elutes in SEC mode as is shown in Fig. 7.36.

When selecting specific solvents as mobile phases, the requirements of the chromatographic and NMR experimental conditions have to be considered. Regarding NMR it is important to ensure the detection of the relevant polymer signals. Figure 7.37 shows ^{1}H-NMR spectra of sample 1 in ethyl acetate, dioxane and chloroform-d. Dioxane and ethyl acetate require solvent suppression. Only one resonance of dioxane and three resonances of ethyl acetate are suppressed. In the case of dioxane the olefinic PI and the isotactic (mm), atactic (mr) and syndiotactic (rr) triads of the α-CH_3 group of PMMA can be detected. The OCH_3 group of PMMA overlaps with dioxane. In the case of ethyl acetate the signals of the olefinic PI, OCH_3 and the atactic (mr) and syndiotactic (rr) triads of the α-CH_3 group are visible.

It is evident from Fig. 7.38 that when using the present chromatographic conditions blends of PI and PMMA can be well separated by LCCC-NMR independently of the chosen critical conditions. Figure 7.38a shows an on-flow experiment at critical conditions of PMMA where PI is eluting in SEC mode. Figure 7.38b illustrates the LCCC-NMR experiment at critical conditions of PI and the SEC mode of PMMA. Both diagrams show a clear separation of the two components.

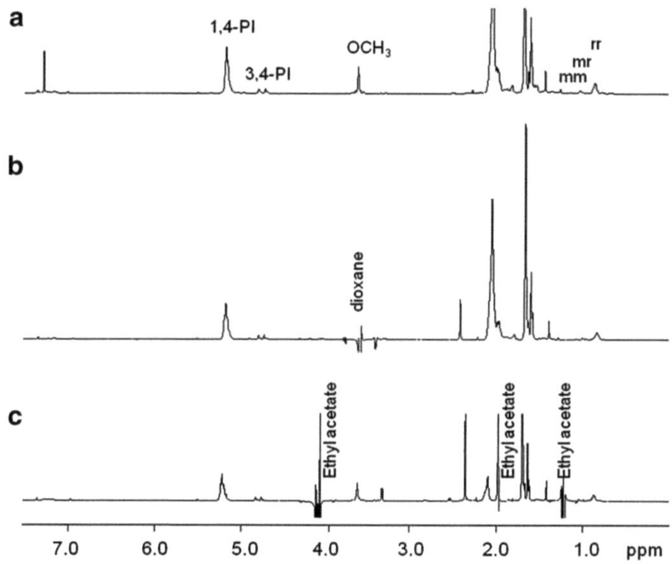

Fig. 7.37 ^{1}H-NMR spectra of PI-b-PMMA (sample 1); (**a**) in $CDCl_3$, (**b**) in dioxane and (**c**) in ethyl acetate (solvent signals of dioxane and ethyl acetate were suppressed with WET) (reprinted with permission from [173]. Copyright (2010) of American Chemical Society)

The investigation of the block copolymers by LCCC-NMR allows the separation of block copolymers and the corresponding homopolymers. Critical conditions can be chosen such as that one polymer block experiences critical conditions while the other elutes in SEC mode. Figure 7.39 shows the on-flow LCCC-NMR experiments of the block copolymer (sample 5) at both critical conditions. The olefinic regions of the 1,2-, 1,4- and 3,4-PI sequences as well as the tacticity of the α-CH_3 group of PMMA are well resolved in both cases. Whereas measurements at critical conditions of PI only need the solvent suppression of one signal (mobile phase is dioxane), critical conditions of PMMA require solvent suppression of three frequencies (mobile phase is ethyl acetate). Dioxane overlaps with the OCH_3 signal of PMMA, however, the α-CH_3 group is visible. The critical conditions of PI are very useful for separating PI homopolymer from the block copolymer. The peak maxima of the copolymer and homopolymer regions are indicated in Fig. 7.39b. Using ethyl acetate as the mobile phase at critical conditions of PMMA, full information of the OCH_3 group and the olefinic region is obtained, but partial overlapping of a 1,2-isoprene signal with the α-CH_3 group of PMMA is observed (see Fig. 7.39a).

In any case, it is possible to calculate the chemical composition of the copolymer at any given retention time for both critical conditions. This is shown for two samples in Fig. 7.40 presenting the individual normalized elution

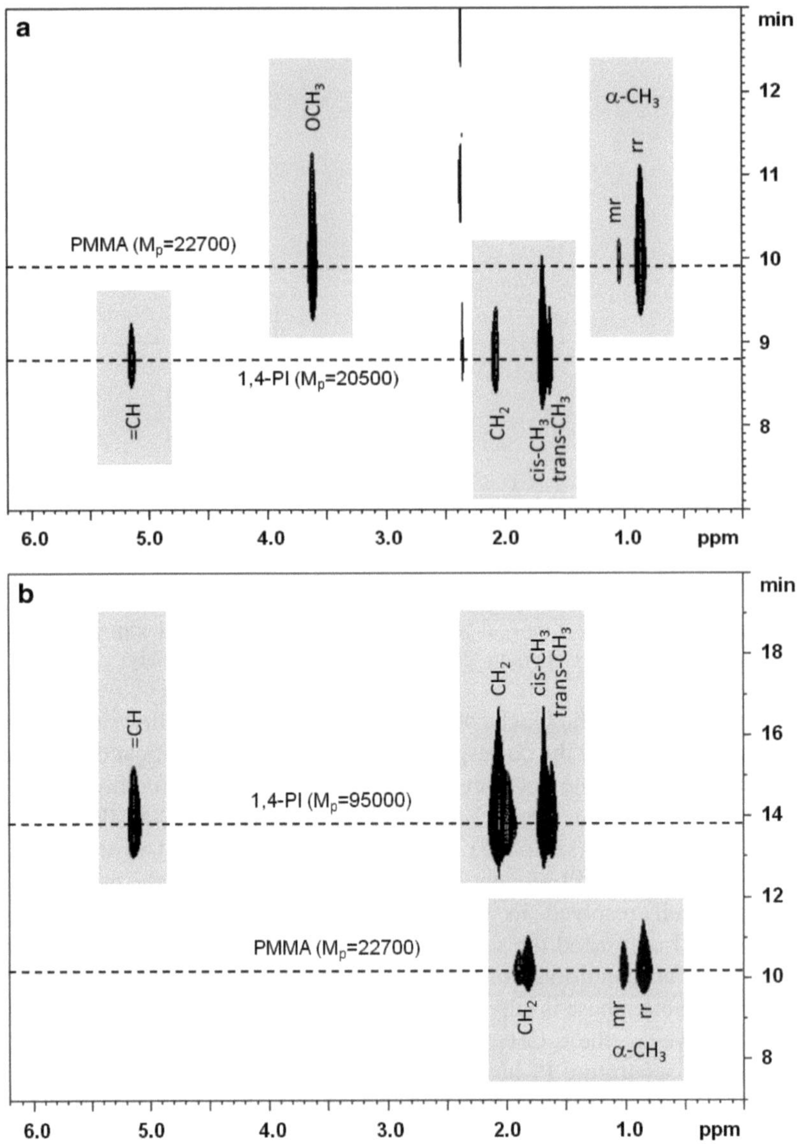

Fig. 7.38 (**a**) LCCC-NMR of a blend of 1,4-PI (M_w = 20.0 kg/mol) and PMMA (M_w = 22.7 kg/mol) at critical conditions of PMMA (ethyl acetate at column temperature T = 10 °C) (10 mg/mL for each polymer, 50 μL injection volume, 0.5 mL/min flow rate) (reprinted with permission from [173]. Copyright (2010) of American Chemical Society). (**b**) LCCC-NMR of a blend of 1,4-PI (M_w = 94.4 kg/mol) and PMMA (M_w = 22.7 kg/mol) at critical conditions of PI (1,4-dioxane at column temperature T = 54 °C) (10 mg/mL for each polymer, 50 μL injection, 0.5 mL/min flow rate) (reprinted with permission from [173]. Copyright (2010) of American Chemical Society)

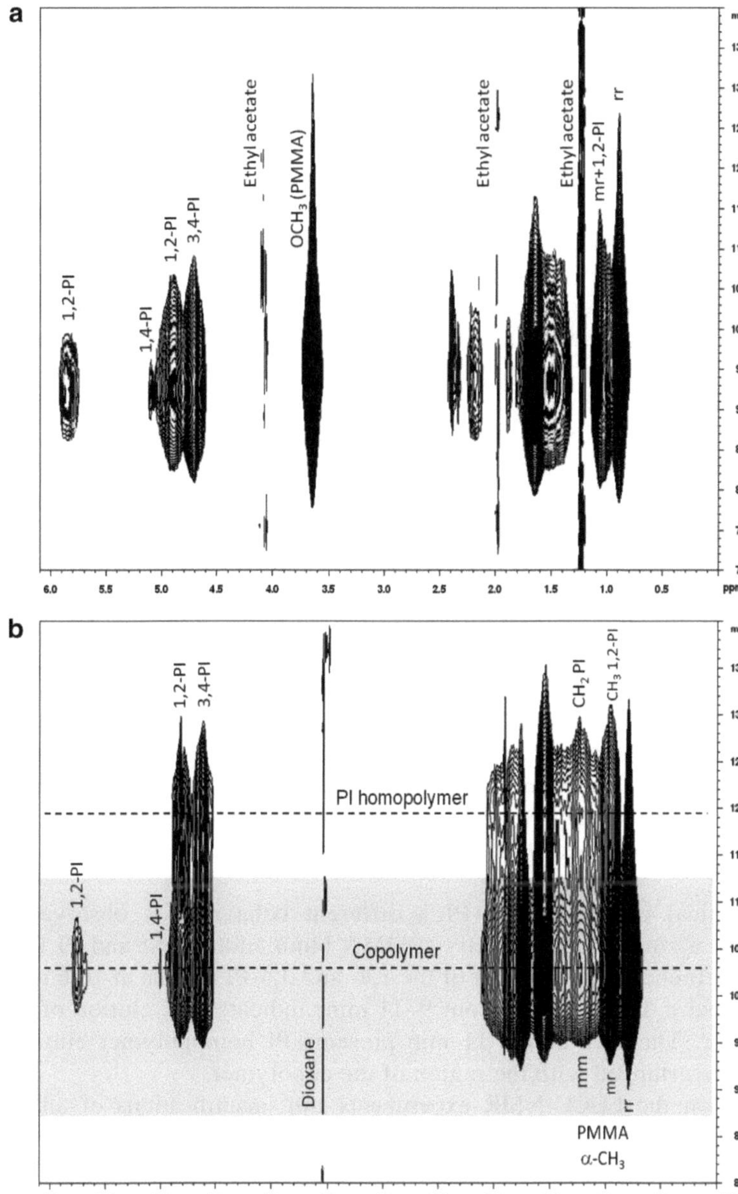

Fig. 7.39 (**a**) LCCC-NMR on-flow run of the PI-*b*-PMMA copolymer sample 5 ($M_w = 52$ kg/mol) at critical conditions of PMMA (30 mg/mL of copolymer, 50 μL injection, 0.5 mL/min flow rate) (reprinted with permission from [173]. Copyright (2010) of American Chemical Society). (**b**) LCCC-NMR on-flow run of the PI-*b*-PMMA copolymer sample 5 ($M_w = 52$ kg/mol) at critical conditions of PI (30 mg/mL of copolymer, 50 μL injection, 0.5 mL/min flow rate) (reprinted with permission from [173]. Copyright (2010) of American Chemical Society)

curves (NMR chromatograms) for the PMMA triads and the PI isomers of samples 5 (predominantly 3,4-PI-b-PMMA) and 2 (predominantly 1,4-PI-b-PMMA). They also show the chemical composition as a function of retention time for the block copolymers at both critical conditions determined from the on-flow experiments.

Figure 7.40a provides monomodal distributions for all monomer units at critical conditions of PMMA. In this case a differentiation between different species is not possible. All curves show almost the same maximum. However, this separation clearly indicates that the sample does not contain PMMA homopolymer. At critical conditions of PMMA, PI homopolymer cannot be detected since the molar masses of the homopolymer and the PI block are similar and, accordingly, they co-elute. Using critical conditions of PI, monomodal distributions for the syndiotactic (rr) and atactic (mr) α-CH_3 groups of PMMA are observed. On the other hand, bimodal distributions for the olefinic 1,2-, 1,4- and 3,4-PI components are found. This is a clear indicator of the existence of a block copolymer fraction (retention time 9–11 min within the grey area) and PI homopolymer fraction (retention time 11–13 min). These NMR chromatograms provide 53.5 mol% block copolymer and 46.5 mol% PI homopolymer related to the total olefinic (PI) content.

Figure 7.40b shows the NMR chromatograms of the individual microstructures as well as the chemical composition of sample 2 for both critical conditions. Sample 2 contains a PI block that is predominantly built of 1,4-PI units. In this case, critical conditions of PMMA provide again monomodal distributions for all microstructures. However, the sample is much more heterogeneous indicated by the shifted elution curves of PI and PMMA. PI elutes first (at the region of higher molar masses), whereas PMMA elutes later (region of lower molar masses) due to the critical conditions of PMMA. As for sample 5, no PMMA homopolymer is visible.

At critical conditions for PI a different behaviour is observed. It also illustrates a strong heterogeneity (PMMA block elutes first and PI later). The main difference is the shoulder of the 1,4- and 3,4-PI elution at 10.6 min (M_p of the PI block). This region (about 9–11 min) indicates the elution of the block copolymer. The area above 11 min presents PI homopolymer elution and is partially overlapped with the region of the copolymer.

Based on the LCCC-NMR experiments full quantifications of all structural features can be conducted. The results provide very detailed information about the chemical composition of the block copolymers, the presence and amount of homopolymers as well as their microstructures. Using the true chemical compositions it is possible to determine the true molar masses of both blocks in the block copolymers. These results are summarized in Tables 7.9 and 7.10.

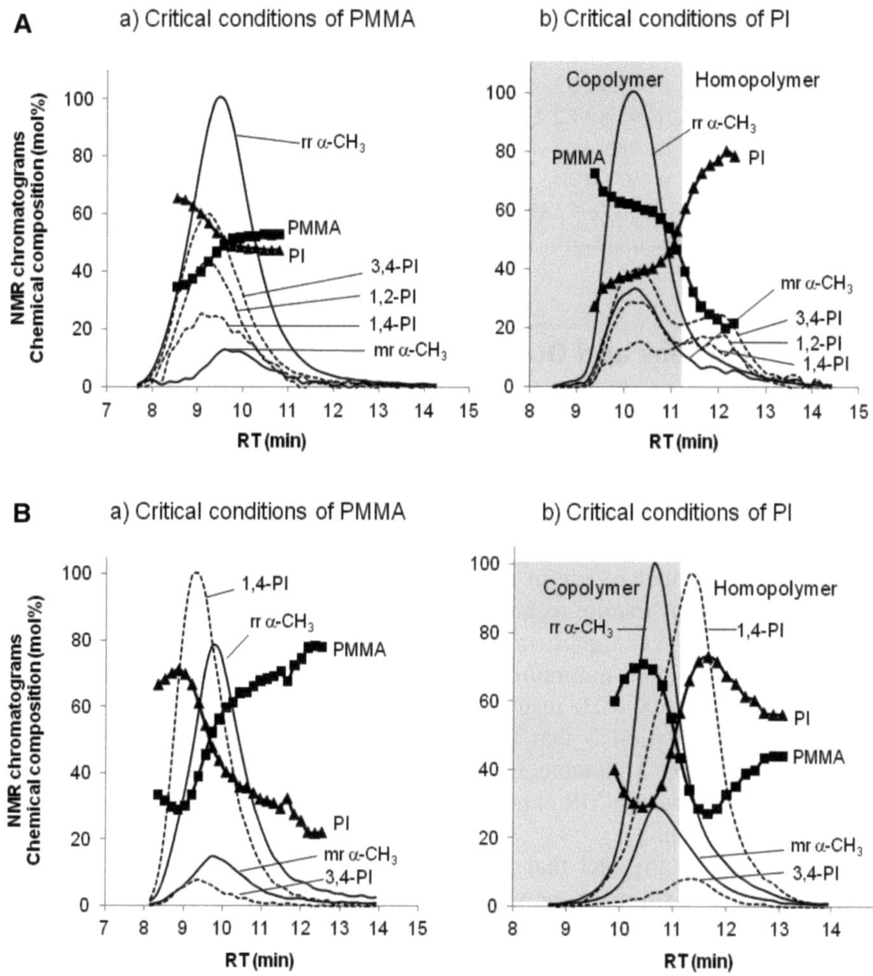

Fig. 7.40 Chemical composition of (**A**) the 3,4-PI-*b*-PMMA copolymer (sample 5) and (**B**) the 1,4-PI-*b*-PMMA copolymer (sample 5) versus retention time (*filled square* = mol % PMMA and *filled triangle* = mol % PI) for critical conditions of (**a**) PMMA and (**b**) PI; lines are representing the NMR projections: *solid lines* = syndiotactic (rr) and atactic (mr) α-CH_3 of PMMA, *dashed lines* = olefinic 1,2-, 1,4- and 3,4-PI (reprinted with permission from [173]. Copyright (2010) of American Chemical Society)

Table 7.9 Molar masses and average chemical compositions of PI-b-PMMA determined by LCCC-NMR at critical conditions of PI, content of PI homopolymer is referred to the total PI

Sample	M_p of PMMA (kg/mol)	M_p of PI (kg/mol)	M_p of PI-b-PMMA (kg/mol)	Isoprene/MMA (mol%)	PI homopolymer (mol%)
1	1.5	8.7	10.2	89.7/10.3	14.9
2	10.9	5.4	16.3	41.9/58.1	3.9
3	83.9	7.3	91.2	11.3/88.7	9.2
4	101.0	6.5	107.5	8.7/91.3	52.4
5	27.1	11.6	38.7	38.7/61.3	46.5

Table 7.10 Microstructures of PI-b-PMMA from LCCC-NMR

Sample	PI (mol%) 1,2 / 1,4 / 3,4	PMMA[a] (mol%) mm / mr / rr
1	0 / 94.9 / 5.1	0 / 13.6 / 86.4
2	0 / 92.8 / 7.2	0 / 24.3 / 75.7
3	0 / 98.9 / 1.1	1.0 / 19.5 / 79.5
4	0 / 97.6 / 2.4	1.1 / 18.4 / 80.5
5	34.2 / 18.5 / 47.3	0 / 25.3 / 74.7

[a]mm = isotactic, mr = heterotactic, rr = syndiotactic triads

7.4 Conclusions and Outlook

As has been demonstrated, the combination of selective separation techniques with powerful spectroscopic detectors enables complex polymers to be analyzed with respect to all possible types of molecular heterogeneity. Chemical composition distribution can be monitored across the molar mass distribution. Steric and functional peculiarities can be detected over the entire molar mass range.

Despite of a number of useful applications, LC-FTIR, LC-MS and LC-NMR cannot be regarded as mature techniques yet. One of the limitations for a broader application of these techniques, in particular of LC-NMR, is the relatively high price of equipment. The minimum requirement for a highly efficient LC-NMR experiment is a 400–500 MHz instrument which is not affordable for every laboratory. The other limitation is that for each technique very specific knowledge is required. To develop a suitable LC-NMR experiment, LC expertise has to be combined with profound NMR experience. This requires an interdisciplinary problem solving approach.

However, due to the fact that polymer structures are becoming increasingly complex and analysis time is one of the key efficiency factors, it is not difficult to predict a bright future for the on-line coupling of liquid chromatography and spectroscopic techniques.

References

1. Housaki T, Satoh K, Nishikida K, Morimoto M (1988) Makromol Chem Rapid Commun 9:525
2. Nishikida K, Housaki T, Morimoto M, Kinoshita T (1990) J Chromatogr 517:209
3. Markovich RP, Hazlitt LG, Smith-Courtney L (1993) In: Provder T (ed) Chromatography of polymers. Characterization by SEC and FFF, ACS Symp. Ser. 521, American Chemical Society, Washington, DC
4. Dhenin V, Rose LJ (2000) Polymer Preprints 41:285
5. DesLauriers PJ, Battiste DR (1995) ANTEC-SPE 53:3639
6. DesLauriers PJ, Rohlfing DC, Hsieh ET (2002) Polymer 43:159
7. DesLauriers PJ (2005) In: Striegel A (ed) Multiple detection in size exclusion chromatography. ACS Symp. Ser. 893, American Chemical Society, Washington, DC
8. Robertson RM, de Haseth JA, Kirk JD, Browner RF (1988) Appl Spectrosc 42:1365

9. Griffiths PR, Conroy CM (1986) Adv Chromatogr 25:105
10. Hellgeth JW, Taylor LT (1986) J Chromatogr Sci 24:519
11. Gagel JJ, Biemann K (1986) Anal Chem 58:2184
12. Gagel JJ, Biemann K (1987) Anal Chem 59:1266
13. Gagel JJ, Biemann K (1988) Microchim Acta 11:185
14. Wheeler LM, Willis JN (1993) Appl Spectrosc 47:1128
15. Willis JN, Dwyer JL, Wheeler LM (1993) Polym Mat Sci 69:120
16. Willis JN, Dwyer JL, Lui MX (1995) In: Proc Int GPC Symp 1994, Lake Buena Vista, FL
17. Polymer Standards Service (Mainz, Germany) webpage: http://www.polymer.de
18. Pasch H (2012) In: Matyjaszewski K, Moeller M (eds) Polymer science: a comprehensive reference, vol 2. Elsevier BV, Amsterdam
19. Pasch H (2000) Adv Polym Sci 150:1–66
20. Pasch H, Esser E, Montag P (1996) GIT Fachz Lab Chromatogr 16:68
21. Adrian J, Esser E, Hellmann G, Pasch H (2000) Polymer 41:2439
22. Willis JN, Wheeler L (1995) In: Provder T, Barth HG, Urban MW (eds) Chromatographic characterization of polymers. Hyphenated and multidimensional techniques. Adv Chem Ser 247, American Chemical Society, Washington, DC
23. Provder T, Kuo C, Whited M, Huddleston D (1995) Proc Int GPC Symp 1994. Lake Buena Vista, FL
24. Cheung PC, Balke ST, Schunk TC (1995) In: Provder T, Barth HG, Urban MW (eds) Chromatographic characterization of polymers. Hyphenated and multidimensional techniques. Adv Chem Ser 247, American Chemical Society, Washington, DC
25. Cheung P, Balke ST, Schunk TC, Mourey TH (1993) J Appl Polym Sci Appl Polym Symp 52:105
26. Kok SJ, Wold CA, Hankemeier T, Schoenmakers PJ (2003) J Chromatogr A 1017:83
27. Esser E, Braun D, Pasch H (1999) Angew Makromol Chem 271:61
28. Pasch H, Rode K, Kloninger C, Esser E (2002) Kautschuk Gummi Kunststoffe 55:9
29. Garcia Romero I, Bashir MA, Pasch H (2005) e-polymers No. 079
30. van Zyl AJP, Graef SM, Sanderson RD, Klumperman B, Pasch H (2003) J Appl Polym Sci 88:2539
31. Pasch H, Siewing A, Heinz LC (2003) Macromol Mat Eng 288:771
32. Garcia Romero I, Pasch H (2005) e-polymers No. 042
33. Graef SM, Brüll R, Pasch H, Wahner UM (2003) e-polymers No. 005
34. Luruli N, Pipers T, Brüll R, Grumel V, Pasch H, Mathot VBF (2007) J Polym Sci Polym Phys 45:2956
35. Albrecht A, Heinz LC, Lilge D, Pasch H (2007) Macromol Symp 257:46
36. Macko T, Schulze U, Brüll R, Albrecht A, Pasch H, Fonagy T, Häussler L, Ivan B (2008) Macromol Chem Phys 209:404
37. Albrecht A, Bruell R, Macko T, Sinha P, Pasch H (2008) Macromol Chem Phys 209:1909
38. Braun D, Esser E, Pasch H (1998) Int J Polym Anal Charact 4:501
39. Tanaka Y, Takeuchi Y, Kobayashi M, Tadokoro H (1971) J Polym Sci 9:43
40. Tanaka Y, Nunogaki K, Adachi J (1988) Rubber Chem Technol 61:36
41. Heinz LC, Siewing A, Pasch H (2003) e-polymers no. 065
42. Kilz P (2002) In: Wu CS (ed) Handbook for size exclusion chromatography and related techniques. Marcel Dekker, New York
43. Kilz P (2002) In: Cazes J (ed) Encyclopedia of chromatography. On-line edition. New York, Marcel Dekker
44. Pasch H, Kilz P (2003) Macromol Rapid Commun 24:104
45. Pasch H, Kilz P (2004) GIT Labor-Fachz 48:25
46. Braun D, Krämer I, Pasch H, Mori S (1999) Macromol Chem Phys 200:949
47. Balogh MP (1997) LC-GC Int 10:728
48. Smits R (1995) LC-GC Int 8:92
49. Niessen WMA, Tinke AP (1995) J Chromatogr A703:37

50. Dole M (1968) J Chem Phys 49:2240
51. Horning EC (1974) J Chromat Sci 12:725
52. Vestal ML, Ferguson GJ (1985) Anal Chem 57:2373
53. Willoughby RC, Browner RF (1984) Anal Chem 56:2626
54. Yergey AL, Edmonds CG, Lewis IAS, Vestal ML (1990) Liquid chromatography/Mass spectrometry—Techniques and applications. Plenum, New York, NY
55. Arpino PJ (1990) Mass Spectrom Rev 9:631
56. Arpino PJ (1992) Mass Spectrom Rev 11:3
57. Behymer TD, Bellar TA, Budde WL (1990) Anal Chem 62:1686
58. Voyksner RD, Smith CS, Knox PC (1990) Biomed Environ Mass Spectrom 19:523
59. The API Book (1989) PE Sciex, Thornhill, Ontario
60. Whitehouse CM, Dreyer RN, Yamashita M, Fenn JB (1985) Anal Chem 57:675
61. Fenn JB, Mann M, Meng CK, Wong SF, Whitehouse CM (1990) Mass Spectrom Rev 9:37
62. Bruins AP, Covey TR, Henion JD (1990) Anal Chem (1987) 59:2642
63. Vouros P, Wronka JW (1991) In: Barth HG, Mays JW (eds) Modern methods in polymer characterization, Chapter 12. New York, Wiley
64. Vestal ML (1984) Science 226:595
65. Vargo JD, Olson KL (1985) Anal Chem 57:672
66. Vestal ML (1983) Mass Spectrom Rev 2:447
67. Vargo JD, Olson KL (1986) J Chromatogr 353:215
68. Leyen DV, Hagenhoff B (1989) J Vac Sci Technol A7:1790
69. Bletsos IV, Hercules DM (1991) Anal Chem 63:1953
70. Lattimer RP, Schulten HR (1985) Int J Mass Spectrom Ion Phys 67:227
71. Rollins K, Scrivens JH, Taylor MJ, Major H (1990) Rapid Commun Mass Spectrom 4:355
72. Fenn JB, Nohmi TJ (1992) J Am Chem Soc 114:3241
73. Kallos GJ, Tomalia DA (1991) Rapid Commun Mass Spectrom 5:383
74. Prokai L, Simonsick WJ (1995) In: Provder T, Barth HG, Urban MW (eds) Chromatographic characterization of polymers. Hyphenated and multidimensional techniques, Chapter 4, Adv. Chem. Ser. 247, American Chemical Society, Washington, DC
75. Prokai L, Simonsick WJ (1993) Rapid Commun Mass Spectrom 7:853
76. Simonsick WJ (1993) Polym Mater Sci Eng 69:412
77. Simonsick WJ, Ross CW (1996) Polymer Prepr 37:286
78. Prokai L, Myung SW, Simonsick WJ (1996) Polymer Prepr 37:288
79. Kemp TJ, Barton Z, Mahon A (1996) Polymer Prepr 37:305
80. Tanaka K, Wadi H, Ido Y, Akita S, Yoshida Y, Yoshida T, Matsuo T (1988) Rapid Commun Mass Spectrom 2:151–153
81. Karas M, Hillenkamp F (1988) Anal Chem 60:2299
82. Beavis RC, Chait BT (1989) Rapid Commun Mass Spectrom 3:233
83. Hillenkamp F, Karas M, Beavis RC, Chait BT (1991) Anal Chem 63:1193A
84. Bahr U, Deppe A, Karas M, Hillenkamp F, Giessmann U (1992) Anal Chem 64:2866
85. Danis PO, Karr DE, Mayer F, Holle A, Watson CH (1992) Org Mass Spectrom 27:843
86. Kahr MS, Wilkins CL (1993) J Am Soc Mass Spectrom 4:453
87. Montaudo G, Montaudo MS, Puglisi C, Samperi F (1994) Rapid Commun Mass Spectrom 8:1011
88. Danis PO, Karr DE, Xiong Y, Owens KG (1996) Rapid Commun Mass Spectrom 10:862
89. Pasch H, Unvericht R (1993) Angew Makromol Chem 212:191
90. Pasch H, Gores F (1995) Polymer 36:1999
91. Braun D, Ghahary R, Pasch H (1996) Polymer 37:777
92. Pasch H, Deffieux A, Ghahary R, Schapacher M, Riquet-Lurbet L (1997) Macromolecules 30:98
93. Weidner SM, Falkenhagen J (2010) In: Li L (ed) MALDI mass spectrometry for synthetic polymer analysis. Wiley, Hoboken, NJ
94. Murray KK, Russell DH (1993) Anal Chem 65:2534

95. Beeson MD, Murray KK, Russell DH (1995) Anal Chem 67:1981
96. Murray KK, Lewis TM, Beeson MD (1994) Anal Chem 66:1601
97. Fei X, Murray KK (1996) Anal Chem 68:3555
98. Fei X, Wei G, Murray KK (1996) Anal Chem 68:1143
99. Laiko VV, Baldwin MA, Burlingame AL (2000) Anal Chem 72:652
100. Daniel JM, Ehala S, Fries SD (2004) Analyst 129:574
101. Available at http://www.apmaldi.com/
102. Available at http://www.kranalytical.co.uk/mass-spec.html
103. Available at http://www.lcpackings.com/products/Instruments/Probot01.html
104. Available at http://www.kromatek.co.uk/MALDI_Fraction_Collector_40.asp
105. Available at http://www.gilson.com/Products/product.asp?pID=116
106. Available at http://www.sunchrom.de/pdf/SunChrom%20SunCollect.pdf
107. Keil O, LaRiche T, Deppe H (2002) Rapid Commun Mass Spectrom 16:814
108. Nielen MWF (1998) Anal Chem 70:1563
109. Grimm R (1997) Int Biotech Lab 58:17
110. Available from: http://www.labconnections.com/products.htm
111. Zhang BY, McDonald C, Li L (2004) Anal Chem 76:992
112. Young JB, Li L (2007) Anal Chem 79:5927
113. Pasch H, Schrepp W (2003) MALDI-TOF mass spectrometry of synthetic polymers. Springer, Berlin
114. Montaudo G, Lattimer RP (2002) Mass spectrometry of polymers. CRC, Boca Raton, FL
115. Pasch H, Rode K (1995) J Chromatogr A699:21
116. Krüger RP, Much H, Schulz G (1996) GIT Fachz Lab 4:398
117. Just U, Krüger RP (1996) In: Auner N, Weis J (eds) Organosilicon chemistry II. Weinheim, VCH
118. Montaudo G (1995) Rapid Commun Mass Spectrom 9:1158
119. Danis PO, Saucy DA, Huby FJ (1996) Polymer Prepr 37:311
120. Montaudo G (1998) In: Proc 11th Int Symp Polym Anal Char, Santa Margerita Ligure, Italy
121. Falkenhagen J, Schulz G, Weidner S (2000) Int J Polym Anal Char 5:549
122. Lee H, Lee W, Chang T, Choi S, Lee D, Ji H, Nonidez WK, Mays JW (1999) Macromolecules 32:4143
123. Montaudo MS, Montaudo G (1999) Macromolecules 32:7015
124. Montaudo MS (2001) Macromolecules 34:2792
125. Falkenhagen J, Weidner S (2009) Anal Chem 81:282
126. Weidner S, Falkenhagen J, Krueger RP, Just U (2007) Anal Chem 79:4814
127. Weidner SM, Falkenhagen J, Maltsev S, Sauerland V, Rinken M (2007) Rapid Commun Mass Spectrom 21:2750
128. Weidner SM, Trimpin S (2010) Anal Chem 82:4811
129. Hart-Smith G, Barner-Kowollik C (2010) Macromol Chem Phys 211:1507
130. Montaudo G, Garozzo D, Montaudo MS, Puglisi C, Samperi F (1995) Macromolecules 28:7983
131. Montaudo MS, Puglisi C, Samperi F, Montaudo G (1998) Macromolecules 31:3839
132. Esser E, Keil C, Braun D, Montag P, Pasch H (2000) Polymer 41:4039
133. Albert K (1995) J Chromatogr 703:123
134. Dorn HC (1984) Anal Chem 56:747A
135. Laude DA, Wilkins CL (1986) Trends Anal Chem 5:230
136. Albert K, Bayer E (1988) Trends Anal Chem 7:288
137. Albert K, Bayer E (1992) In: Patonay G (ed) HPLC Detection: Newer Methods. New York, VCH Publishers
138. Hofmann M, Spraul M, Streck R, Wilson ID, Rapp A (1993) Labor Praxis 10:36
139. Smallcombe HS, Pratt LP, Keifer PA (1995) J Magn Reson 117A:295
140. Ogg RJ, Kingsley PB, Taylor JS (1994) J Magn Reson B104:1
141. Albert K, Braumann U (1994) Anal Chem 66:3042
142. Albert K, Braumann U, Streck R, Spraul M, Ecker R (1995) Fresenius J Anal Chem 352:521
143. Albert K (1997) J Chromatogr A785:65

144. Braumann U, Händel H, Albert R (1995) Anal Chem 67:930
145. Wu N, Peck TL, Webb AG, Magin RL, Sweedler JV (1994) J Am Chem Soc 116:7929
146. Wu N, Peck TL, Webb AG, Magin RL, Sweedler JV (1994) Anal Chem 22:3849
147. Sweedler JV (1995) In: Proceedings of the 7th International Symposium on High Performance Capillary Electrophoresis, Würzburg, Germany
148. Hatada K, Ute K, Okamoto Y, Imanari M, Fujii N (1988) Polym Bull 20:317
149. Hatada K, Ute K, Kashiyama M, Imanari M (1990) Polym J 22:218
150. Ute K (1998) In: Proc 11th Int Symp Polym Anal Char, Santa Margerita Ligure, Italy
151. Ute K, Niimi R, Matsunaga M, Hatada K, Kitayama T (2001) Macromol Chem Phys 202:3081–3086
152. Krämer I, Pasch H, Händel H, Albert K (1999) Macromol Chem Phys 200:1734
153. Pasch H, Hiller W (1996) Macromolecules 29:6556
154. Pasch H, Hiller W, Haner R (1998) Polymer 39:1515
155. Schlotterbeck G, Pasch H, Albert K (1997) Polym Bull 38:673
156. Hiller W, Brüll A, Argyropoulos D, Hoffmann E, Pasch H (2005) Magn Reson Chem 43:729–735
157. Kitayama T, Janco M, Ute K, Niimi R, Hatada K (2000) Anal Chem 72:1518–1522
158. Krämer I, Hiller W, Pasch H (2000) Macromol Chem Phys 201:1662–1666
159. Hiller W, Sinha P, Pasch H (2007) Macromol Chem Phys 208:1965
160. Hiller W, Sinha P, Pasch H (2009) Macromol Chem Phys 210:605
161. Hiller W, Pasch H, Macko T, Hoffmann M, Ganz J, Spraul M, Braumann U, Streck R, Mason J, van Damme F (2006) J Magn Res 183:309
162. Hatada K, Kitayama T (2004) NMR spectroscopy of polymers. Springer, Berlin
163. Hatada K, Ute K, Kitayama T, Yamamoto M, Nishimura T, Kashiyama M (1989) Polym Bull 23:549
164. Shepherd L, Chen TK, Harwood HJ (1979) Polym Bull 1:445
165. Segre AL, Ferruti P, Toja E, Danusso F (1969) Macromolecules 2:35
166. Matsuzaki K, Uryu T, Osada K, Kawamura T (1974) J Polym Sci C 12:2873
167. Randall JC (1977) Polymer sequence determination, carbon-13 method. Academic, New York, NY
168. Chen TK, Gerken TA, Harwood HJ (1980) Polym Bull 2:37
169. Sato H, Saito K, Miyashita K, Tanaka Y (1981) Makromol Chem 182:2259
170. Tanaka Y, Sato H, Saito K, Miyashita K (1980) Makromol Chem Rapid Commun 1:551
171. Sato H, Tanaka Y, Hatada K (1982) Makromol Chem Rapid Commun 3:175
172. Eisenbeiss F, Dumont E, Henke H (1978) Angew Makromol Chem 71:67
173. Hiller W, Pasch H, Sinha P, Wagner T, Thiel J, Wagner M, Muellen K (2010) Macromolecules 43:4853

Polyolefin Analysis by Multidimensional Liquid Chromatography

8

The synthesis and characterization of polyolefins continues to be one of the most important areas for academic and industrial research considering the fact that polyolefins constitute about 60% of the total polymer market. One consequence of the development of new "tailor-made" polyolefins is the need for new and improved analytical techniques for the analysis of polyolefins with respect to molar mass, molecular topology and chemical composition distribution. Very frequently, polyolefins exhibit multiple distributions, e.g. long chain branching and molar mass distribution (MMD) in low-density polyethylene (LDPE) or chemical composition distribution (CCD) and MMD in linear low-density polyethylene (LLDPE), copolymers and polymer blends.

A number of important methods are well established that provide average information on the molecular structure of polyolefins, the most prominent ones being FTIR and NMR spectroscopy for the average chemical composition and microstructure, viscometry and light scattering for the average molar mass, and thermal analysis for glass transition, melting and crystallization temperatures and enthalpies. Most polymer fractionation methods operate in dilute polymer solutions requiring good solubility of all polymer components in the solvent that is used for the fractionation procedure. The majority of technically important polyolefins are semicrystalline materials with melting points above 100 °C. They are not soluble in most of the typical organic solvents. Typically, the polyolefin must be heated above its melting temperature to be soluble and, therefore, specific high boiling point solvents are required in polyolefin analysis. Another problem is the fact that polyolefins tend to undergo thermo-oxidative degradation. This of course has to be prevented by suitable measures when dissolving the sample. Typically, polyolefins are dissolved at temperatures between 130 and 160 °C. To prevent degradation, stabilizers and antioxidants are added to the solvent. The most common solvents for polyolefin fractionations are 1,2,4-trichlorobenzene (TCB), 1,2-dichlorobenzene, decaline, and in some cases cyclohexane.

There is a group of techniques available based on the different crystallization behaviour of semicrystalline polyolefins in dilute solution for obtaining information about CCD. The crystallization behaviour of polyolefins is determined by the

H. Pasch and B. Trathnigg, *Multidimensional HPLC of Polymers*, Springer Laboratory, DOI 10.1007/978-3-642-36080-0_8, © Springer-Verlag Berlin Heidelberg 2013

molecular structure including the type of monomer, the copolymer composition and the molecular size.

Very early in polyolefin research the potential of crystallization behaviour as an analytical tool was recognized. The major techniques in this category are temperature rising elution fractionation (TREF), crystallization analysis fractionation (CRYSTAF) and crystallization elution fractionation (CEF). All these techniques are based on differences in their crystallizability as a function of temperature but different experimental approaches are used to get the final results which will be discussed in detail as follows. The main disadvantage of crystallization based techniques is the very long analysis times and research is mainly focused on decreasing analysis times along with other aspects like better resolution of different fractions. Another disadvantage is the fact that only the crystallizable part of a material can be fractionated. The amorphous part is obtained as a bulk fraction. In the early days of TREF, the analysis time was around 100 h per sample. Recent improvements in TREF allow analysis of a sample in 3–4 h while the development of CRYSTAF reduced this analysis time to 100 min. The latest development in crystallization based techniques—CEF—allows analyzing a sample in 30 min. Overviews on the fundamentals, the different experimental approaches and a number of important applications are given in [1–8].

For the molar mass analysis of polyolefins, high-temperature SEC is routinely used [9, 10]. The corresponding high-temperature chromatography instruments have been commercially available since 1964. The instruments may be equipped with different detectors such as refractive index, viscometer, light scattering or infrared. Since HT-SEC for instrument producers is rather a niche market there are only a few types of instruments that are most commonly used including the Agilent PL-GPC 220 [11], the Malvern/Viscotek HT-GPC [12] and the GPCIR of Polymer Char [13].

HT-SEC is a reliable, precise, and fast method to measure the molar mass averages, the polydispersity index and the complete molar mass distribution (MMD) of polyolefins. Depending on the complexity of the sample to be analyzed there are several techniques available that mainly differ by the added detectors and calibration options [14] including (1) conventional HT-SEC with a concentration detector, (2) HT-SEC-light scattering, (3) HT-SEC-viscometry. Frequently a triple-detector SEC technology is used, where three on-line detectors are used together in a single SEC system. In addition to the concentration detector, an on-line viscometer and a MALLS instrument are coupled to the SEC (TriSEC). With TriSEC, absolute molar mass determination is possible for polymers that are very different in chemical composition and molecular conformation. The usefulness of the TriSEC approach has been demonstrated in a number of applications [15–19].

8.1 Coupled HT-SEC-FTIR

For a detailed analysis of olefin copolymers or polyolefin blends it is important to determine the CCD in addition to the MMD. The bulk chemical composition of polyolefins can be determined quantitatively by FTIR or NMR spectroscopy. Dual information on the chemical composition as a function of molar mass can be obtained when HT-SEC is directly coupled to these spectroscopic methods.

The coupling of high-temperature LC with FTIR is an important technique because of the robustness, the simplicity and the cost effectiveness, LC-FTIR is the method of choice in most applications as compared to the more costly LC-NMR. The hyphenation of LC and FTIR can be realized in two ways: (1) on-line mode via a flow cell, (2) off-line mode via a solvent elimination interface. The typical characteristics of the two approaches are discussed in Sect. 7.1. As has been discussed there, a major limitation of all flow-through cells is the limited selection of solvents/mobile phases that exhibit sufficiently large spectral windows for high-sensitivity measurements. One of the few very fortunate cases is the SEC-FTIR analysis of polyolefins. In this case 1,2,4-trichlorobenzene is used as the mobile phase which is sufficiently transparent in the range of 2,700–3,000 cm^{-1} that is used for polyolefin detection. Alternatively, o-dichlorobenzene (ODCB) or tetrachloroethylene may be used. As has been shown by DesLauriers and others, the compositional heterogeneity (short chain branching, SCB) in polyolefins can be analyzed sensitively by on-flow SEC-FTIR [9–15]. Chromatograms are generated from ratio-recorded transmittance spectra where the spectrum of the pure mobile phase is used as background. Typical sample concentrations are 1–3 mg/mL and rather large injection volumes of 400–1,000 μL are used for sufficient signal-to-noise ratio. In the case of low density materials branching is determined as the levels of methyl (2,958 cm^{-1}) and methylene endgroups (2,928 cm^{-1}) [9, 10, 12]. For high density materials with low degrees of branching multivariable statistical techniques are preferred [14].

A typical analytical result is shown in Fig. 8.1 comparing Ziegler-Natta catalyzed ethylene-1-hexene resins with high and low comonomer levels [15]. The degree of branching is given as "branches per 1,000 total carbons". Similar approaches can be used for other polymers provided that a spectral window is available for selective detection of the polymer species. Piel et al. [16] have recently significantly increased the signal-to-noise ratio in SEC-FTIR after application of a bandpass filter instead of a steel mesh attenuator and by changes in data processing. The signal obtained with the bandpass filter was almost four times higher than that with the steel mesh attenuator. They used the proposed method for the determination of short-chain branching. Apart from SEC other fractionation techniques can also be applied such as analytical temperature rising elution fractionation (A-TREF).

A rather broad applicability of FTIR as a detector in liquid chromatography can be achieved when the mobile phase is removed from the sample prior to detection. In this case the sample fractions are measured in pure state without interference from solvents. This situation is realized when the LC Transform system is used, see Sect. 7.1 for more details. As an example for this approach, the analysis of a blend of two EPDM copolymers with different molar masses and chemical compositions is presented in Fig. 8.2 [17]. The FTIR spectrum of an EPDM copolymer is given in Fig. 8.2a. The propylene percentage is determined from the absorption peak at 1,378 cm^{-1}, while the ethylidene norbornene is determined from the peak at

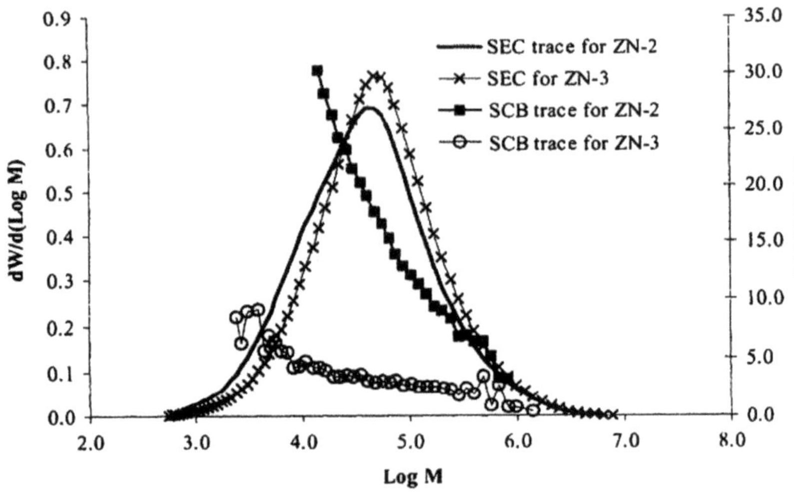

Fig. 8.1 SEC-FTIR analysis of LLDPE, comparison of comonomer incorporation in Ziegler-Natta catalyzed ethylene-1-hexene resins using high (ZN-2) and low (ZN-3) comonomer levels (reprinted with permission from [15]. Copyright (2004) of American Chemical Society)

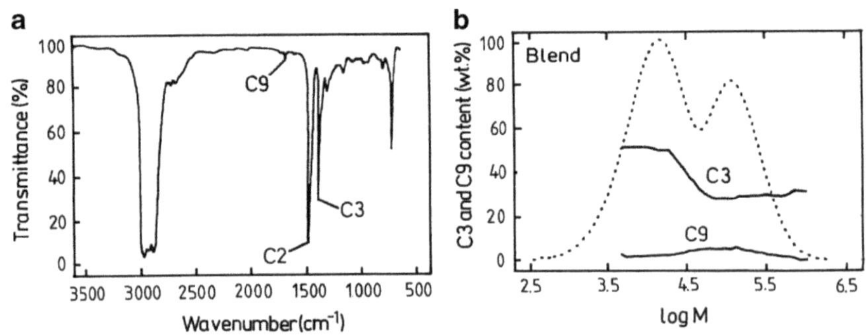

Fig. 8.2 FTIR spectrum of an EPDM copolymer (**a**) and HT-SEC/FTIR analysis of the blend of two EPDM copolymers (**b**) (reprinted from [8] with permission of Springer Science + Business Media)

1,690 cm^{-1}. The percentage of the two monomers across the molar mass axis is given in Fig. 8.2b. As can be seen clearly, the propylene content of the higher molar mass copolymer is lower [17].

Using this experimental set-up, a multitude of different materials can be analyzed, including α-olefin copolymers, and polyolefin blends. In addition to the analysis of macromolecular components, the technique can be used for the detection and quantification of additives.

Over the past few years a number of applications for the analysis of olefin copolymers have been published that make use of the LC-Transform system. These include the SEC-FTIR analysis of ethylene-vinyl acetate copolymers [18], ethylene-methyl methacrylate copolymers [19, 20], ethylene-styrene copolymers [21], high density polyethylene (HDPE) and polypropylene (PP) [22]. A number of studies used SEC-FTIR for monitoring the thermo-oxidative degradation of polyolefins [23–27] and a combination of TREF and SEC-FTIR to investigate the complex structure of olefin copolymers [28, 29].

The challenges for further improvement in sensitivity are to overcome the loss of IR sensitivity in the reflectance mirrors of the optics module and to deposit the effluents in rather narrow tracks on the substrate. The configuration of the DiscovIR-LCTM interface which was recently commercialized by Spectra Analysis Inc. (Marlborough, MA, USA), accounts for the energy loss in the optics module by using IR microscopy [30]. The instrument is a single unit that eliminates the solvent from the eluate received from the LC system and deposits the chromatogram as a track, which is a function of retention time. The track is scanned with a built-in FTIR microscope in real time. The deposition occurs under high vacuum and low temperatures (−140 to 100 °C), which protects the compounds from oxidation. The deposition matrix is ZnSe and allows measurements in the transmission mode. A number of applications have been presented, however, none for the analysis of polyolefins.

8.2 Coupled HT-SEC-^{1}H-NMR

Another most exciting new tool for the analysis of complex polyolefins is the direct coupling of high-temperature liquid chromatography and ^{1}H-NMR. Such equipment became available only recently when a high-temperature flow-through NMR probe was introduced by Bruker. The construction and experimental setup of the LC-NMR coupling is described in detail by Hiller et al. [31]. In brief, the NMR flow probe can operate at temperatures up to 150 °C. The probe has an active flow cell with a volume of 120 μL. It is a dual inverse ^{1}H/^{13}C probe with pulse field gradients. A stop-flow valve was developed as an interface for the SEC and the NMR. The valve is a two position device and guides the flow either from the SEC to the NMR or directly to the waste, see Fig. 8.3. This setup allows on-flow experiments, automatic stop-flow experiments and time-slicing.

To evaluate the capabilities of the system, a polymer blend comprising PE and PMMA homopolymers and a PE-PMMA copolymer was prepared and analysed. The molar masses of PE, PMMA and the copolymer were M_n = 1,100 g/mol, M_n = 263,000 g/mol and M_n = 10,600 g/mol, respectively. The experiments were performed with TCB as the mobile phase. WET suppression was applied to the intrinsic solvent signals, i.e. the three aromatic proton signals were suppressed.

Figure 8.4 shows the on-flow run of the blend as a corrected contour plot by subtracting signals that correspond to impurities of the solvent. In the SEC system the elution of the blend components is in the order of decreasing molar mass. This elution order can be clearly seen in the SEC-NMR contour plot. The spectra of the early eluting fractions show signals for MMA but not for ethylene. In contrast, the late eluting fractions exhibit signals for ethylene but not for MMA and can be

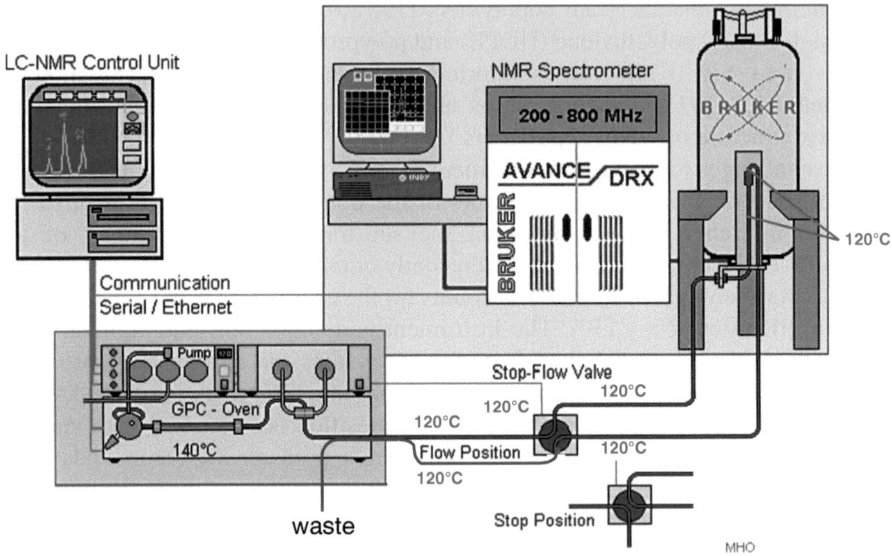

Fig. 8.3 Experimental set-up of the high temperature SEC-NMR (SEC: 130 °C; LC probe, stop-flow valve and transfer lines: 120 °C) (reprinted from [31] with permission of Elsevier)

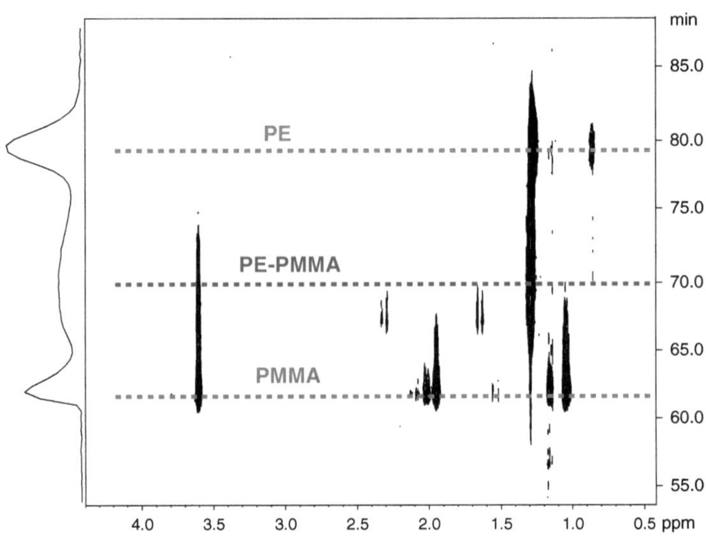

Fig. 8.4 SEC-NMR (400 MHz) on-flow run (corrected) of a PE-PMMA-copolymer blend at 130 °C in TCB. (flow rate 0.5 mL/min, concentration 2 mg/mL of each polymer, 300 μL injection volume, 5 Waters columns, 24 scans per FID, 1.24 s repetition delay) (reprinted from [31] with permission of Elsevier)

assigned to PE. Between the two homopolymers, the elution of the copolymer can be measured by detecting signals for both MMA and ethylene. Figure 8.4 also shows the vertical projections taken from the sum of the NMR signals. It can be used as the chromatogram which also indicates three separated peaks.

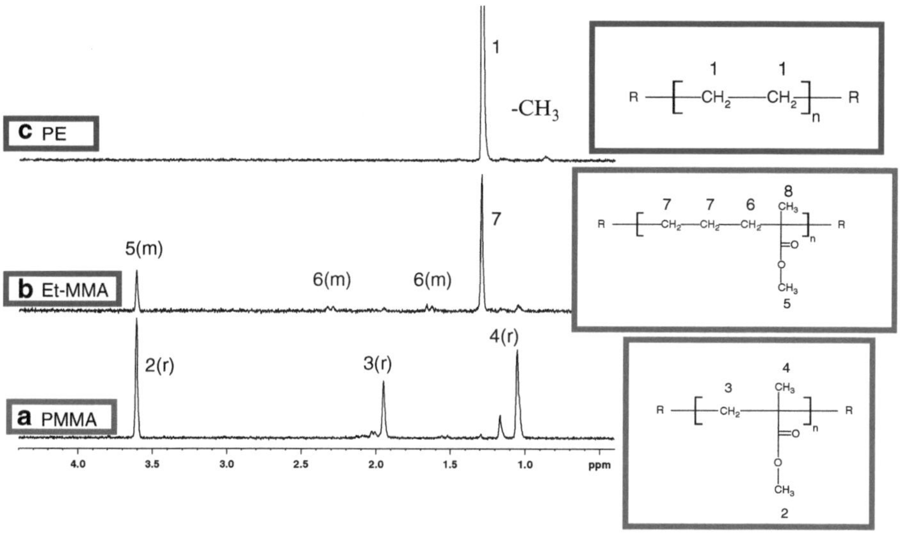

Fig. 8.5 ^{1}H traces of the on-flow run of Fig. 8.4: (**a**) PMMA (RT = 60.5 min); (**b**) PE-PMMA copolymer (RT = 66.0 min); (**c**) PE 1100 g/mol (RT = 79.4 min).

Figure 8.5 shows the different traces of the on-flow experiment. These traces clearly indicate the different components of the blend. The signals of PMMA (a) correspond to syndiotactic species of this homopolymer. The second trace (b) contains the copolymer. It is a block copolymer in which MMA is mainly isotactic. The third trace contains only the PE component. It even shows the CH_3 end group at 0.86 ppm. However, the signal-to-noise of the CH_3 group is not sufficient for a precise molar mass calculation.

In the second experiment, the CCD of the PE-PMMA copolymer was investigated by using on-flow and stop-flow experiments. The distributions of the different structural moieties corresponding to MMA and ethylene can be seen and correlated with the corresponding molar masses.

Recently, a new cryoprobe for high-temperature NMR has been introduced. This cryoprobe enables a dramatic increase of signal-to-noise ratio. Using this cryoprobe it is even possible to perform ^{13}C NMR analyses on a small quantity of a material with a reasonable acquisition time for sample concentrations as low as 0.9–3.2 mg/mL. These concentrations are significantly lower than the concentrations usually used in ^{13}C NMR with a conventional probe [32, 33]. Cong et al. [34] used this cryoprobe and collected fractions from 20 chromatographic runs. After evaporation of the mobile phase NMR measurements were performed. The new cryoprobe enabled determination of the content of octene in the collected copolymer fractions thus demonstrating the practical applicability and the excellent improvement of detectability of polyolefins in NMR.

8.3 High-Temperature 2D Liquid Chromatography

As has been discussed in the previous chapters, HPLC is an important tool for the fast separation of complex polymers with regard to chemical composition. HPLC separations can be achieved via different mechanisms, including adsorption–desorption and precipitation–redissolution. In gradient HPLC, frequently precipitation and adsorption processes are combined.

Until recently, standard HPLC methods for polymers, e.g. gradient chromatography or LCCC, were limited to ambient or slightly elevated temperatures [35, 36]. The majority of published HPLC separations were conducted at operating temperatures of a maximum of 80 °C. These temperatures are too low for the dissolution of polyolefins, which require at least 120 °C for dissolution due to their mostly semi-crystalline nature. It was, therefore, a challenge to develop HPLC methods for the separation of polyolefins that operate at temperatures of 120 °C and higher.

A major breakthrough was achieved by Macko et al. with the successful separation of PE, PP and EP copolymers using silica-based interactive stationary phases [37–39]. An overview on the elution behaviour of different polyolefins on a number of interactive stationary phases was given in [40] (Table 8.1). In 2004, as a joint development of Polymer Laboratories, Ltd. (Church Stretton, England) and the group of Pasch and Macko, the first instrument that combines both high operation temperatures and the necessary requirements for gradient HPLC was introduced [41]. This (first of its kind) pioneering instrument contained a high pressure gradient pump for either running a binary solvent gradient or pumping of a single solvent (SEC) or a mixture of two solvents at constant composition (for HPLC), see Fig. 8.6.

Mobile phase changes were accomplished via a multi-solvent management system. The chromatograph was equipped with a robotic sample handling system, which enabled sample preparation and injection at temperatures up to 220 °C. For fast column and mobile phase screening, a column switching valve inside the column compartment enabled the successive use of up to six different columns (or five columns and a reference capillary for direct injection into the detector). The choice of detectors for high-temperature HPLC of polyolefins and their copolymers is very limited. The present instrument contained a high-temperature differential refractive index (RI) detector for isocratic elution (e.g., SEC and LCCC) and an ELSD for gradient and isocratic elution modes. The ELSD was attached to the chromatograph via a heated transfer line.

This instrument was subsequently used to develop a number of important methods for the separation of complex polyolefins. A detailed discussion of the major developments in the field has been presented recently by Pasch et al. [8]. LCCC has been used for the separation of polyethylene-polystyrene blends [43]. Critical conditions for PMMA at a temperature of 140 °C have been also identified and the separation of ethylene-methyl methacrylate block copolymers using

Table 8.1 Overview of elution behaviour of isotactic PP and linear PE on different column packings (reprinted from [40] with permission of Wiley-VCH)[a]

Microporous sorbents (zeolites)										
Solvent ↓	SH-300		Silicalite		CBV-780		CP814E		MCM-41	
	Si/Al = 150		Si/Al = 400		Si/Al = 40		β Si/Al = 12.5		Pores 44 Å	
	Pores 5–6 Å		Pores 5–6 Å		Pores 7–12 Å	Mesopores 40–400 Å	Pores 5.6–7.5 Å		→	
Polymer →	PP	PE	PP	PE	PP	PE	PP	PE	PP	PE
TCB	E	**PR**	E	E	E	E	E	**FR**	E	E
Decalin	E	**FR**	E	**PR**	E	**PR**	E	**FR**	E	E
Tetralin	E	**PR**	E	E	E	E	E	**FR**	E	E
1,4-Dimethylbenzene[b]	E	**FR**	E	E	E	E	E	**FR**		
1,2-Dichlorbenzene	E	**FR**	E	E	E	E	E	**FR**		
1,3-Dichlorobenzene	E	**PR**	E	E	E	E	E	**FR**		
1,3,5-Trimethylbenzene	E	**FR**	E	**PR**	E	E	E	**FR**	E	E
1,1,2,2-Tetrachlorethane[b]	E	**FR**	E	E	E	E	E	**FR**	E	E
Tetrachlorethylene[c]	E	**FR**	E	E	E	E	E	**FR**	E	E
Diphenylether	E	**FR**	E	E	**FR**	E	E	**FR**		
Cyclohexanol					**PR**	**PR**				
n-Decanol	E	E	E	E	**PR**	**PR**	E	E	E	E
n-Dodecanol					**PR**	**PR**				
Diphenylmethane					**PR**	**PR**				
Macroporous sorbents										
Solvent ↓	Silica gel		Aluminium oxide		Zirconium oxide		Hydroxy-appatite		Florisil	
	Pore size 100 Å		Pore size 200 Å		Pore size 150 Å		Pore size 300 Å		Pore size 300 Å	
Polymer →	PP	PE	PP	PE	PP	PE	PP	PE	PP	PE
1,1,2,2-Tatrachloroethane	**PR**	**PR**	**PR**	**PR**	**PR**	**PR**	**PR**	**PR**	**PR**	**PR**
1,2,3-Trichloropropane	**PR**	**PR**	**PR**	**PR**	**PR**	**PR**	**PR**	**PR**	**PR**	**PR**
1,1,2,2-Trichloropropane	E	E	E	E	E	E	E	E	E	E

[a]Symbols: *FR* full retention of polymers on sorbent, *PR* partial retenion of polymers on sorbent, *E* full elution of polymers from sorbent
[b]Measured at 110 °C
[c]Measured at 135 °C. Other data measured at 140 °C

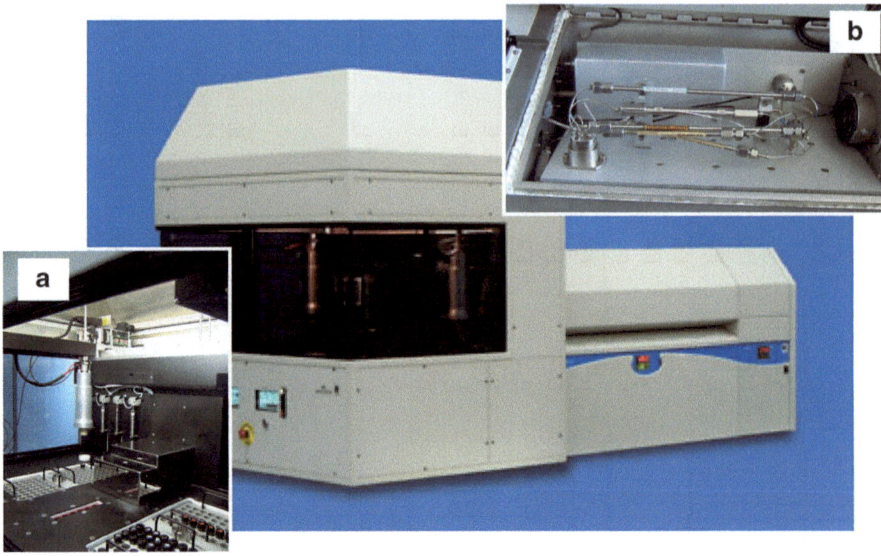

Fig. 8.6 Polymer Labs HT-HPLC instrument with sample robot (**a**) and column switching valve (**b**) (reprinted from [42] with permission of Elsevier)

high-temperature LCCC has been accomplished [20]. Using a solvent gradient of ethyleneglycol monobutylether (EGMBE)–TCB on silica gel, a baseline separation of PE and PP was achieved [44]. In this case PE was completely precipitated on the column with the initial mobile phase, while PP eluted in the size exclusion mode. When the content of TCB in the mobile phase was increased by performing a gradient the precipitated polyethylene was eluted. As was shown, for the first time blends of different polyolefins were separated quantitatively over a wide range of concentrations by liquid chromatography at 140 °C. Moreover, EP copolymers were separated into a propylene-rich part and an ethylene-rich part [45]. This chromatographic approach was also applied to the separation of ethylene/propylene copolymers [46] and for the separation of various polyolefins with regard to the chemical composition of the components [47].

HT-HPLC based on adsorption–desorption was used to separate random ethylene-vinyl acetate copolymers according to chemical composition. On silica gel as the stationary phase and using decaline-cyclohexanone as the eluent full separation of copolymers of different compositions was achieved, see Fig. 8.7. In addition, the homopolymers PE and PVAc were well separated from the copolymers. This was the first time that a chromatographic system was available that separates olefin copolymers irrespective of crystallinity and solubility over the entire range of compositions. The components of the mobile phase are solvents for both PE and PVAc. The non-polar solvent, decalin, supports adsorption of PVAc on the silica gel, while the polar solvent, cyclohexanone, enables desorption and elution of the adsorbed polymer sample from the column [48]. In a next step, this highly selective type of copolymer separation was coupled to FTIR spectroscopy to

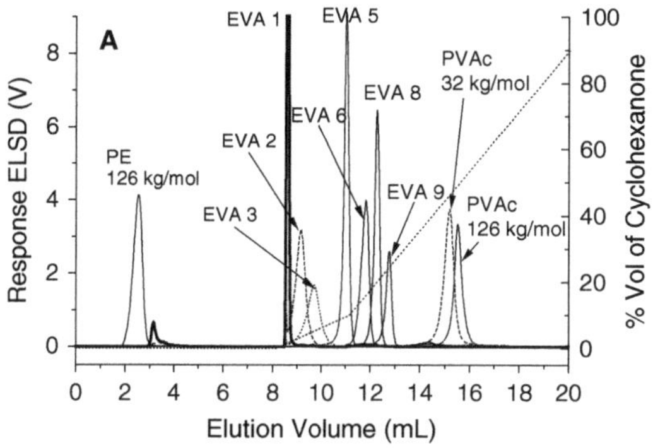

Sample	M_w [kg/mol]	VA [wt.%]
EVA 1	330	5
EVA 2	270	12
EVA 3	330	14
EVA 5	144	28
EVA 6	430	45
EVA 8	370	60
EVA 9	267	70

Fig. 8.7 Overlay of the chromatograms of EVA copolymers; stationary phase: Polygosil 1000; mobile phase: gradient decalin/cyclohexanone (*dotted line*); temperature: 140 °C; detector: ELSD; sample solvent: decalin (TCB for the PVAc standards) (reprinted with permission from [48]. Copyright (2007) of American Chemical Society)

analyze the CCD of the samples. For the HPLC-FTIR coupling the LC Transform interface system was used [49].

In a ground-breaking study Macko and Pasch found that a specific carbon-based stationary phase—Hypercarb [50]—enables highly selective separations of polyolefins. Hypercarb was originally developed by Knox and coworkers [51] and had been used in HPLC analysis of small molecules; it was, however, never applied to separate synthetic polymers. Macko et al. found that porous carbon Hypercarb adsorbs linear PE from 1-decanol as the mobile phase at 160 °C [52–54]. The retained polymer was desorbed from the column using a linear gradient from 1-decanol to TCB. Moreover, this HPLC system separated isotactic, atactic and syndiotactic PP from each other, see Fig. 8.8. It was shown further that the same chromatographic system separates ethylene/hexene and propene/1-alkene copolymers according to their chemical compositions [55, 56].

Macko et al. demonstrated the usefulness of the approach for ethylene–propylene copolymers [57] and copolymers of propylene with different tacticities [58]. Moreover, terpolymers of ethylene, propylene and a diene monomer (EPDM) were separated [59]. It was found that both comonomers, ethylene and diene, are adsorbed. On the other hand, adsorption of EP, ethylene-butene (EB), ethylene-hexene (EH), ethylene-octene (EO) or ethylene/1-decene copolymers depends linearly on the average content of ethylene [60].

It is known that adsorption of polymers is a function of temperature [61]. This phenomenon has been applied to the separation of synthetic polymers by Lochmüller [62] for poly(ethylene glycol) and by Chang [63] for polystyrene. Very recently, Cong et al. described experimental conditions that enable the application of temperature changes to the separation of polyolefins [34]. The separation was achieved by the interaction of the polyolefin with a graphite surface in a thermodynamically good solvent for PE. The solvent used was o-dichlorobenzene and the commercially available Hypercarb column was applied. This method is now termed ‘high-temperature temperature-gradient interaction

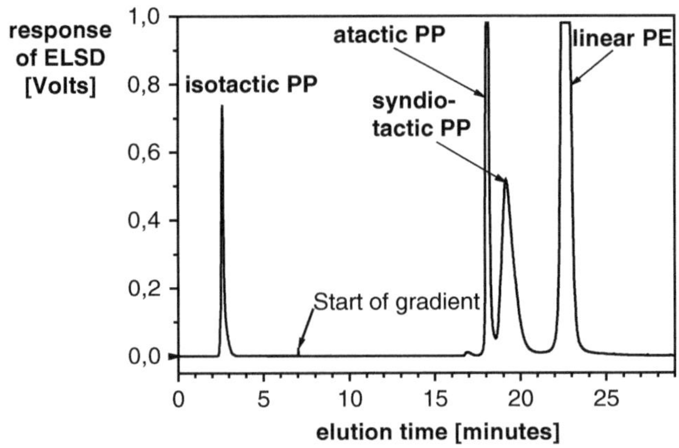

Fig. 8.8 Separation of a blend of isotactic, syndiotactic and atactic PP and linear PE; stationary phase: Hypercarb; mobile phase: gradient 1-decanol/TCB; temperature: 160 °C; detector: ELSD (reprinted with permission from [53]. Copyright (2009) of American Chemical Society)

chromatography' (HT-TGIC), see scheme of the experimental protocol in Fig. 8.9. A typical copolymer separation is presented in Fig. 8.10. In contrast to HT-HPLC, HT-TGIC uses a single solvent as the mobile phase enabling the detection of the polymers with different detectors, including a light scattering detector, refractometer, infrared or viscosimetric detector.

The different techniques of high-temperature interaction chromatography have been coupled to SEC in comprehensive 2D-LC setups. In 2009 the first commercial HT-2D-LC system has been introduced into the market by PolymerChar (Valencia, Spain) that enabled the on-line coupling of HT-HPLC and HT-SEC. In this instrument, isocratic and solvent gradient separations can be conducted in the first dimension to provide information on the chemical composition of olefin copolymers and polyolefin blends. A photograph of the instrument is shown in Fig. 8.11. It comprises a separate sample dissolution and injection module, a solvent delivery module and the chromatographic unit containing two separate column ovens for the HPLC and the SEC columns. The instrument is equipped with RI, IR and ELSD detectors with options to add a MALLS or viscometer detector.

The first results on 2D-LC for polyolefins were published by Ginsburg et al. [64, 65] and Roy et al. [66]. Roy et al. [66] applied a separation system that was previously described by Macko et al. [52, 53, 55]. This system was applied to the separation of ethylene/1-octene copolymers regarding chemical composition and molar mass. Ginzburg et al. [64] used gradient HPLC coupled to SEC to separate blends of PP stereoisomers, ethylene/propylene rubbers, ethylene/norbornene copolymers and ethylene/1-hexene copolymers, all at an operating temperature of 160 °C using a stationary phase of Hypercarb and a mobile phase of 1-decanol-TCB. The 2D contour diagram (composition vs. molar mass) for one example is shown in Fig. 8.12a. This is a most convincing application that gives a clear idea of the capabilities of HT-2D-LC. In a similar experiment a complex mixture of PE and PPs with different

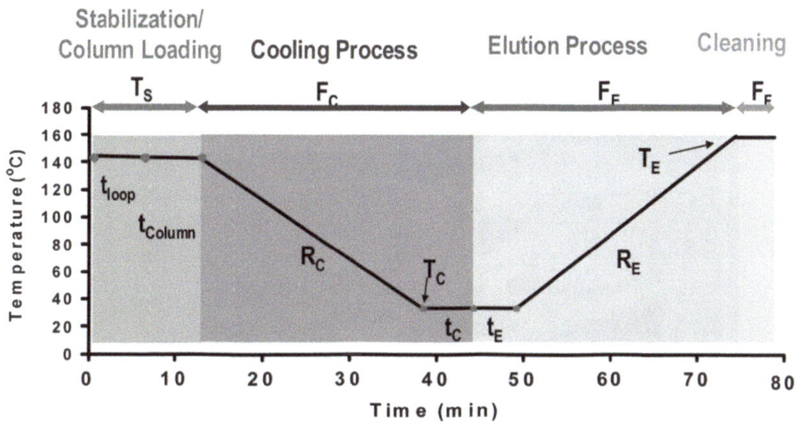

Fig. 8.9 Schematic of HT-TGIC experimental set-up, the definition of each variable is described in Table 5 (reprinted with permission from [34]. Copyright (2011) of American Chemical Society)

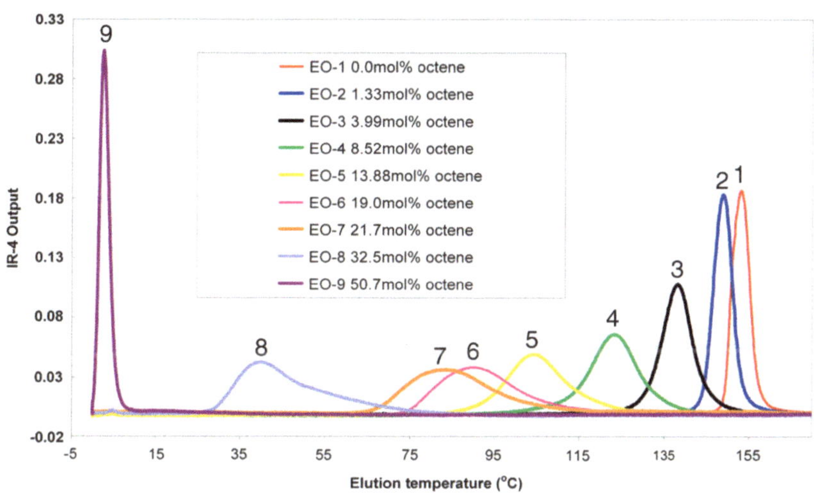

Fig. 8.10 HT-TGIC chromatograms of ethylene-octene random copolymers, Hypercarb column, experimental conditions see Fig. 8.9 (reprinted with permission from Ref. [34]. Copyright (2011) of American Chemical Society)

tacticities has been separated, see Fig. 8.12b. Both axes in the contour plot may be calibrated, as recently illustrated for the HT-2D-LC separation of EVA copolymers by Ginzburg et al. [64]. The SEC separation was calibrated with PE standards, while the HPLC separation was calibrated with EVA copolymers with a known content of VA. Moreover, the coupling of HPLC with SEC where TCB is used as the mobile phase enables the application of RI, IR, VIS or LS detectors. This was demonstrated recently for the 2D-LC separation of EP and EO copolymers [67]. Molar masses of the polymers eluting from the

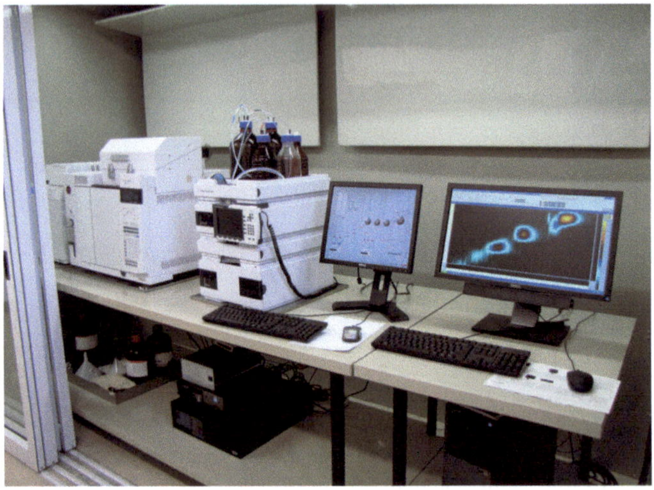

Fig. 8.11 High-temperature 2D-LC system of PolymerChar (Valencia, Spain)

2D-LC system were calculated on the basis of signals from the IR and LS detectors.

8.3.1 Analysis of Polypropylenes by Tacticity and Molar Mass [65]

Aim

Polyolefins are, similar to other synthetic polymers, complex with regard to different parameters of molecular heterogeneity. Polypropylenes exhibit a MMD and a distribution regarding tacticity. Due to the chiral center of propylene, different catalysts and polymerization conditions may result in the formation of isotactic (iPP) and syndiotactic (sPP) polymer chains or subunits. When units of different tacticity are distributed along the polymer chain, the material is atactic (aPP). The average tacticity of PP can be analysed by FTIR or NMR spectroscopy. In FTIR spectroscopy the information relates to a global % tacticity while NMR provides the types and concentrations of tactic triads or pentads depending on the technical parameters of the spectrometer. Quantitative information on the composition of a single polymer chain cannot be obtained, nor can a blend of e.g. iPP, sPP and aPP differentiated from a PP containing different tactic units. It has been shown in a number of applications that the Hypercarb stationary phase exhibits a remarkable selectivity towards different polyolefin structures [47, 53, 59]. This stationary phase shall now be used for the separation of PP according to tactivity. The molar mass of the different tactic polymers shall be analyzed by on-line coupled SEC.

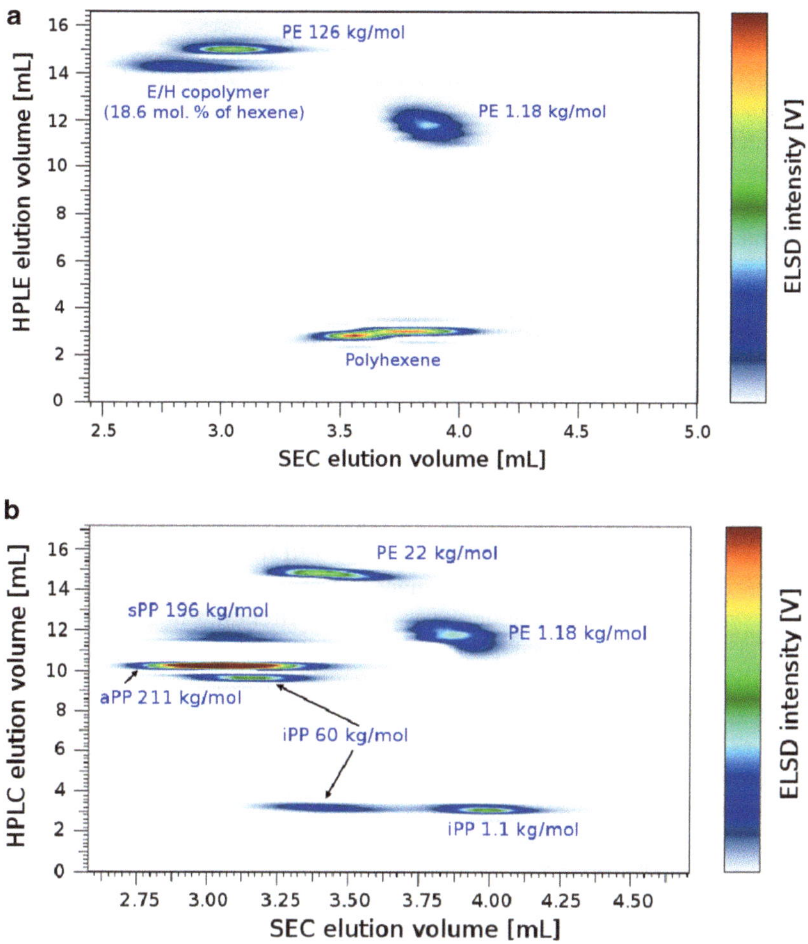

Fig. 8.12 Contour diagrams of the HT-2D-LC separation of (**a**) a blend of PE, poly-1-hexene and an ethylene-1-hexene and (**b**) a blend of PE and PPs with different tacticities; stationary phase: Hypercarb (first dimension) and PL Rapide H (second dimension); mobile phase: gradient 1-decanol/TCB (first dimension) and TCB (second dimension); temperature: 160 °C; detector: ELSD (reprinted from [65] with permission of Elsevier)

Materials

- **Calibration Standards**. Linear PE and PP standards from PSS GmbH (Mainz, Germany)
- **Polymers**. sPP with M_w 196 kg/mol from Sigma–Aldrich (Munich, Germany), aPP with M_w 211 kg/mol from LyondellBasell (Ferrara, Italy), iPP with M_w 45 kg/mol from the University of Stellenbosch (South Africa).

Equipment

- **Chromatographic System**. All experiments were realized using a prototype chromatographic system for HT-2D-LC constructed by PolymerChar (Valencia, Spain), comprising an autosampler, two separate ovens, valves and two pumps equipped with vacuum degassers (Agilent, Waldbronn, Germany). One oven was used for thermostatting the SEC column, while the second one, where the injector and a switching valve were housed, was used to thermostat the HPLC column. The coupling of HT-HPLC and HT-SEC was achieved by using an electronically controlled eight-port valve EC8W (VICI Valco instruments, Houston, Texas, USA) equipped with two 200 μL loops. From the moment of injection into the HPLC column (50 μL injection loop), the 8-port valve was switched every 2 min in order to inject 200 μL of effluent from the HPLC into the SEC column. 2D-LC system handling was done with software provided by Polymer Char (Valencia, Spain). WinGPC-Software v. 7.0 (Polymer Standards Service, Mainz, Germany) was used for data acquisition and evaluation.
- **Columns**. Chromatograph 1: Hypercarb column packed with porous graphite particles with the following parameters: column size 250 × 4.6 mm i.d., average particle size diameter 5 μm, surface area of 120 m^2/g and pore size of 250 Å (Thermo Scientific, Dreieich, Germany). Chromatograph 2: PL Rapide H, 150 × 7.5 mm (Polymer Laboratories, Church Stretton, England).
- **Mobile Phase**. Chromatograph 1: Linear gradient 1-decanol to TCB starting with 100 % of 1-decanol for 40 min, the volume fraction of TCB was linearly increased to 100 % within 80 min and then held constant for 80 min. The flow rate was 0.1 mL/min. Chromatograph 2: TCB with a flow rate of 2.5 mL/min.
- **Detectors**. ELSD PL-ELS 1000 (Polymer Laboratories, Church Stretton, England). The following parameters were set on the ELSD: air flow rate 1.5 L/min, nebulizer temperature 160 °C, evaporation temperature 260 °C.
- **Column Temperature**. 160 °C
- **Sample Concentration**. 2–3 mg/mL. All samples are dissolved in 1-decanol.
- **Injection Volume**. 50 μL (first dimension).

Preparatory Investigations

The separation in the first dimension was conducted according the method published by Macko and Pasch [53] using Hypercarb as the stationary phase and a solvent gradient of decanol-TCB. The separation is according to tacticity of PP and chemical composition separating PP and PE. In preliminary investigations it has been found that iPP elutes in two peaks. The first peak elutes in decanol before the start of the gradient while the second peak elutes with the solvent gradient. To investigate this phenomenon in more detail, iPP with different molar masses was analyzed by HT-2D-LC, see Fig. 8.13.

The contour plots prove that the portion of iPP that elutes in the gradient has in all cases a larger molar mass than the portion that elutes in 1-decanol. Moreover, the higher the molar mass of the injected iPP standard, the larger is the portion that elutes in the gradient. The standard with M_w 350 kg/mol is almost completely retained and elutes mostly with the gradient. At present it is not quite clear what the reason for the elution behaviour is. This should be considered in future investigations.

Separations

The separation of a blend of iPP, sPP, aPP and PE is presented in Fig. 8.14. As was expected all components are perfectly separated from each other. Their molar masses are different as is proven by the different elution volumes in the second dimension. It can be seen clearly, that aPP and sPP have significantly higher molar masses than iPP and PE.

Evaluation

A most important and difficult topic in 2D-LC is calibration of the second dimension to determine the molar masses of the separated species. In a comprehensive 2D-LC set-up, two chromatographic modes (HPLC and SEC) are on-line coupled. This means that the polymer sample is introduced into the SEC column in a mixed solvent via an automated switching valve. In the present case the composition of the mixed solvent changes from pure 1-decanol to 1-decanol/TCB. Because the hydrodynamic volume of a macromolecule depends on the solvent, this may affect the calibration of the SEC. In order to study the influence of the injection solvent on the behaviour of macromolecules in SEC the PE and iPP standards were individually analyzed by SEC as stand alone. The sample solvent for PE and iPP was either 1-decanol or TCB. Figure 8.15 shows an overlay of two SEC calibration curves constructed for iPP and PE standards. As can be seen, the calibration curves obtained with iPP standards corresponding to the different injection solvents are different. On the other hand, the two curves obtained with PE standards overlap over almost the entire elution range except in the low molar mass region. It is, therefore, important to investigate the calibration behaviour of different polyolefins in detail. An even more complex situation is encountered when the calibration standards are injected into the first dimension and undergo the entire 2D-LC separation. In this case, very scattered data have been obtained as is seen in Fig. 8.15c that cannot be explained at present.

8.3.2 Analysis of Ethylene-Vinyl Acetate Copolymers [64]

Aim

Copolymers of ethylene and vinyl acetate (EVA) are commercially important products. Depending on their comonomer content, these materials are used in the production of films, foams or hot melt adhesives. As in all other copolymers EVA may be distributed with regard to the molar mass, the chemical composition and the branch length. A comprehensive characterization of these polymers being distributed in more than one compositional feature is essential for optimization of

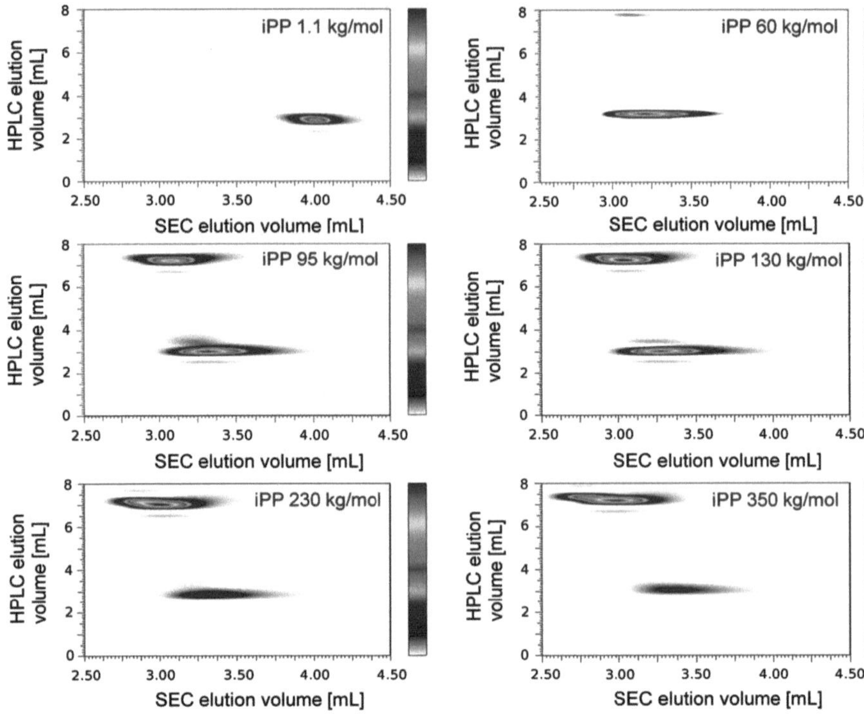

Fig. 8.13 2D-LC plots of iPP samples with different molar masses, first dimension: Hypercarb, mobile phase: gradient decanol-TCB; second dimension: PL Rapide, TCB; detector: ELSD, column temperature 160 °C (reprinted from [65] with permission of Elsevier)

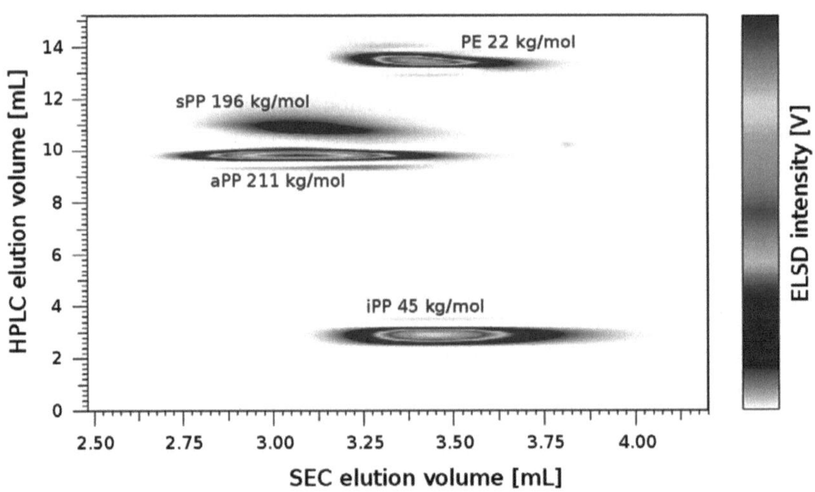

Fig. 8.14 2D-LC plot of a blend of iPP, sPP, aPP and PE with different molar masses, experimental conditions see Fig. 8.13 (reprinted from [65] with permission of Elsevier)

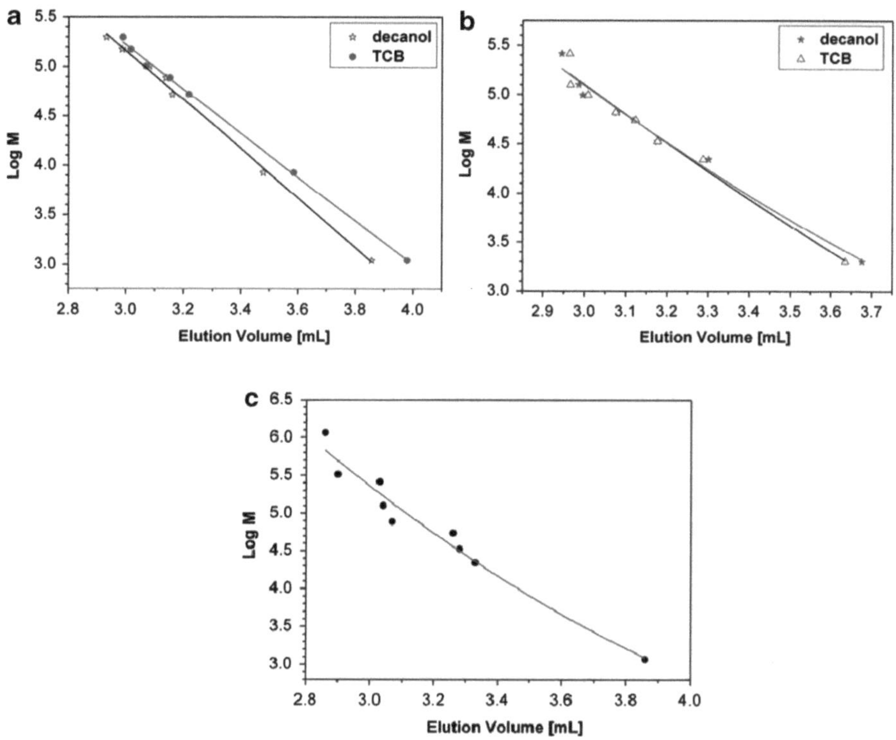

Fig. 8.15 2D-LC calibration curves for iPP (**a**) and PE (**b**) obtained by injection of the samples in the second dimension and for PE (**c**) obtained by injection in the 2D-LC (reprinted from [65] with permission of Elsevier)

the synthesis and fundamental understanding of key structure–property relationships.

EVA with low vinyl acetate (VA) content is a semicrystalline material that can be separated according to composition by TREF. EVA copolymers containing 9–42 wt% VA were analyzed and it was found that copolymers with VA contents higher than 20 wt% are fully amorphous and thus cannot be separated by TREF or CRYSTAF. It is, therefore, the aim of the present application to separate EVA copolymers over the entire comonomer concentration range by HT-HPLC. The molar mass information shall be obtained by on-line coupled SEC.

Materials

- **Calibration Standards**. Linear PE standards from PSS GmbH (Mainz, Germany).
- **Polymers**. EVA copolymers were obtained from Exxon-Mobil Chemical (Meerhout, Belgium) and Bayer (Leverkusen, Germany). Their characteristics were as follows: M_w (kg/mol)/PDI/VA(mol%): ESCORENE 0019 (Exxon Mobil) 197.5/3.1/6.5; LEVAPREN 450 (Bayer) 377.9/8.1/20; LEVAPREN 800 HV (Bayer) 224.6/4.1/57.

Equipment

- **Chromatographic System**. All experiments were realized using a prototype chromatographic system for HT-2D-LC constructed by PolymerChar (Valencia, Spain), comprising an autosampler, two separate ovens, valves and two pumps equipped with vacuum degassers (Agilent, Waldbronn, Germany). One oven was used for thermostatting the SEC column, while the second one, where the injector and a switching valve were housed, was used to thermostat the HPLC column. The coupling of HT-HPLC and HT-SEC was achieved by using an electronically controlled eight-port valve EC8W (VICI Valco instruments, Houston, Texas, USA) equipped with two 200 μL loops. From the moment of injection into the HPLC column (50 μL injection loop), the 8-port valve was switched every 2 min in order to inject 200 μL of effluent from the HPLC into the SEC column. 2D-LC system handling was done with software provided by Polymer Char (Valencia, Spain). WinGPC-Software v. 7.0 (Polymer Standards Service, Mainz, Germany) was used for data acquisition and evaluation.
- **Columns**. Chromatograph 1: Perfectsil 300, 250 mm × 4.6 mm i.d., average particle size diameter 5 μm (MZ Analysentechnik, Mainz, Germany). Chromatograph 2: PL Rapide H, 150 × 7.5 mm (Polymer Laboratories, Church Stretton, England).
- **Mobile Phase**. Chromatograph 1: linear gradient TCB-cyclohexanone. The flow rate was 0.1 mL/min. Chromatograph 2: TCB with a flow rate of 2.5 mL/min.
- **Detectors**. ELSD PL-ELS 1000 (Polymer Laboratories, Church Stretton, England). The following parameters were set on the ELSD: air flow rate 1.5 L/min, nebulizer temperature 160 °C, evaporation temperature 260 °C.
- **Column Temperature**. 150 °C.
- **Sample Concentration**. 2 mg/mL. All samples are dissolved in TCB.
- **Injection Volume**. 50 μL (first dimension).

Preparatory Investigations

Having appropriate chromatographic systems is the main prerequisite for the realization of 2D-LC separation. As has been shown previously, EVA copolymers could be separated according to their VA content on bare silica using TCB and cyclohexanone as components of the mobile phase [48]. The separation is based on the full adsorption of EVA from TCB and a subsequent controlled desorption by a TCB-cyclohexanone solvent gradient.

Separations

The contour plot in Fig. 8.16 shows the 2D-LC separation of a blend of the homopolymers PVAc and PE and three EVA copolymers. The gradient separation is represented along the *y*-axis whereas the elution along the *x*-axis corresponds to the SEC separation. As can be seen, the individual samples elute in the order of their polarity. The first eluting spot can be assigned to PE and the last one to PVAc which are the least and most polar component, respectively. Between these the three EVA copolymers elute. Only two EVA copolymers (6.5 and 20 mol% of VA) are not

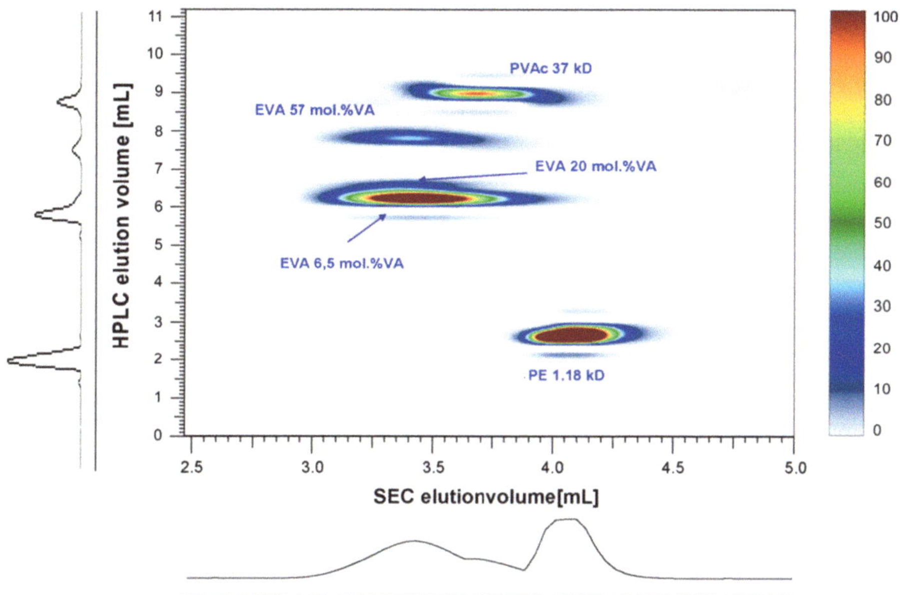

Fig. 8.16 2D-LC plot of a blend of PVAc, PE and three EVA copolymers, first dimension: Perfectsil, mobile phase: gradient TCB-cyclohexanone; second dimension: PL Rapide, TCB; detector: ELSD, column temperature 150 °C (reprinted from [64] with permission of Elsevier)

baseline separated, but the presence of two components with different chemical compositions as well as with different molar masses can be concluded. The small narrow part in the contour plot eluting between 5.6 and 6.0 mL is an artifact produced by the WinGPC software.

Evaluation

In 2D-LC typically only the second dimension providing the molar mass information is calibrated. In the present application, however, both dimensions shall be calibrated. A calibration of the HPLC requires knowledge of the delay volume of the system, i.e. when a given gradient reaches the detector.

The delay volume is the sum of a void volume and a dwell volume of the corresponding system. Here the void volume was considered as the volume of the component that is not retained by the stationary phase while the dwell volume is the volume of liquid contained in the system between the point where the gradient is formed and the injector. The dwell and the void volume were determined by a modified procedure proposed by Bashir et al. for HPLC [68]. We have found previously that the dependence between the elution volume and the average chemical composition of EVA copolymers in gradient HPLC is linear. The obtained relationship is depicted in Fig. 8.17 together with the SEC calibration curve. As the delay volume of the system was determined, the content of cyclohexanone in the mobile phase could be related to the elution volume and the dependence between the VA content and the elution volume applied to the 2D contour plot.

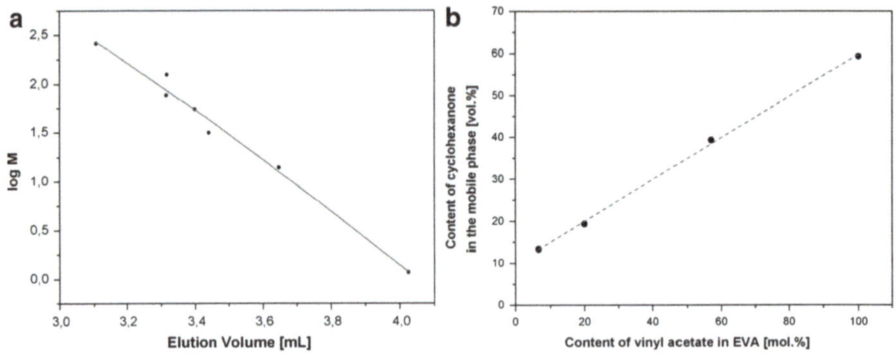

Fig. 8.17 Molar mass calibration curve for PE (**a**) and chemical composition calibration curve for VA content obtained by injection in the 2D-LC (**b**) (reprinted from Ref. [64] with permission of Elsevier)

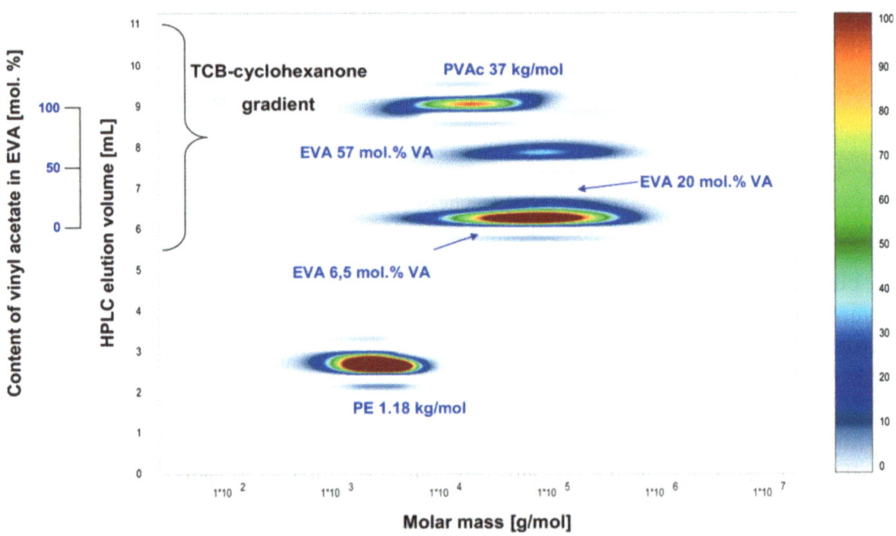

Fig. 8.18 2D-LC plot obtained from the original data in Fig. 8.16 after calibration of both dimensions (reprinted from [64] with permission of Elsevier)

Consequently, the *x*- and *y*-axis of the contour plot were converted and thus a new (quantitative) contour plot was obtained, see Fig. 8.18.

8.4 Conclusions and Outlook

Polyolefins belong to the most important synthetic polymeric materials in all spheres of human activities ranging from packaging and construction to computer science and medicine. Similar to other polymeric materials, polyolefins are distributed in their molecular properties and in-depth analysis of these properties is required using the most sophisticated analytical methods. The classical techniques for chemical composition analysis of polyolefins are based on the crystallization behaviour of different components of these materials. These techniques are only applicable for the crystalline part of the sample and the amorphous part is obtained as a bulk fraction. Nevertheless, these techniques are still the analytical workhorse in most polyolefin research laboratories. There is a number of recent advancements in these techniques that enable the reduction of analysis time, better resolution and mathematical modelling etc.

A fascinating new development in column-based chromatographic techniques for polyolefin analysis is high-temperature interaction chromatography. In contrast to crystallization-based techniques, IC can address the whole sample irrespective of whether it is crystalline or amorphous. The use of gradient HT-HPLC, LCCC at high temperatures above 120 °C, HT-HPLC based on precipitation–redissolution or adsorption–desorption for chemical composition analysis of polyolefins have been reported in recent years. These methods are a major breakthrough in the field of chemical composition analysis of polyolefins. They overcome the drawbacks of other techniques used previously for chemical composition analysis as they address both the amorphous and the crystalline part of the sample. The ultimate recent development in polyolefin analysis is coupling of HT-HPLC with online SEC. This fascinating development leads to the MMD of the sample as a function of its chemical composition. 2D-HT-HPLC is a major advancement in polyolefin analysis and promises to be the future for research-oriented polyolefin laboratories. The most recent step regarding hyphenation of 2D-HT-HPLC is the coupling with infrared and light scattering detectors [67].

All column-based separation methods can be coupled to information-rich detectors such as FTIR and NMR. It has been demonstrated recently, that HT-SEC and ^{1}H-NMR can be coupled on-flow to provide information on the chemical composition as a function of molar mass.

To summarize, all techniques used for polyolefin characterization have advantages and disadvantages. Some information can be obtained more reliably from one technique and some other from other techniques. One has to decide on the problems to be addressed using a given technique. Nevertheless, 2D-HT-HPLC seems to be one major technique to be used for polyolefin analysis in the future due to its ability to provide MMD as a function of CCD of the sample which is not possible by other approaches.

References

1. Wild L, Ryle T, Knobeloch D, Peat IR (1982) J Polym Sci Polym Phys 20:441
2. Wild L (1991) Adv Polym Sci 98:1
3. Monrabal B (1994) J Appl Polym Sci 52:491
4. Monrabal B (1996) Macromol Symp 110:81
5. Monrabal B, Blanco J, Nieto J, Soares JBP (1999) J Polym Sci Polym Chem 37:89
6. Monrabal B, Mayo N, Cong R (2012) Macromol Symp 312:115
7. Li Pi Shan C, Miller M, Lee D, Zhou Z (2012) Macromol Symp 312:1

8. Pasch H, Malik I, Macko T (2012) Adv Polym Sci. doi: 10.1007/12_2012_167
9. Housaki T, Satoh K, Nishikida K, Morimoto M (1988) Makromol Chem Rapid Commun 9:525
10. Nishikida K, Housaki T, Morimoto M, Kinoshita T (1990) J Chromatogr A 517:209
11. Markovich RP, Hazlitt LG, Smith-Courtney L (1993) In: Provder T (ed) Chromatography of polymers. Characterization by SEC and FFF, ACS Symp Ser 521. American Chemical Society, Washington, DC
12. Dhenin V, Rose LJ (2000) Polym Prepr 41:285
13. DesLauriers PJ, Battiste DR (1995) ANTEC-SPE 53:3639
14. DesLauriers PJ, Rohlfing DC, Hsieh ET (2002) Polymer 43:159
15. DesLauriers PJ (2004) Measuring compositional heterogeneity in polyolefins using SEC/FTIR spectroscopy. In: Striegel A (ed) Multiple detection in size exclusion chromatography, ACS Symp Ser 893. American Chemical Society, Washington, DC
16. Piel C, Albrecht A, Neubauer C, Klampfl CW, Reussner J (2001) Anal Bioanal Chem 400:2607
17. Tackx P, Bremmers S (1997) Proc ISPAC-10, Toronto, p 42
18. Albrecht A, Bruell R, Macko T, Malz F, Pasch H (2009) Macromol Chem Phys 210:1319
19. Albrecht A, Bruell R, Macko T, Sinha P, Pasch H (2008) Macromol Chem Phys 209:1909
20. Heinz LC, Graef S, Macko T, Bruell R, Balk S, Keul H, Pasch H (2005) e-polymers no. 54
21. Macko T, Schulze U, Bruell R, Albrecht A, Pasch H, Fonagy T, Haeussler L, Ivan B (2008) Macromol Chem Phys 209:404
22. Verdurmen-Noel L, Baldo L, Bremmers S (2001) Polymer 42:5523
23. de Goede S, Bruell R, Pasch H, Marshall N (2003) Macromol Symp 193:35
24. de Goede S, Bruell R, Pasch H, Marshall N (2004) e-polymers no. 012
25. de Goede E, Mallon P, Pasch H (2010) Macromol Mat Eng 295:366
26. de Goede E, Mallon P, Pasch H (2012) Macromol Mat Eng 297:26
27. de Goede E, Mallon P, Pasch H (2011) Macromol Mat Eng 296:1018
28. Graef S, Bruell R, Pasch H, Wahner UM (2003) e-polymers no. 005
29. Luruli N, Pipers T, Bruell R, Grumel V, Pasch H, Mathot VBF (2007) J Polym Sci Polym Phys 45:2956
30. Kearney T, Dwyer JL (2008) Am Lab 40:8
31. Hiller W, Pasch H, Macko T, Hoffmann M, Ganz J, Spraul M, Braumann U, Streck R, Mason J, van Damme F (2006) J Magn Res 183:290
32. Zhou Z, Kuemmerle R, Stevens JC, Redwine D, He Y, Qiu X, Cong R, Klosin J, Montanez N, Roof G (2009) J Magn Res 200:328
33. Zhou Z, Stevens JC, Klosin J, Kuemmerle R, Qiu X, Redwine D, Cong R, Taha A, Winniford B, Chauvel P, Montanez N (2009) Macromolecules 42:2291
34. Cong R, de Groot AW, Parrott A, Yau W, Hazlitt L, Brown R, Miller MD, Zhou Z (2011) Macromolecules 44:3062
35. Berek D (2000) Prog Polym Sci 25:873
36. Chang T (2003) Adv Polym Sci 163:1
37. Macko T, Pasch H, Kazakevich YV, Fadeev AY (2003) J Chromatogr A 988:69
38. Macko T, Pasch H, Denayer JF (2003) J Chromatogr A 1002:55
39. Macko T, Pasch H, Milonjic SK, Hiller W (2006) Chromatographia 64:183
40. Macko T, Bruell R, Zhu Y, Wang Y (2010) J Sep Sci 33:3446
41. Heinz LC, Macko T, Williams A, O'Donohue S, Pasch H (2006) The Column (electronic journal) Feb 13
42. Pasch H (2012) Chromatography. In: Matyjaszewski K, Moeller M (eds) Polymer science: a comprehensive reference, vol 2. Elsevier BV, p 33
43. Heinz LC, Macko T, Pasch H, Weiser MS, Mülhaupt R (2006) Int J Polym Anal Charact 11:47
44. Heinz LC, Pasch H (2005) Polymer 46:12040
45. Albrecht A, Heinz LC, Lilge D, Pasch H (2007) Macromol Symp 257:46
46. Weiser MS, Thomann Y, Heinz LC, Pasch H, Mülhaupt R (2006) Polymer 47:4505
47. Dolle V, Albrecht A, Brüll R, Macko T (2011) Macromol Chem Phys 212:959

48. Albrecht A, Brüll R, Macko T, Pasch H (2007) Macromolecules 40:5545
49. Pasch H, Albrecht A, Bruell R, Macko T, Hiller W (2009) Macromol Symp 282:71
50. http://www.interscience.be/promotiesites/hypersil/topics/promotiesites/hypersil/nieuws/hypercarb_technical.pdf
51. Gilbert MT, Knox JH, Kaur B (1982) Chromatographia 16:138
52. Macko T, Pasch H, Wang Y (2009) Macromol Symp 282:93
53. Macko T, Pasch H (2009) Macromolecules 42:6063
54. Macko T, Brüll R, Wang Y (2009) Polym Prepr (Am Chem Soc Div Polym Chem) 50:228
55. Macko T, Brüll R, Alamo RG, Thomann Y, Grumel V (2009) Polymer 50:5443
56. Macko T, Brüll R, Wang Y, Thomann Y (2009) The Column (electronic journal) 4:15
57. Macko T, Brüll R, Wang Y, Coto B, Suarez I (2011) J App Polym Sci 122:3211
58. Macko T, Cutillo F, Bussico V, Brüll R (2010) Macromol Symp 298:182
59. Chitta R, Macko T, Brüll R, van Doremaele G, Heinz LC (2011) J Polym Sci Polym Chem 49:1840
60. Macko T, Brüll R, Alamo RG, Stadler FJ, Losio S (2011) Anal Bioanal Chem 399:1547
61. Lipatov YS, Sergeeva LM (1974) Adsorption of polymers. Wiley, New York, NY
62. Lochmüller CH, Moebus MA, Liu QC, Jung C, Elomaa M (1996) J Chromatogr Sci 34:69
63. Lee HC, Chang T (1996) Polymer 37:S747
64. Ginsburg A, Macko T, Dolle V, Bruell R (2010) J Chromatogr A 1217:6867
65. Ginsburg A, Macko T, Dolle V, Bruell R (2011) Eur Polym J 47:319
66. Roy A, Miller MD, Meunier DM, de Groot AW, Winniford WL, van Damme FA, Pell RJ, Lyons JW (2010) Macromolecules 43:3710
67. Lee D, Miller MD, Meunier DM, Lyons JW, Bonner JM, Pell RJ, Li Pi Chan C, Huang T (2011) J Chromatogr A 1218:7173
68. Bashir MA, Brüll A, Radke W (2005) Polymer 46:3223

9 Conclusions and Future Trends

More than half a century has passed since the introduction of size exclusion chromatography. Yet this 'mature' method of polymer fractionation remains an important topic of research in Analytical Polymer Science. Over the last 20 years a tremendous growth in SEC capabilities has been observed due to the fact that (1) novel and more efficient stationary phases have been developed and (2) SEC has become a part of multi-detector systems where concentration detectors are coupled to detectors that are sensitive to molar mass and chemical composition. The most recent major achievements in this field are the direct coupling of SEC with FTIR and NMR spectroscopy as well as mass spectrometry. Most remarkably, such couplings are available now even in high-temperature liquid chromatography for the analysis of semicrystalline polyolefins.

It took significant time for chromatography practitioners to recognize that SEC (even in multi-detector systems) cannot address all polymer distributions. While it is a perfect tool for the quantitative analysis of molecular size (molar mass) distributions, it fails when it comes to chemical composition, functionality or topology distributions. It was the fundamental research of Belenkii, Glöckner, Mori and others that lead to the development of polymer fractionation methods that are highly selective regarding chemical composition and functionality. Balke introduced the term 'chromatographic cross-fractionation' and took the first step into two-dimensional liquid chromatography.

Today multidimensional tools for comprehensive polymer analysis are becoming increasingly routine methods that are used for the analysis of random and segmented copolymers, polymers with complex topologies, polymer blends and polymers with complex functional groups. In addition to isothermal methods (e.g. SEC, LCCC, solvent gradient HPLC) temperature changes have been found to be useful parameters for the optimization of chromatographic separations (e.g. TGIC). The development of such methods is still progressing and the limits in polymer separations have not been reached yet. Very recently it has been shown that comprehensive 2D-LC can be coupled on-flow to ^{1}H-NMR to provide information on molecular size, functionality and endgroup topology, see Fig. 9.1 [1]. In another development, the separation capabilities of LCCC have been investigated. It was

H. Pasch and B. Trathnigg, *Multidimensional HPLC of Polymers*, Springer Laboratory,
DOI 10.1007/978-3-642-36080-0_9, © Springer-Verlag Berlin Heidelberg 2013

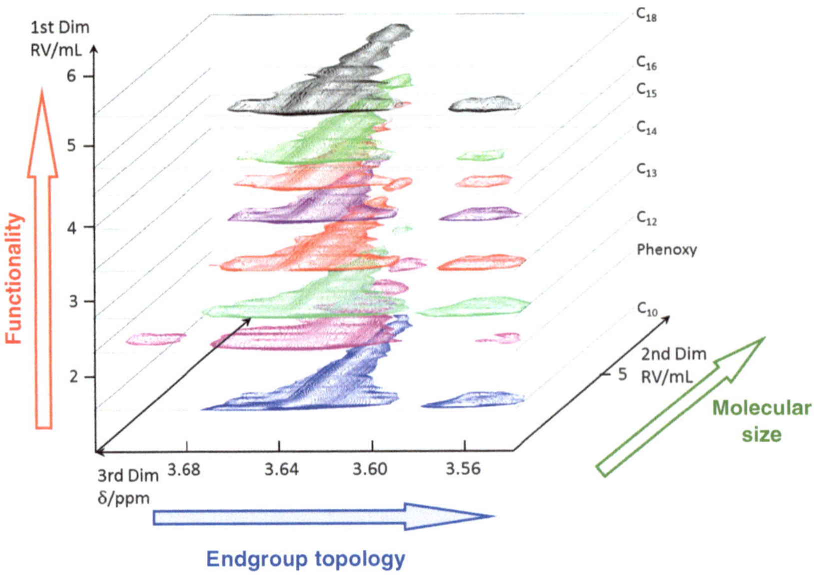

Fig. 9.1 Three-dimensional analysis of polyethylene oxides by online 2D-LC-NMR coupling (reprinted with permission from [1]. Copyright (2012) American Chemical Society)

shown that this method is sufficiently selective to separate deuterated and protonated polymers of similar molar masses, chemical compositions and functionalities, the deuteration being the only molecular difference [2] (Fig. 9.2).

Column-based liquid chromatography reaches its limits when polymer samples with very high molar masses (>1,000 kg/mol) or polymer nanocomposites are targeted. Due to high viscosity and very strong shear forces in the column the sample may degrade thus changing the size and the chemical composition of the macromolecules. As an alternative (and complementary) technique field flow fractionation has become increasingly popular [3]. This channel-based family of methods uses external fields for the separation of complex samples regarding different molecular parameters. Asymmetric field-flow fractionation (AF4) is based on differences in the diffusion coefficient and uses a cross-flow. The separation is mainly based on molecular size. Thermal FFF (ThFFF) uses thermal fields and is based on the normal and thermal diffusion coefficients. Separation in this case takes place regarding size and chemical composition. The schematic set-up of an AF4 instrument in which FFF and SEC separations can be conducted alternatively is shown in Fig. 9.3.

Finally, significant progress has been achieved recently in high-temperature fractionation and analysis of complex polyolefins. In addition to introducing high-temperature HPLC, TGIC and 2D-LC (see Chap. 8), molecular parameters can now be coupled to thermal properties. This is the first step towards linking molecular properties to processing properties/conditions and ultimately to more advanced structure–property correlations. It has been shown that fractions from

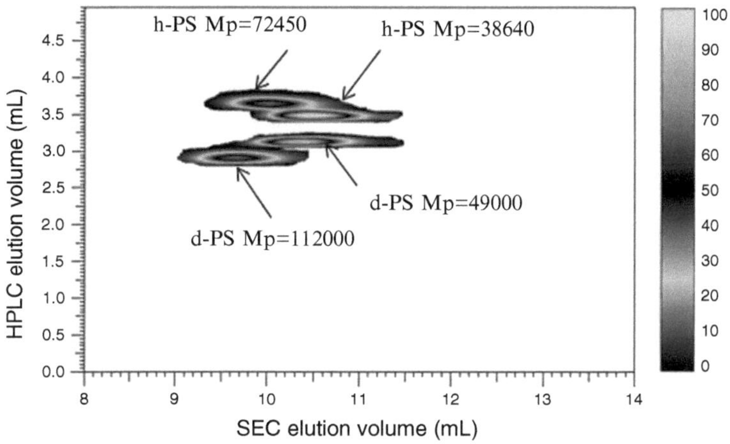

Fig. 9.2 Two-dimensional contour plot of a blend of d-PS (M_p = 112,000) + d-PS (M_p = 72,450) + h-PS (M_p = 38,640) + h-PS (M_p = 72,450), all molar masses in g/mol. *d-PS* deuterated polystyrene, *h-PS* protonated polystyrene (reprinted from [2] with permission of Elsevier)

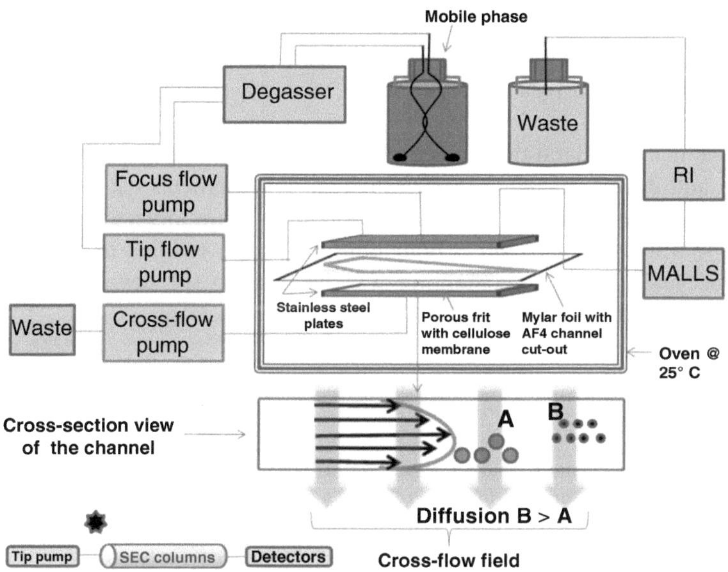

Fig. 9.3 AF4 instrumentation with a cross-section view of the channel at the *bottom* and insert of how the SEC columns are connected to the tip pump when the FFF channel is not in use (reprinted with permission from [4]. Copyright (2012) American Chemical Society)

chromatographic separations can be directly analyzed by fast scanning calorimetry (HPer DSC, Flash DSC) to provide crystallization and melting temperatures of chromatographic fractions (SEC, HPLC) [5, 6]. The experimental set-up of such complex analyses is presented schematically in Fig. 9.4.

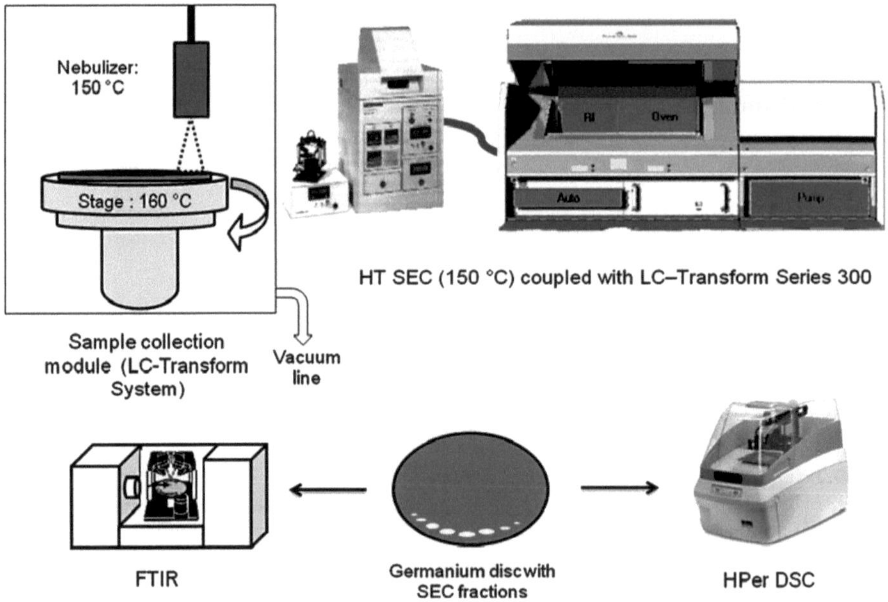

Fig. 9.4 Combination of polymer fractionation with FTIR and DSC analysis (reprinted with permission from [5]. Copyright (2012) American Chemical Society)

In conclusion, separation science of polymers is a rapidly progressing field of research and development. New instrumental developments can be expected that will trigger more advanced method developments. One molecular parameter that can not be fully addressed yet is branching. It is expected that in the forthcoming couple of years significant attention will be paid to the development of more advanced methods for branching analysis.

References

1. Hiller W, Hehn M, Sinha P, Raust J, Pasch H (2012) Macromolecules 45:7740
2. Sinha P, Harding G, Maiko K, Pasch H (2012) J Chromatogr A 1265:95
3. Messaud FA, Sanderson RD, Runyon JR, Williams SKR, Otte T, Pasch H (2009) Progr Polym Sci 34:351
4. Makan AC, Otte T, Pasch H (2012) Macromolecules 45:5247
5. Cheruthazhekatt S, Pijpers TFJ, Harding GW, Mathot VBF, Pasch H (2012) Macromolecules 45:2025
6. Cheruthazhekatt S, Pijpers TFJ, Harding GW, Mathot VBF, Pasch H (2012) Macromolecules 45:5866

Index

H. Pasch and B. Trathnigg, *Multidimensional HPLC of Polymers*, Springer Laboratory,
DOI 10.1007/978-3-642-36080-0,

Printed by Libri Plureos GmbH
in Hamburg, Germany